Embedda

World Class Designs

Newnes World Class Designs *Series*

Analog Circuits: World Class Designs
Robert A. Pease
ISBN: 978-0-7506-8627-3

Embedded Systems: World Class Designs
Jack Ganssle
ISBN: 978-0-7506-8625-9

Power Sources and Supplies: World Class Designs
Marty Brown
ISBN: 978-0-7506-8626-6

For more information on these and other Newnes titles, visit: **www.newnespress.com**

Embedded Systems

World Class Designs

Jack Ganssle

with

Stuart Ball
Arnold S. Berger
Keith E. Curtis
Lewin A. R. W. Edwards
Rick Gentile
Martin Gomez
John M. Holland
David J. Katz
Chris Keydel
Jean LaBrosse
Olaf Meding
Robert Oshana
Peter Wilson

ELSEVIER

AMSTERDAM • BOSTON • HEIDELBERG • LONDON
NEW YORK • OXFORD • PARIS • SAN DIEGO
SAN FRANCISCO • SINGAPORE • SYDNEY • TOKYO
Newnes is an imprint of Elsevier

Newnes

Newnes is an imprint of Elsevier
30 Corporate Drive, Suite 400, Burlington, MA 01803, USA
Linacre House, Jordan Hill, Oxford OX2 8DP, UK

⊗ Recognizing the importance of preserving what has been written, Elsevier prints its
books on acid-free paper whenever possible.

Library of Congress Cataloging-in-Publication Data
Embedded systems: world class designs/Jack Ganssle with Stuart Ball ... [et al.].
    p. cm. – (World class designs)
    Includes bibliographical references and index.
    ISBN-13: 978-0-7506-8625-9 (pbk. : alk. paper)   1. Embedded computer systems–Design
and construction.   I. Ganssle, Jack G.   II. Ball, Stuart R., 1956–
    TK7895.E42E648 2007
    004.16–dc22

                                                                    2007038488

British Library Cataloguing-in-Publication Data
A catalogue record for this book is available from the British Library.

ISBN: 978-0-7506-8625-9

For information on all Newnes publications
visit our Web site at www.books.elsevier.com

Contents

About the Editor

Jack Ganssle (Chapters 5, 6, 7, and 8) is the author of *The Firmware Handbook, Embedded Systems*. He has written over 500 articles and six books about embedded systems, as well as a book about his sailing fiascos. He started developing embedded systems in the early '70s using the 8008. He's started and sold three electronics companies, including one of the bigger embedded tool businesses. He's developed or managed over 100 embedded products, from deep-sea navigation gear to the White House security system ... and one instrument that analyzed cow poop! He's currently a member of NASA's Super Problem Resolution Team, a group of outside experts formed to advise NASA in the wake of Columbia's demise, and serves on the boards of several high-tech companies. Jack now gives seminars to companies world-wide about better ways to develop embedded systems.

About the Authors

Stuart Ball, P.E., (Chapter 1) is an electrical engineer with over 20 years of experience in electronic and embedded systems. He is currently employed with Seagate Technologies, a manufacturer of computer hard disc drives.

Arnold S. Berger (Chapter 2) is a Senior Lecturer in the Computing and Software Systems Department at the University of Washington-Bothell. He received his BS and PhD degrees from Cornell University. Dr. Berger has held positions as an R&D Director at Applied Microsystems Corporation, Embedded Tools Marketing Manager at Advanced Micro Devices and R&D Project Manager at Hewlett-Packard. Dr. Berger has published over 40 papers on embedded systems. He holds three patents.

Keith E. Curtis (Chapter 3) is the author of *Embedded Multitasking*. He is currently a Technical Staff Engineer at Microchip, and is also the author of Embedded Multitasking. Prior and during college, Keith worked as a technician/programmer for Summit Engineering. He then graduated with a BSEE from Montana State University in 1986. Following graduation, he was employed by Tele-Tech Corporation as a design and project engineer until 1992. He also began consulting, part time, as a design engineer in 1990. Leaving Montana in 1992, he was employed by Bally Gaming in Las Vegas as an engineer and later the EE manager. He worked for various Nevada gaming companies in both design and management until 2000. He then moved to Arizona and began work as a Principal Application Engineer for Microchip.

Lewin A. R. W. Edwards (Chapter 4) is the author of *Embedded System Design on a Shoestring*. He hails from Adelaide, Australia. His career began with five years of security and encryption software at PC-Plus Systems. The next five years were spent developing networkable multimedia appliances at Digi-Frame in Port Chester, NY. Since 2004 he has been developing security and fire safety devices at a Fortune 100 company in New York. He has written numerous technical articles and three embedded systems books, with a fourth due in early 2008.

Rick Gentile (Chapter 10) is the author of Embedded Media Processing. Rick joined ADI in 2000 as a Senior DSP Applications Engineer, and he currently leads the Processor Applications Group, which is responsible for Blackfin, SHARC and TigerSHARC processors. Prior to joining ADI, Rick was a Member of the Technical Staff at MIT Lincoln Laboratory, where he designed several signal processors used in a wide range of radar sensors. He has authored dozens of articles and presented at multiple technical conferences. He received a B.S. in 1987 from the University of Massachusetts at Amherst and an M.S. in 1994 from Northeastern University, both in Electrical and Computer Engineering.

Martin Gomez (Chapter 6) is a contributor to The Firmware Handbook. He currently works at Aurora Flight Sciences Corporation, as the manager of the Aerial Robotics Group. He has 24 years of experience in the Aerospace field. Martin holds a BS in Aerospace Engineer, an M.Eng. in EE, and an MS in Applied Physics. He can be reached at MLG28@cornell.edu.

John M. Holland (Chapter 9) is a well-known pioneer of mobile robotics, founding Cybermotion in 1984, the first company to successfully manufacture and sell commercial mobile robot units to customers such as the U.S. Department of Energy, U.S. Army, Boeing, NASA, General Motors and many others. An electrical engineer by training, John holds six U.S. patents and is the author of two previous books, including the foundational book Basic Robotics Concepts (Howard Sams, 1983). He has written and lectured extensively and is an irreverent and outspoken futurist.

David J. Katz (Chapter 10) is the author of Embedded Media Processing. He has over 15 years of experience in circuit and system design. Currently, he is the Blackfin Applications Manager at Analog Devices, Inc., where he focuses on specifying new convergent processors. He has published over 100 embedded processing articles domestically and internationally, and he has presented several conference papers in the field. Previously, he worked at Motorola, Inc., as a senior design engineer in cable modem and automation groups. David holds both a B.S. and M.Eng. in Electrical Engineering from Cornell University.

Chris Keydel (Chapter 5) has been involved with the various aspects of embedded hardware and software design since 1994, starting his career as an R&D engineer for In-circuit Emulators. As a director and tutor of the Embedded Systems Academy,

Christian supervises new class development and teaches and consults clients on embedded technologies including CAN and CANopen, Embedded Internetworking, Real-Time Operating Systems, and several microcontroller architectures as well as the associated development tools. He is a frequent speaker at the Embedded Systems Conferences and the Real-Time and Embedded Computing Shows and co-authored the book "Embedded Networking with CAN and CANopen" by Annabooks/RTC Group. Keydel holds an MSEE and an MSCS from the University of Karlsruhe, Germany.

Jean LaBrosse (Chapter 11) is author of MicroC/OS-II and Embedded Systems Building Blocks. Dr. Labrosse is President of Micrium whose flagship product is the Micrium µC/OS-II. He has an MSEE and has been designing embedded systems for many years.

Olaf Meding (Chapter 5) is a senior software development engineer with more than 15 years experience in all aspects of software analysis, design, development and configuration management in ISO 9001, FDA, and NASA regulated environments. He started his career as a software developer using the Forth programming language working on a large telephony client server project. Since then he served as chief architect and software developer of the Biomass Production System (BPS), a NASA project studying gravitational effects on plant growth in space. And he designed the control software for a TomoTherapy machine that will revolutionize the radiotherapy delivery process for cancer treatment. Olaf currently works for Bruker-AXS where he uses Trolltech's Qt toolkit and the Python programming language to control state of the art X-ray diffraction instruments.

Robert Oshana (Chapter 12) is the author of DSP Software Development Techniques. He has over 25 years of experience in the real-time embedded industry in both embedded application development as well as embedded tools development. He is currently director of engineering for the Development Technology group at Freescale Semiconductor. Rob is also a Senior Member of IEEE and an adjunct at Southern Methodist University. He can be contacted at: robert.oshana@freescale.com

Dr. Peter Wilson (Chapter 13) worked for many years as a Senior Design Engineer in industry with Perranti plc (Edinburgh, Scotland) and as an EDA technical specialist with Analogy, Inc. (Beaverton, Oregon) before joining the Department of Electronics and Computer Science in 1999 at the University of Southampton, UK, where he is currently Senior Lecturer in Electronics. He is also a consultant for Integra Design Ltd. in various aspects of embedded systems including design and modeling with VHDL, Verilog-AMS, and VHDL-AMS.

Preface

At least one bird species uses a twig to dig insects out of holes in trees. Some primates open hard shells with branches used as clubs. The Attini ant is quite literally a farmer, cultivating patches of fungus and exchanging fungi species with other ant colonies. Clearly Homo sapiens is not the only species that uses tools or manipulates the environment to serve our ends. Four billion years of evolution has produced creatures whose fitness for their place in an ecological niche must be augmented by using and building things.

But tools are surely one of the defining marks of our species. In 10,000 years we've gone from the wheel to the microprocessor, each invention a means to acquire wealth and ultimately improve our lot in life.

We engineers are today's inventors, the creators of many of the products that improve and change the life of billions of our fellow Earth-dwellers. Most of the planet's population must now be aware of the microprocessor; surely billions of lives have been touched by embedded products in one form or another. Recently I talked with a company building tiny 5 KW microprocessor-controlled generators for use in the smallest villages in Nepal. Spacecraft spot beams dump TV and other communications products to all but the most remote African communities. Poverty and remoteness no longer isolate societies from the computer age.

In our rush to build products, we forget to look at the impact of our creations. Do you make routers? That project might seem to be just a very sophisticated way of shoving packets around, but in fact it's much more. As a primary component of the Internet it quite literally puts food in hungry bellies. savethechildren.org and a hundred other aid groups solicit donations online and help current and potential donors see in near real-time how a few dollars quite literally saves lives.

Do you build automotive electronics? Your device might be critical to the health of the electric vehicle market, reducing emissions, asthma, and emphysema.

A webcam lets distant grandparents connect with their descendants, the NC milling machine reduces the cost of all sorts of consumer products, a DVD player substitutes a tiny hunk of plastic for hundreds of feet of videotape built from the most toxic of chemicals. The amount of good done by our products far exceeds what we imagine, especially when we're caught up in the drama and frustration of getting the silly thing to market on schedule.

And yet there are only a half million of us producing these smart products. A mere handful of engineers whose impact has been quite profound. Perhaps this has always been true of the engineering profession. When Roebling designed the Brooklyn Bridge, cars didn't exist; could he have dreamed of the vast numbers of people who would use his concrete and steel edifice to earn a living, feed their children, and pay the mortgage? Could De Lessups foresee the Arabian Gulf becoming the world's major source of energy and his Suez Canal one of the primary ways to deliver this resource? Did Jack Kilby and Robert Noyce anticipate how integrating a few transistors onto a single chip meant devices could be smart, energy requirements reduced, emissions curtailed...and of course create employment for the 500,000 of us?

We're an odd and almost invisible breed. Mention computer and the average person thinks of a PC. Yet of the 9 billion processors made every year only tens of millions go to the desktop market. That's pretty close to zero percent. The rest are for embedded systems. Tell your neighbor you're a computer designer and he'll immediately target you forever for answers about his problems with spreadsheet macros and sporadic system crashes. Embedded? Hey, is that why my car goes 100,000 miles between tune-ups? Ah, my new refrigerator uses half the energy of the older one...you mean there's a computer in there that manages temperature control, cuts my expensive electrical bill, and so helps me send the kids through college?

Resource Hunters

We embedded folks include quite a disparate range of people, with skill sets that range widely. In the early days of the business all developers were EEs, who both designed the hardware and wrote all of the code. Most worked in isolation; few projects required more than a couple of people. The high cost of memory at the time had a self-limiting effect on program size: 8k of code could easily be written by a single engineer in very little time.

Moore's Law drove, and continues to drive, the price of transistors practically to zero, so the limits to program size have largely disappeared. Systems with over a million lines of code are quite common today. This technological change has created a corresponding one in us; now more and more developers have little to do with the hardware. Quite a few EEs leave college and embark in careers in firmware. Though embedded systems are the last place where hardware and software intertwine to form a unified whole, there's quite a gulf between the people developing each component. The circuit design of systems in many cases is becoming more complex despite an ever-increasing suite of tools and off-the-shelf chips. In many cases hardware engineering now looks a lot like software development as the EEs struggle with VHDL equations and SystemC code that compiles to a hardware description language.

But one characteristic that both disciplines combine is a continuous search for new ideas, time-saving products, and canned solutions to common problems. Wise developers are resource-hunters.

Engineering is the art of solving problems, along the way producing products. It does not require us to invent everything. The best engineers are those who seek out and use solutions that already exist, saving their employers money and speeding the product to market.

Many EEs get a dozen magazines a month, most free, all containing valuable information. No one can consistently, month after month, read all that material, so the typical engineer flips through the publications, perhaps noting an article or two of interest to read later. The ads get as much attention as editorial content. "Hey, TI now has a DSP part that does 1.6 bflops!" Though you may not need that today, you're building a wetware database—a file in your brain that's every bit as useful and important as the coolest link site on the Net. A year later, when a colleague asks about available DSPs you may respond, "Gee, I dunno but something makes me think TI does this. Check them out."

Even as far back as 1890 "The Electrical Engineer" was a weekly publication engineers read to stay on top of the latest technology, which may have been breaking news in the exciting field of insulating wires. But for 100+ years there's been an EE tradition of keeping up, noting resources, and finding useful commercial products.

This is much less common in the software world, particularly for firmware. After all, this is a very new business. One survey suggested that the average firmware developer reads just one technical magazine per month, and a single tech book per year.

Once engineers strove to acquire and save all relevant knowledge. In fact, a system called the "EEM File" helped catalog the data in file cabinets. This is no longer possible due to the sheer proliferation of information, but resources like this *World Class Designs* series give immediately useful ideas and even code and schematics that can be incorporated into your work, today.

This book is both a service and a tribute to the working embedded developer. It's a service in that it brings resources and ideas we need together. It's a tribute in that it supports the intriguing mix of people needed to make today's complex products.

And that's what it all comes down to—the people doing it and those whose lives we benefit. A poll on embedded.com showed that 65% of developers felt they were building products important to the world, that affected people's lives in positive and meaningful ways.

Few other professions can make such a noble claim.

—Jack Ganssle

CHAPTER 1

Motors

Stuart Ball

Many engineers tell me that circuit diagrams always grab their attention. That's probably why EDN's "Design Ideas" remains that magazine's most popular section. Stuart Ball's chapter on motors is filled with schematics, and even cross-sectional diagrams of motors and related components. This is the ideal introduction to motor control for newbies and is an important refresher for experienced developers.

A surprising number of projects use various flavors of motors, and very few of those employ a simply driven motor that freewheels without some sort of feedback.

Stuart starts off with a complete description of the frequently misunderstood stepper motor, a device that uses multiple windings, each generally driven by a sequence of pulses created by the software. They require power—much more than generated by a logic element's CMOS outputs, and so he gives useful circuit diagrams that beef up the computer-generated signals. The circuit descriptions are so complete that you don't need an EE degree to completely understand how they work. Stuart also covers commercial ICs that do the amplification work.

Tougher and sometimes-ignored issues like half-stepping, resonance and even microstepping get full treatment. Chopper control, too, is covered. This subject is gaining in importance due to increasing need for precise control and quick operation while maintaining minimum power consumption.

While stepper motors are everywhere, simple DC versions fill in the low-cost and the high-torque ends. But they lack any sort of precise position control. Stuart describes in detail how to control these frustrating beasts, and includes a nice circuit that uses magnetic braking. Both brush and brushless devices are covered. He gets drive circuits implemented using discrete semiconductors as well as specialized driver ICs.

Finally, there's a treatise about encoders which are often coupled with motors to provide position feedback. Omitted from the discussion is one interesting fact, that some encoder companies will build custom versions that let you have nonlinear pulse spacing. Sometimes one must translate a real-world non-linear sort of motion into equal steps for software processing, and it can be cheaper to use a special encoder rather than implement complex software at high processing speeds.

—Jack Ganssle

Motors are key components of many embedded systems because they provide a means to control the real world. Motors are used for everything from the vibrator in a vibrating pager to moving the arm of a large industrial robot. All motors work on the same principles of electromagnetism, and all function by applying power to an electromagnet in some form or another. We won't spend our time on magnetic theory here. Instead, we will look at the basic motor types and their applications in embedded systems.

1.1 Stepper Motors

Stepper motors come in three flavors: permanent-magnet, variable-reluctance, and hybrid. Figure 1-1 shows a cross-sectional view of a variable-reluctance (VR) stepper motor. The VR stepper has a soft iron rotor with teeth and a wound stator. As current is applied to two opposing stator coils (the two "B" coils in the figure), the rotor is pulled into alignment with these two coils. As the next pair of coils is energized, the rotor advances to the next position.

The permanent magnet (PM) stepper has a rotor with alternating north and south poles (Figure 1-2). As the coils are energized, the rotor is pulled around. This figure shows a single coil to illustrate the concept, but a real stepper would have stator windings surrounding the rotor. The PM stepper has more torque than an equivalent VR stepper.

The hybrid stepper essentially adds teeth to a permanent magnet motor, resulting in better coupling of the magnetic field into the rotor and more precise movement. In a hybrid stepper, the rotor is split into two parts, an upper and lower (Figure 1-3). One half is the north side of the magnet and one is the south. The teeth are offset so that when the teeth of one magnet are lining up with the mating teeth on the stator, the teeth on the other magnet are lining up with the grooves in the stator (in the side view in Figure 1-3, the tops of the teeth are crosshatched for clarity). Some hybrid steppers have more than one stack of magnets for more torque.

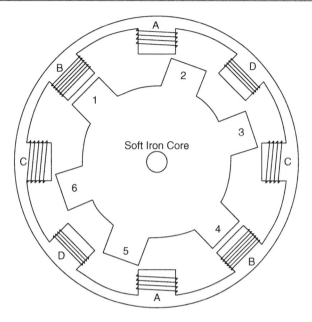

Figure 1-1: Variable-Reluctance Stepper.

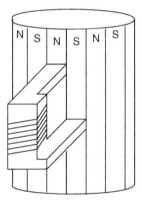

Figure 1-2: Permanent Magnet Stepper.

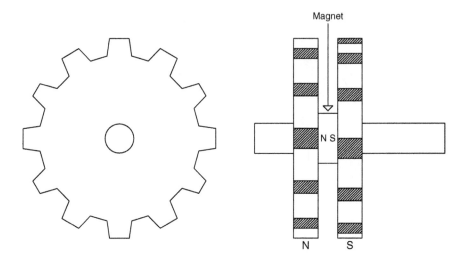

Figure 1-3: Hybrid Stepper.

1.1.1 Bipolar versus Unipolar

All steppers work by creating a rotating magnetic field in the stator, to which the rotor aligns itself. There are two types of stator winding methods for stepper motors: bipolar and unipolar. Bipolar windings use field coils with no common connections. The coils must be driven independently to reverse the direction of motor flow and rotate the motor. Unipolar motors use coils with centertaps. The centertap is usually connected to the positive supply, and the individual coils are grounded (through transistors) to drive the motor. Figure 1-4 shows the difference between bipolar and unipolar motors. Each time the field is changed in a bipolar motor or a different coil is turned on in a unipolar motor, the motor shaft steps to the next rotation position. Typical step sizes for a stepper are 7.5° or 15°. A 7.5° stepper will have 360/7.5 or 48 steps per revolution. The step size depends on the number of rotor and stator teeth.

1.1.2 Resonance

When a stepper motor rotates, it aligns the rotor with the magnetic field of the stator. In a real motor, the rotor has some inertia and is moving when it reaches the ideal alignment, so it overshoots the final position. Because it is now out of alignment with the magnetic field, it "bounces" back and overshoots in the other direction. This continues, with smaller oscillations, until the rotor finally stops. Figure 1-5 illustrates this. The frequency at which the rotor oscillates depends on the motor characteristics (rotor mass and construction, for instance) and the load. If the motor is connected to a load that looks like a flywheel (a mechanical shutter in an optical system, for example), resonance may be more of a problem than it is with an unloaded motor. A load with a lot of friction, such as a belt-driven pulley, has a damping effect that will reduce resonance (unless the belt is connected to a flywheel).

Many stepper motors exhibit a sudden loss of torque when operating at certain step rates. This occurs when the step rate coincides with the oscillation frequency of the rotor. The torque can change enough to cause missed steps and loss of synchronization. There may be more than one band of step rates that cause this effect (because the motor has more than one resonant frequency). In a design that uses only one step rate, these frequency bands (usually fairly narrow) can be avoided by simply picking a step rate that is not a problem.

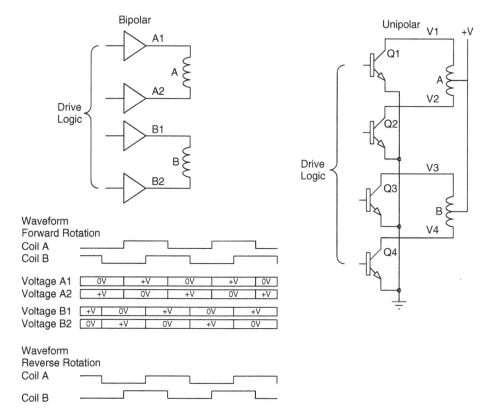

Figure 1-4: Bipolar versus Unipolar Operation.

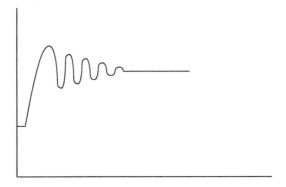

Figure 1-5: Step Motor Ringing.

In a design in which the step rate has to vary, the system may need to be characterized to identify the problem frequencies. The software may then need to avoid operating the motor at these step rates. When accelerating a stepper up to a particular speed, the software may have to accelerate rapidly through these problem areas (Figure 1-6). This is particularly true if the acceleration ramp is fairly slow, which would otherwise cause the step rate to spend some time in the resonance area.

1.1.3 Half-Stepping

As was already mentioned, the rotor in a stepper motor aligns itself to the magnetic field generated by applying voltage to the stator coils. Figure 1-7 shows a simple stepper with a single pair of rotor poles and two stator coils. Say that coil A is energized, and the rotor aligns itself to magnet A with the north pole up (position 1), as shown in the figure. If coil A is turned off and B is energized, the rotor will rotate until the north pole is at position 3. Now if coil B is turned off and coil A is energized but in the reverse direction of what it was before, the rotor will go to position 5. Finally, if coil A is turned off and coil B is energized with the reverse of its original polarity, the rotor will move to position 7. This sequence is called *one-phase-on drive*.

Say that instead of energizing one magnet at a time, we energize coils A and B at the same time. The rotor will move to position 2, halfway between magnets A and B. If we then reverse the current through coil A, the rotor will move to position 4. If we reverse B, the rotor moves to position 6, and, finally, if we reverse A again the rotor moves to position 8. Each of these methods generates a full step of the rotor (in this case, 45° per step), but the actual position is different for the two drive methods. If we combine the two, we can half-step the rotor:

A+, B off: position 1

A+, B+: position 2

A off, B+: position 3

A−, B+: position 4

In this simple example, half-stepping permits a step angle of 22.5°, as opposed to 45° for a full step. The same principle applies to a real motor with several rotor teeth. A motor with a 15° full step can be half-stepped in 7.5° increments.

Figure 1-8 shows all three drive methods. Half-stepping provides smoother rotation and more precise control. It is important to note, though, that for the positions where only one phase is energized (positions 1, 3, 5, 7), the coils need more current to get the same torque. This is because there is only one coil (electromagnet) pulling the rotor.

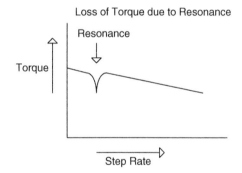

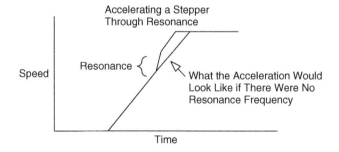

Figure 1-6: Step Motor Resonance.

Switching from two coils to one coil reduces the torque by approximately 30%, so two coils have about 140% of the torque of a single coil. You can compensate for this loss of torque by increasing the coil current by 140% when driving a single coil.

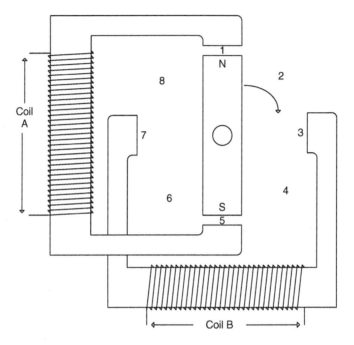

Figure 1-7: Half-Stepping.

1.1.4 Microstepping

If you examine the drive waveform for half-stepping a motor, you can see that it looks like a pair of digitized sine signals, offset by 90°. When the rotor is at position 1, coil A is at the maximum voltage and coil B is at minimum voltage. At position 3, coil A is off and coil B is at maximum voltage. For half-stepping, each coil has three possible drive values: positive drive, off, and negative drive.

If the rotor is at position 1 and coil B is energized slightly, the rotor will rotate toward position 3. If the current through coil A is gradually decreased as the current through coil B is increased, the rotor will slowly move toward position 3, where it ends up when the current in coil A is zero and the current in coil B is maximum. If coil A and B are driven with sine signals that are offset by 90°, the motor will rotate smoothly. Figure 1-9 shows the discrete drive waveform with the equivalent sine/cosine drive and the corresponding rotor positions. A stepper can actually be driven this way.

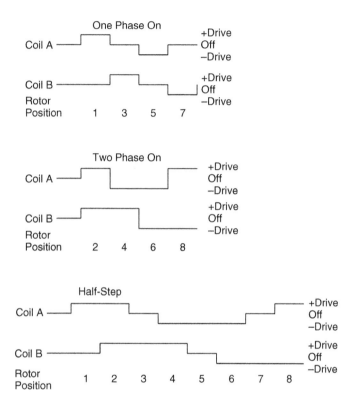

Figure 1-8: Half-Step Drive Waveforms.

If the drive signals are generated from a DAC, the motor can be moved to discrete points between the normal step or half-step positions. This is called *microstepping*. It provides finer control of shaft position, but at the expense of more expensive analog drive circuitry. The actual resolution obtainable by microstepping depends on the resolution of the DAC, the torque of the motor, and the load. For instance, say the motor is very close to position 2 and you want to microstep it to position 2. If the load is too large, you may find that you have to apply more torque than you wanted to move it, and then it may overshoot the position and stop in the wrong place.

If you do need to perform small steps, you can use a bigger motor that can overcome the load. In some cases, this may be a lower-cost solution than other possibilities, such as a

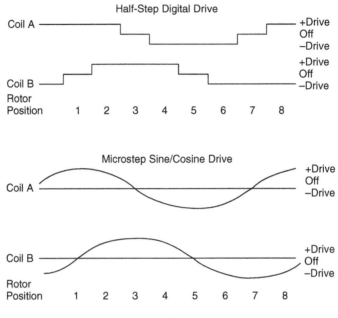

Figure 1-9: Microstepping.

geared DC motor. Microstepping also reduces resonance problems because the motor does not receive discrete steps, so the mechanical ringing is less likely to occur. In a real application, a high-precision DAC is not usually needed because the stepper will not respond to very small changes in the drive waveform. Typical microstep increments are 1/3 to 1/16 of a full step. In other words, using a 10-bit DAC to microstep a stepper motor will not provide any practical advantage over using an 8-bit DAC.

1.1.5 Driving Steppers

The coils of a bipolar stepper are typically driven with an H-bridge circuit. Figure 1-10 shows a circuit that will drive both coils in a two-coil bipolar stepper. This circuit consists of a pair of N-channel MOSFETs and a pair of P-channel MOSFETs for each coil. When input "A" is high, transistors Q1 and Q3 are turned on and current flows from the positive supply, through Q1, through the motor winding, through Q3, and to ground. When "A" is low and "B" is high, Q2 and Q4 are on and current through the motor winding is reversed. The circuit for the other coil works the same way.

The diodes, D1-D8, protect the transistors against the coil flyback voltage when the transistors are turned off. The motor shaft is rotated by applying drive voltage to each input in the proper sequence.

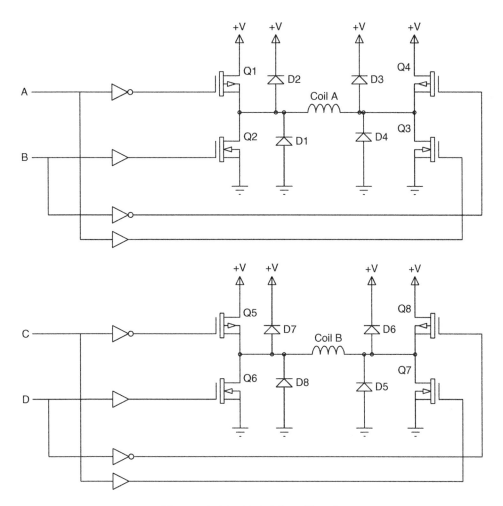

Figure 1-10: H-Bridge Circuit.

1.1.6 Cross-Conduction

One common problem for designers who want to build their own H-bridge circuits from discrete transistors is cross-conduction, also known as shoot-through. This is the condition that occurs when the upper and lower transistors on the same side of the coil turn on at the same time. In the example in the previous section, this would be transistors Q1 and Q2 or Q3 and Q4. If Q1 and Q2 turn on at the same time, there will be a very low impedance between the supply voltage and ground—effectively a short. This usually destroys one or both transistors. In a high-power circuit, the results can be quite dramatic, with blue sparks and pieces of transistor flying across the room.

Shoot-through can be caused (again going back to the same example) by bringing inputs "A" and "B" high at the same time. As shown in Figure 1-11, it can also be caused by bringing one input high while simultaneously taking the other input low. If one of the transistors in the bridge turns off a little more slowly than the others turn on, the result will be momentary shoot-through. It may not be enough to destroy the part, but over time it can cause premature failure. Or, worse, the problem may show up only at high or low temperatures, making failures that only happen in the field.

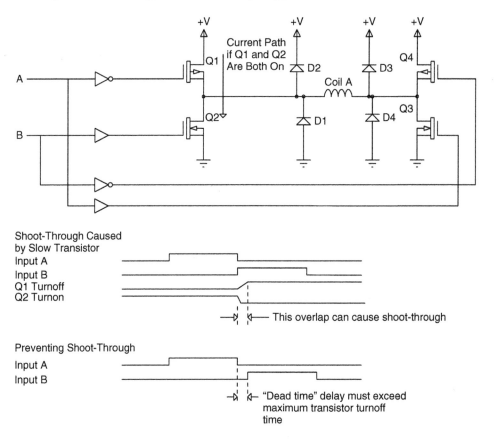

Figure 1-11: Shoot-Through.

The usual method to avoid shoot-through is to introduce a short delay between turning off one side of the H-bridge and turning on the other. The delay must be long enough to allow both transistors to turn off before the other pair turns on.

I saw a design once (Figure 1-12) that used optocouplers to provide isolation between the motor-control circuitry and the driving circuitry. The problem was that optocouplers have a wide variation in turn-on/turn-off times. In production, the only way to make the circuit work reliably was to hand-select optocouplers that had similar characteristics. If the operating temperature varies widely, it is possible that a circuit like this can fail in the field.

If you drive an H-bridge directly from the port outputs of a microcontroller, be sure to take power-up conditions into account. Until they are initialized, the port bits of most microcontrollers are floating. Depending on whether the H-bridge logic sees this condition as logical "1" or "0," it can turn on both sides of the bridge and cause shoot-through. Be sure everything comes up in a safe condition and add pull-ups to the port pins if necessary. If the H-bridge drive inputs cannot be guaranteed during power-up, use a power supply for the stepper motor that has the ability to be disabled with a shutdown input.

Keep the motor power off until everything on the control side is stable. It may be tempting to depend on the microprocessor getting out of reset and getting its port bits set to the right state before the motor power supply comes up to a high enough voltage to do any damage. This is a risky approach, as a faulty processor may never get the ports set up right. If you use an emulator for debugging, there may be a considerable delay between applying power and getting the ports set up correctly. And what happens if you turn the power on but you forget to plug the emulator in? You could destroy the entire

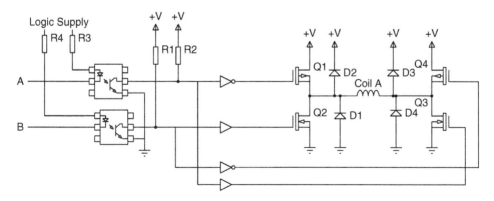

Figure 1-12: Shoot-Through Caused by Optoisolator Delay.

prototype setup. This can be a real problem if there is only one of them. The safest route is to ensure that the power-up state of the processor can't do any damage.

Shoot-through can also be caused by the driver transistors themselves. Figure 1-13 shows one half of an H-bridge driver constructed with MOSFET transistors. MOSFETs have a fairly high capacitance between the gate terminal and both of the other terminals (source and drain). In the figure, the capacitance is represented by the capacitance C, between the gate and drain of Q2. This capacitance is usually on the order of a few tens of picofarads for a typical MOSFET used in a motor application.

If transistor Q1 turns on to apply voltage to one side of the motor (the transistor opposite Q2, not shown, on the other side of the bridge would turn on as well), there will be a voltage spike at the junction of the drains of Q1 and Q2. This voltage spike will be coupled to the gate of Q2 by the capacitance C. If the impedance of the device driving the gate of Q2 is high enough, the voltage spike may be enough to turn on Q2 and cause shoot-through. Remember that the voltage on the motor may be 24 V, 36 V, or more, and the gate of Q2 may need only a few volts to turn on. So even if the signal is

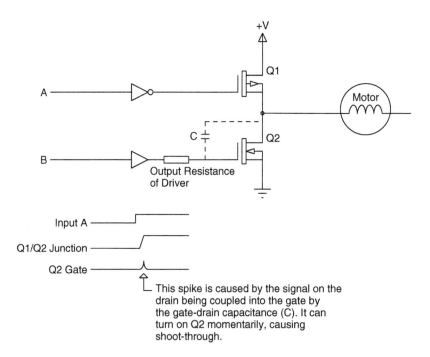

Figure 1-13: Shoot-Through Caused by MOSFET Capacitance.

significantly attenuated, it still may be able to turn on the MOSFET. This problem can be minimized by ensuring that the impedance of the driver is low; if a series resistor is used to limit current flow into the gate in case of transistor failure, make the value as small as possible. Minimize trace lengths between the MOSFET and the driver device.

1.1.7   Current Sensing

Many designs need to sense the current through the stepper motor coils. The usual method for doing this is to place a small-value precision resistor in series with the ground lead of the driver circuit (Figure 1-14). When the motor is turned on, the current through the winding must pass through the sense resistor to reach ground. This develops a voltage across the resistor that can be amplified and sensed with an opamp amplifier. The amplifier output can be connected to an ADC so it can be read by a microprocessor, or it can connect to one side of a comparator for digital detection of an overcurrent condition.

To avoid stealing excessive power from the motor winding, the sense resistor is usually small, on the order of 1Ω or less. Even a 1Ω resistor will take a watt in a motor drive circuit that uses one amp. This is a watt of power that is wasted as heat. Generally, you want to make the sense resistor as small as possible without making sensing difficult. International Rectifier makes a series of MOSFETs known as SENSEFETs with an extra pin that mirrors a fraction of the transistor current. This can be used for current sensing.

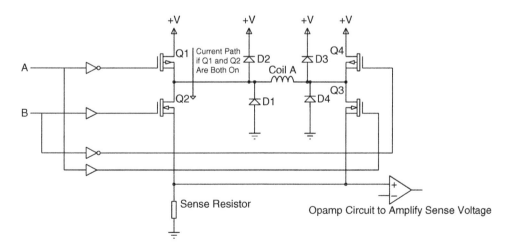

Figure 1-14: H-Bridge Current Sensing.

1.1.8 Motor Drive ICs

There are a number of ICs that can control and drive stepper motors. The L6201 from SGS-Thompson is a typical part. The L6201 can drive motors up to 5 A with supply voltages up to 48 V. The L6201 includes internal flyback protection diodes and includes a thermal shutdown that turns the motors off if the part overheats. The L6201 is available in DIP, SMT, and multiwatt packages.

The LMD18200 from National is another motor driver IC. This part includes a pin that provides a thermal warning when the device is about to overheat. Unlike the L6201, the LMD18200 does not require a sense resistor in the ground connection of the driver transistors. Instead, the LMD18200 has a separate pin that mirrors the current in the H-bridge. This pin (CURRENT SENSE OUTPUT in Figure 1-15) typically carries $377\,\mu A$ per amp of current in the bridge. If a motor winding draws 2 amps, and a 4.99 K resistor is connected from the current sense pin to ground, then the voltage developed across the resistor will be:

$$377 \times 10^{-6} \times 2 \times 4990 = 3.76\,V$$

The current sense output pin can be connected directly to an ADC or comparator input.

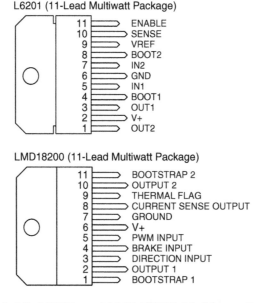

Figure 1-15: L6201 and LMD18200 Multiwatt Packages.

1.1.9 Chopper Control

Torque in a stepper motor is controlled by adjusting the current through the windings. Because the winding is an inductor, applying voltage to the coil doesn't cause the current to change instantly (Figure 1-16). As the current in the coil increases, torque increases. So, if we want to have a particular torque, it takes a while to get there once voltage is applied. However, as shown in Figure 1-16, if we operate at a higher voltage (V2 in the figure), we get to the original torque value much more quickly because the current increases along an exponential curve. The problem is that we end up with too much current in the winding because the current keeps climbing past the torque we wanted.

One way of generating torque faster is to use a higher drive voltage to get fast current buildup, but turn off the voltage to the coil when the current reaches the desired value. The chopper circuit in Figure 1-17 illustrates a way to do this. The voltage from the sense resistor (amplified if necessary) is applied to one input of a comparator. The other side of the comparator connects to a reference voltage that sets the drive current.

A chopper oscillator, typically operating from 20 kHz to 200 kHz (depending on the motor and driver characteristics) sets a flip-flop. The output of the flip-flop enables the H-bridge outputs. When the flip-flop output is low, the H-bridge is disabled, even if one of the control inputs is high.

When voltage is applied to the coil and the current builds to the desired level, the voltage across the sense resistor becomes greater than the comparator reference, and the comparator output goes low. This turns off the flip-flop and disables the H-bridge until the next oscillator pulse occurs. As long as the current is less than the desired level, the H-bridge will remain enabled.

The circuit shown in Figure 1-17 illustrates the concept. In practice, the comparator reference voltage could be fixed, or it could come from a microprocessor-controlled DAC. This would permit software control of the current and therefore the torque. This would allow a stepper motor to be used in an application with varying loads, as long as the microprocessor knows approximately what the load is. It could also be used to compensate for the torque variation between a single-coil and two-coil drive when half-stepping, or to generate the varying signals needed for microstepping.

The chopping frequency has to be high enough to be significantly greater than the maximum step rate of the motor, but low enough that the transistors can respond. If the

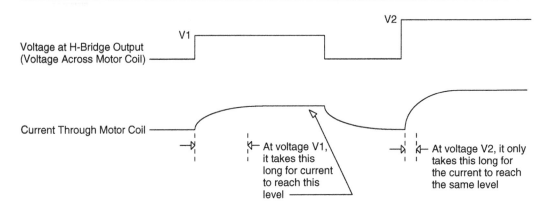

Figure 1-16: Coil Current as a Function of Supply Voltage.

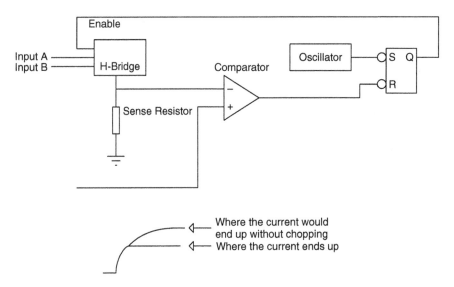

Figure 1-17: Chopper Control of Coil Current.

chopping frequency is too high, the drive transistors will spend too much time in the linear region (during the turn-on and turn-off times) and will dissipate significant power.

The chopper oscillator and comparator could be eliminated and this entire function could be performed in software. A regular interrupt at the chopping frequency would be used as a time base. Each time the interrupt occurred, the microprocessor would examine the sense resistor voltage (via an ADC) and either enable or disable the H-bridge. Of course, the processor must be able to service interrupts at the chopping frequency, which would limit that frequency in a practical design. Using a microprocessor just to chop a single motor would probably be overkill, but it might be cost-effective to use a single microprocessor to control several motors if all motors were chopped with the same clock.

1.1.10 Control Method and Resonance

Stepper motors driven with constant current drive (chopped or analog) are more likely to have resonance problems at low step rates. Using half-stepping or microstepping can usually overcome these problems. Of course, going from a simple on-off H-bridge to a DAC-controlled microstepping scheme is a large step in system complexity.

Steppers that are driven with constant voltage are more likely to have resonance problems at higher step rates. Half-stepping and microstepping will not solve these problems. However, a load with a significant damping effect (such as a high-friction load) reduces resonance effects overall. If your application calls for high step rates and a load that doesn't provide much damping, use constant current drive and half-stepping or microstepping to avoid low-frequency resonance problems. What is a high step rate? It depends on the motor, but will generally be in the range above 200 to 500 steps/sec.

1.1.11 Linear Drive

If you don't want to use chopping to get a constant current drive, you can use a circuit like that shown in Figure 1-18. In this circuit, a power opamp, capable of controlling the current required by the motor coils, drives the top of the coil. The voltage across the

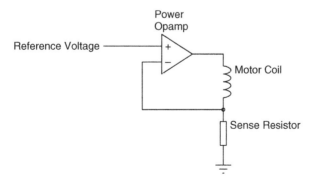

Figure 1-18: Linear Constant-Current Drive.

sense resistor (amplified if necessary) drives the inverting input of the opamp. The opamp will attempt to keep the motor current equal to the reference voltage.

A circuit like this is electrically quieter than the chopper, but it is much less efficient. The power opamp will dissipate considerable power because it will carry the same current as the motor coil and will usually have a significant voltage drop. The power dissipated by the opamp at any time is given by where V is the supply voltage, Vm is the motor coil voltage, and I is the coil current.

A linear drive like this requires a negative supply voltage. It is possible to build a bridge driver using two opamps that operates from a positive supply and works like the H-bridge, driving one side of the coil positive or negative with respect to the other.

The L297 (Figure 1-19) from SGS-Thompson is a stepper-controller IC. It provides the on-off drive signals to an H-bridge driver such as the L6201 or to a discrete transistor driver design. The L297 controls current in the motor windings using chopping. It has an internal oscillator, comparators, and chopping logic. The oscillator frequency can be set by using an external resistor/capacitor or an external clock. The chopping clock is also used to time turn-on and turn-off of the phases to prevent shoot-through.

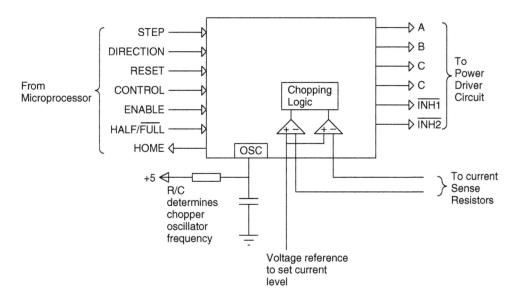

Figure 1-19: SGS-Thompson L297.

The L297 provides four phase outputs (ABCD) and two inhibit outputs for chopping (INH1, INH2). An open-collector HOME signal goes low when the L297 phase outputs are at the home position (ABCD = 0101). The L297 can control a stepper in half or full steps.

1.2 DC Motors

Figure 1-20 shows a cross-section of a DC motor, sometimes referred to as a permanent magnet DC (PMDC) motor. A DC motor consists of a permanent magnet stator and a wound rotor. Connection to the rotor windings is made with brushes, which make contact with a commutator that is affixed to but insulated from the shaft. When power is applied, the rotor rotates to align its magnetic field with the stator. Just as the field is aligned, the commutator sections that had been in contact with the brushes break contact and the adjacent commutator sections make contact. This causes the polarity of the windings to reverse. The rotor then tries to align its new magnetic field with the stator. The rotor rotates because the brushes keep changing the winding polarity. The example shown in Figure 1-20 has four rotor arms, four brushes, and four commutator contacts. Some high-performance DC motors do not use wound rotors, but instead print the rotor winding as traces on a printed circuit. This provides a very low-inertia motor, capable of high acceleration.

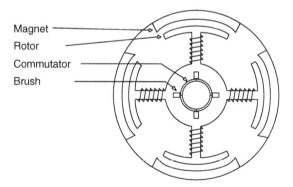

Magnet
Rotor
Commutator
Brush

Figure 1-20: Cross-Section of PMDC Motor.

DC motors do not lose synchronization as stepper motors do. If the load increases, the motor speed decreases until the motor eventually stalls and stops turning. DC motors are typically used in embedded systems with position encoders that tell the microprocessor what the motor position is. Encoders will be covered in detail later in this chapter.

A DC motor is typically driven with an H-bridge, like a bipolar stepper. However, a DC motor requires only one bridge circuit, because there are only two connections to the motor windings. DC motors will typically operate at higher speeds than equivalent stepper motors.

1.2.1 Driving DC Motors

Like steppers, DC motors can be driven with an on-off chopped H-bridge or by an analog driver such as a power opamp. However, where a stepper motor typically uses an analog drive or chopped PWM signal to control motor current, the DC motor driver design does not usually depend on current control. Instead, the DC motor controller provides sufficient current to meet a particular acceleration curve (as measured by the encoder feedback). If the motor has a larger-than-normal load, then the driver circuit will increase the current to force the motor to the correct speed. In other words, the DC motor controller increases or decreases the current to maintain a particular speed. Speed is monitored, not motor current. DC motor control circuits do sometimes sense current in the H-bridge, but it is usually to detect an overcurrent condition, such as occurs when the motor stalls.

Figure 1-21 shows a typical DC motor operation with two different loads. The motor accelerates to a constant speed, runs for a certain time, then decelerates back to a stop.

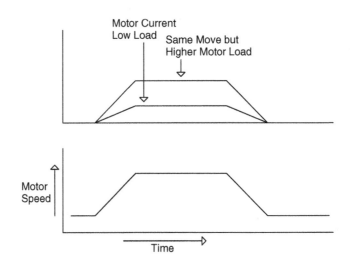

Figure 1-21: DC Motor Operation with Different Loads.

With light loading, the motor current profile is lower than with higher loading. However, the controller applies sufficient current to the motor to produce the required speed/time curve regardless of motor load. For this reason, DC motors are usually better for applications with large load variations.

One feature of DC motors is the ability to brake them. If you manually turn the shaft of a DC motor, you get a small generator. If you short the terminals of a DC motor, it becomes difficult to turn the shaft because of the electromotive force (EMF) the motor generates when it turns. If you short the motor terminals while the motor is running, it quickly comes to a halt. This is called *dynamic braking*.

Figure 1-22 shows the H-bridge we've looked at before, but with a modification. Here, we have separated the motor control inputs so we can turn each transistor on and off separately. If we take inputs "A" and "D" high at the same time, transistors Q1 and Q3 both turn on and the motor turns in one direction. If "B" and "C" are both high, the other pair turns on and the motor turns in the opposite direction.

Now, suppose the motor is turning and inputs "B" and "D" go low, then inputs "A" and "C" are both driven high. This turns on transistors Q1 and Q4. One side of the motor will be more positive than the other; let's say it is the left side for this example. Current will flow from the positive supply, through Q4, through the motor winding, through D2, and back to the positive supply. The motor is effectively shorted out by Q4 and D2. This will stop the motor quickly. If the right side of the motor is the positive one, the current will flow through Q1 and D3. If we drive inputs "B" and "D" high instead of "A" and "C," we get the same effect, with the current flowing through Q3/D1 or Q2/D4.

Many motor H-bridge ICs include braking capability. These include the L6201 and LMD18200. The L6201 has two inputs to control the two halves of the bridge circuit. If both inputs are brought to the same level (high or low), the driver will brake the motor. The LMD18200 has a separate input signal for braking.

Braking can be used to stop a motor quickly, or to hold it in position. One limitation on dynamic braking as a holding force is that there will be no braking until the EMF generated by the motor exceeds the forward drop of the diode in the braking circuit.

There are ICs that provide a motor drive subsystem for DC motors; we will examine this subject after covering brushless DC motors and encoders.

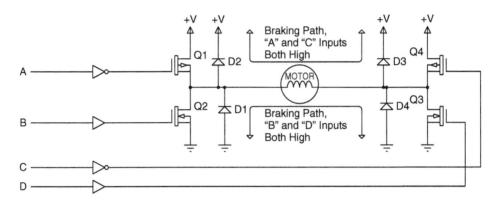

Figure 1-22: DC Motor Braking.

1.3   Brushless DC Motors

Figure 1-23 shows a cross-section of a brushless DC motor. This looks very much like a stepper motor, and in fact a brushless DC motor works much the same way. The stator in this motor consists of three coils (A1/A2, B1/B2, and C1/C2). The coils are connected in a three-phase arrangement, with a common center point. A brushless DC motor is more efficient than a brushed DC motor of the same size. This is because the coils in a brushless DC motor are attached to the case (instead of to the rotor), so it is easier to get the heat generated in the windings out of the motor.

A brushless DC motor functions essentially as a DC motor, but without the brushes. Instead of mechanical commutation, the brushless DC motor requires that the drive electronics provide commutation. A brushless DC motor can be driven with a sine signal, but is more typically driven with a switched DC signal. Figure 1-24 illustrates both drive waveforms. For sinusoidal drive, the current can be controlled with a chopper circuit, or a linear drive can be used. Because the coil positions are 120° apart, the sinusoidal drive waveforms for the coils are 120° apart. The sum of the currents in the three coils is 0. For the switched DC waveform, there are always two phases on (one high, one low), and the third phase is floating (off).

Note that if you use a sinusoidal drive, the driver does not need a negative supply; the sinusoid can swing between ground and a positive voltage (or for that matter, between two different positive voltages). If the drive goes from 0 V to 5 V, when all three coils are at the same voltage there is no current flowing. So the midpoint between the two drive voltages (in this case, 2.5 V) can be picked as a "virtual ground."

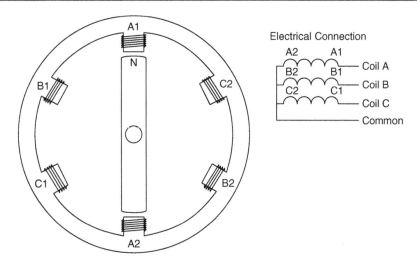

Figure 1-23: Brushless DC Motor.

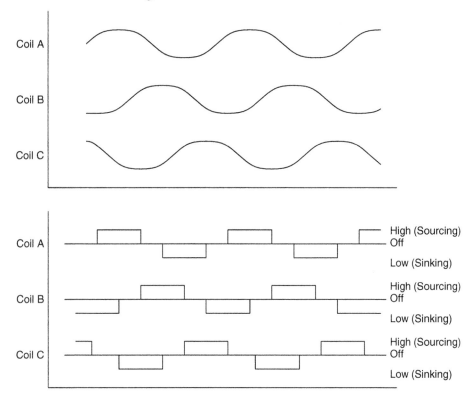

Figure 1-24: Brushless DC Motor Waveform.

For digital drive, the driver circuitry for a brushless DC motor is simpler than for a stepper or brushed DC motor. Because each phase is either high, low, or off (high impedance), an H-bridge is not needed. Instead, the driver circuitry can just be a totem pole output. Figure 1-25 illustrates how two MOSFETs can be used to drive a brushless DC motor. The inputs to this circuit could come from a controller IC or a microprocessor. Note that flyback protection diodes are needed in this circuit.

A brushless DC motor usually has at least one Hall-effect sensor (and more typically three) to indicate position. However, it is possible to drive a brushless DC motor without any sensors. If you look at the digital drive waveforms in Figure 1-24, you will see that there are always two phases that are on (either positive or negative drive) and one that is off. The moving rotor will generate a voltage in the coil that is not driven. This voltage will cross zero once during the OFF period, and can be sensed to indicate the rotor

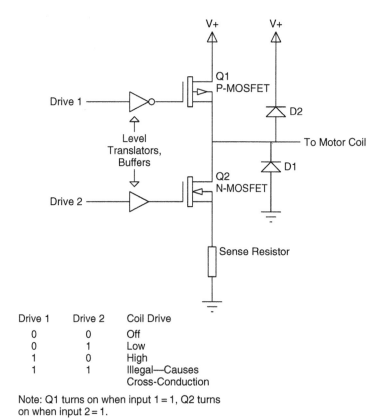

Drive 1	Drive 2	Coil Drive
0	0	Off
0	1	Low
1	0	High
1	1	Illegal—Causes Cross-Conduction

Note: Q1 turns on when input 1 = 1, Q2 turns on when input 2 = 1.

Figure 1-25: Brushless DC Motor Drive.

position. Note that the voltage being measured is the voltage across the unused coil—in other words, the difference between the coil connection and the common connection point for all the coils.

Figure 1-26 shows a sensorless drive configuration for a brushless DC motor. This circuit brings the common connection point of the three motor coils back to the ADC circuitry as a reference. This is not always necessary; however, this technique can reduce the noise in the measurement. If the common point cannot be used as a reference, it could be connected to a fourth ADC channel and the value subtracted from the sensed coil in software. If the common point isn't brought out of the motor, you can calculate its value in software if the microprocessor is powerful enough. If the processor isn't powerful enough to perform the calculation in real time, you can calculate the values and put them in a lookup table.

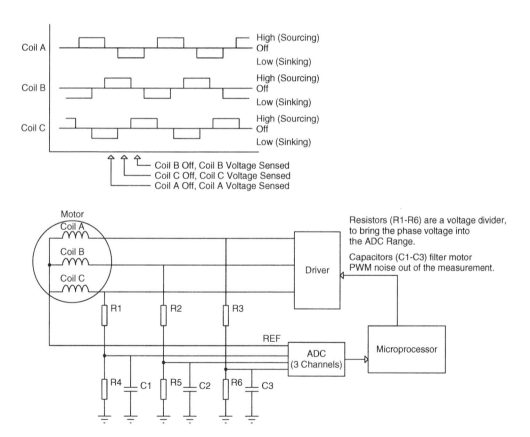

Figure 1-26: Sensorless Brushless DC Motor Drive.

When using the sensorless technique with a microprocessor, you will find that there are noise spikes on the sensed coil when the transistors switch on and off. You can filter this out with capacitors on the sense line, as shown in Figure 1-26, or you can just ignore the samples from the sensed winding during this interval.

There are a number of brushless DC motor drivers that can take advantage of sensorless, EMF-based position sensing. The Philips TDA5140 will drive motors up to about 8A and can use either a sensor or sensorless driving.

1.3.1 Encoders

PMDC and brushless DC motors are usually used in embedded systems with an encoder attached to the shaft. This provides feedback to the microprocessor as to motor position. A typical encoder is shown in Figure 1-27. In this scheme, four magnets are placed around the shaft of the motor and a Hall-effect sensor is placed on the case. The Hall-effect sensor will produce four pulses per revolution of the motor shaft.

Four pulses per rotation of the motor shaft is sufficient to regulate motor speed for a low-resolution application such as a cooling fan. If the motor is geared, so that it takes many revolutions of the motor shaft to produce one revolution of the (geared) output shaft, then this type of encoder is also suitable for more precise applications. However, for cases where you need accurate information about the position of the motor shaft within a single rotation, an optical encoder is normally used.

Figure 1-28 shows a simple optical encoder. A glass disk is printed with opaque marks, 16 in this example. The glass disk is attached to the motor shaft and a slotted optical switch straddles the edges of the disk. Every time an opaque spot passes through the slotted switch, the phototransistor turns off and a pulse is generated. This encoder will produce 16 pulses for every rotation of the motor shaft. The controller can count pulses to determine the angle of the motor shaft and the number of revolutions.

This simple encoder has one major drawback, common to the simple Hall-effect encoder—how do you tell which way the motor is turning? Figure 1-29 shows a practical encoder arrangement that provides direction information. This encoder still uses a glass disk with opaque stripes, but now there are two slotted switches, located next to each other. The opaque stripe is wider than the distance between the switches. As the opaque stripe moves under switch A, the output (channel A) goes high. As the opaque stripe moves under switch B, that output (channel B) goes high. As the motor shaft continues to rotate, the stripe clears switch A and its output goes low, followed by switch B.

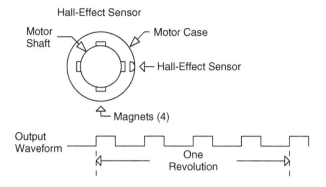

Figure 1-27: Hall-Effect Motor Shaft Encoder.

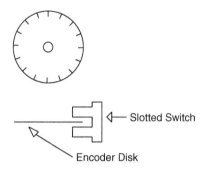

Figure 1-28: Simple Motor Encoder Glass Disk.

If the motor reverses direction, switch B is covered first, followed by switch A. So this two-channel encoder (called a *quadrature* encoder) provides information on position, speed, and direction. Typical encoders of this type produce between 50 and 1000 pulses per revolution of the motor shaft.

Encoders are also available with an index output, which uses a third encoder and a single opaque stripe closer to the center of the disk. As shown in Figure 1-29, there is a single index stripe, so only one pulse is produced per revolution of the shaft. This allows the system to know the absolute starting position of the motor shaft, for cases in which this is important.

Figure 1-30 shows the pattern for a section of an absolute encoder. The absolute encoder encodes the opaque stripes in a binary fashion so that the absolute position is always known. Of course, this requires as many slotted switches and stripe rings as there are bits of resolution. The figure shows the outer four rings; an encoder with 6 rings would require 6 switches and would divide one revolution into 64 unique codes. An encoder that provides 1024 unique positions would require 10 switches and 10 concentric rings on the encoder disk. Absolute position encoders are extremely expensive. Their primary use is in systems where the position of the motor shaft needs to be known at power-up.

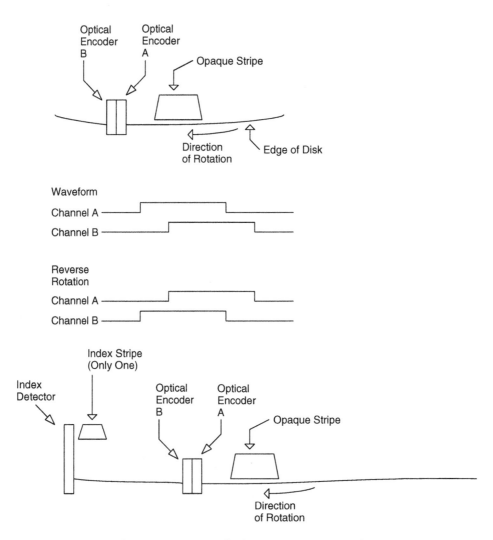

Figure 1-29: Practical Quadrature Encoder.

1.3.2 DC Motor Controller ICs

There are ICs that are designed for the control of DC motors. The LM628/LM629 from National Semiconductor are typical devices. Figure 1-31 shows how these two devices would work in a system. The LM628 has an 8-bit or 12-bit output word (selectable) for driving the motor through an analog interface using a DAC. The LM629 has PWM outputs for driving a motor, using PWM, through an H-bridge. Both parts use a similar microprocessor interface. There is an 8-bit data bus, READ and WRITE signals, a chip

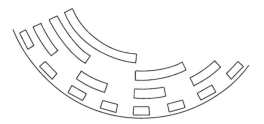

Figure 1-30: Absolute Position Shaft Encoder.

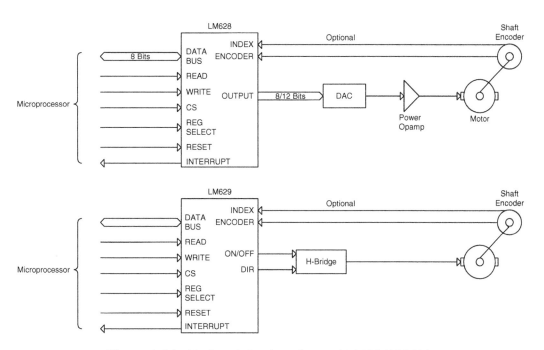

Figure 1-31: National Semiconductor LM628/LM629.

select, a reset, and a register select signal. The LM628/9 also provides an interrupt output to the microprocessor. The motor interface includes the output (PWM or DAC) and an input for a two-channel quadrature encoder. There is also an input for an index pulse from the encoder if the encoder provides one; this input is optional and need not be used.

When connected to a DAC and power opamp (LM628) or an H-bridge driver (LM629), the LM628/9 provides a complete motor control subsystem. The microprocessor issues a series of commands such as "move to position *x* with acceleration *y*," and the LM628/9 will execute a trapezoidal move, accelerating the motor to a particular speed, holding that speed, then decelerating the motor to a stop at the right position. (The "position" is a count of encoder pulses, maintained in a 32-bit register.)

The LM628/9 uses two addresses. One address is a command address and the other is for data. A command sequence starts with an 8-bit command opcode, written to the command register by the microprocessor. This is followed by anywhere from 0 to 14 bytes of data, either read from or written to the data register. The commands for the LM628/9 are as follows:

Table 1-1

Command	Opcode	Data following
Reset	00	None
Select 8-bit DAC output	05	None
Select 12-bit DAC output	06	None
Define home	02	None
Set index position	03	None
Interrupt on error	1B	2 bytes, written
Stop on error	1A	2 bytes, written
Set breakpoint, absolute	20	4 bytes, written
Set breakpoint, relative	21	4 bytes, written
Mask interrupts	1C	2 bytes, written
Reset interrupts	1D	2 bytes, written
Load filter parameters	1E	2 to 10 bytes, written
Update filter	04	None
Load trajectory	1F	2 to 14 bytes, written
Start motion	01	None
Read signals register	0C	2 bytes, read
Read index position	09	4 bytes, read
Read desired position	08	4 bytes, read
Read real position	0A	4 bytes, read
Read desired velocity	07	4 bytes, read
Read real velocity	0B	2 bytes, read
Read integration sum	0D	2 bytes, read

The LM628/9 index input is intended for use with an encoder that provides an index output. The LM628/9 can capture the encoder position count and store it in a separate register when the index pulse occurs. However, the index input does not have to be connected to an encoder output. I have used the LM628/9 index input to indicate other conditions. For instance, in one system we had a rotating carousel that was connected to the motor shaft via a gearbox. It took many revolutions of the motor to produce one revolution of the carousel. We did not need to know when the motor shaft reached a specific position, but we did need to know when the carousel reached its home position. So the sensor (slotted switch) that indicated when the carousel was at home was connected to the index input.

One caution if you use this technique: the LM628/9 responds to the index input when both the encoder channels are low, so the sensor output has to be low while both encoder channels are low. To avoid multiple index capture events from a single sensor input signal, be sure the index input to the LM628/9 occurs for only one encoder cycle, regardless of how long the actual sensor input lasts. In the actual application, a small CPLD handled the index inputs for multiple LM629s. Figure 1-32 shows how the timing worked.

The interrupt output can be asserted for any combination of various conditions, including a breakpoint, index pulse, wraparound, position error, or command error. The software determines which conditions generate an interrupt, by setting a mask byte in the LM628/9. The interrupt output is level sensitive and true when high. When using the LM628/9 motor controller, there are some software considerations:

- The position registers in the device have a limited size: 32 bits for the LM628/9. This means that if enough forward movements are made, or if the motor continuously rotates, the registers will eventually overflow. The software must take this condition into account. This is especially true if the software uses, say, 64-bit math. It would be possible, in software, to add an offset to a current position and get an answer that is greater than 32 bits: for example, C017B390 (hex) plus 40000000 (hex) results in a result larger than 32 bits and cannot be stored in the LM628/9 registers.

- When using the index input, the LM629 will capture the count. This becomes, in effect, the "zero" or "home" position of the motor, and all moves are relative to that position. However, the 32-bit position counter is not reset by the index. So the software must offset moves from the index position.

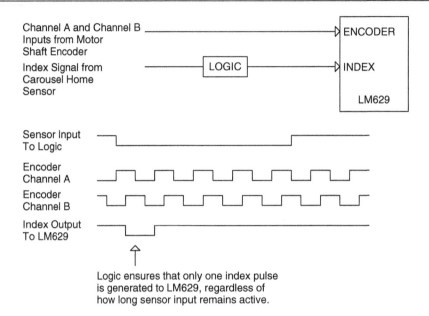

Figure 1-32: LM628/LM629 Index Timing.

- The fact that the LM628/9 uses two addresses (command and data) means that there is the potential for a race condition. If an interrupt occurs in the middle of a command sequence and the ISR also communicates with the LM628/629, the original command will be corrupted. An example would be an interrupt that notifies the processor that the index pulse has occurred. If the ISR reads the index position, and the interrupt happens in the middle of another command, the non-ISR code will get garbage data. Figure 1-33 illustrates this. To avoid this condition, the software should disable interrupts around non-ISR code (or interruptible ISR code) that accesses the LM628/9.

These restrictions are typical and are not unique to the LM628/9. There are other motor controller ICs available, and all have their quirks. The MC2300 series from Precision Motion Devices (PMD) is a two-chip set that can control up to four brushless DC motors. These parts can control two-phase or three-phase brushless motors and can provide several motion profiles. The MC2300 can provide a digital word for a DAC/amplifier driver, or PWM outputs for an H-bridge.

The MC2100 series, also from PMD, is a two-chip set for brushed DC motors. Like the MC2300, the MC2100 parts support one to four motors, have 32-bit position registers, and support multiple types of motion profiles. Both of the PMD devices are based on a fast DSP that performs the actual motor manipulation.

The Agilent HCTL-1100 is a single-motor controller with a 24-bit encoder counter and PWM or 8-bit digital outputs. The HCTL-1100 does not use an address- and data-register scheme, but instead multiplexes the address signals with 6 of the 8 data lines.

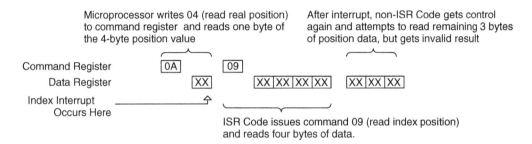

Figure 1-33: LM628/LM629 Interrupt Timing.

1.3.3 Software Controllers

In some cases, a DC motor might be directly controlled by a microcontroller, using software, instead of using an off-the-shelf controller such as the LM628. Reasons for this include the following:

- **Cost**: An off-the-shelf controller must be coupled with a microprocessor anyway, so why not do away with the controller and just use the processor?

- **Simplicity**: In an off-the-shelf controller, you pay for all the generalized functionality that the part provides. If you need only slow speeds, simple controls, and limited features, you may be able to implement them in software.

- **Flexibility**: You can design the control algorithms to your requirements, instead of just modifying PID parameters. You can also make very deep position registers, 64 or 128 bits for specialized applications.

- **Custom design**: If your system has special requirements, such as special sensors or a move-to-stop-and-apply-pressure for *x* milliseconds, you can implement this because you will develop and control the algorithms.

If you decide to roll your own controller, there are a few things to consider. The processor has to be fast enough to keep up with whatever processing demands are required. This means also servicing encoder interrupts in a timely fashion. In a software-based controller, the encoder on a DC motor typically connects to one or more interrupt inputs. Figure 1-34 illustrates this. One method of handling interrupts is to let one channel ("A" in the figure) generate an edge-sensitive interrupt to the microcontroller. When the interrupt occurs, the microcontroller reads the state of the other encoder channel ("B" in the figure). If channel B is low, motor motion is forward, and if "B" is high, motion is reversed. For forward motion, the software-maintained position register would be incremented, and for backward motion the register would be decremented.

As shown in Figure 1-34, if there is enough latency between the rising edge of channel "A" and the state of the ISR, channel B may have changed states and the wrong result will be calculated by the firmware. If you implement a motor controller with a system like this, be sure that your interrupt latency never allows this condition to occur, even at maximum motor speed.

It is a good idea to make the interrupt a timer input if one is available. The timer can be set one count before rollover, and the encoder input will cause the timer to roll over and generate an interrupt. If an interrupt is missed, the timer count will be 0001 instead of 0000 (for a timer that increments starting from FFFF) and the missed interrupt can be detected. The system as shown in Figure 1-34 will have only 1/4 the resolution of a typical system using a motor controller IC, because it captures new position information on only one encoder edge (rising edge of "A") instead of on all four edges. You could compensate for this by using an encoder with more lines, but that could cost as much as a motor controller IC. You can double the resolution of this circuit by connecting both encoder channels to interrupts on the microcontroller. Most microcontrollers permit you to read the state of an interrupt input as if it were a port pin. When an interrupt occurs, the software reads the state of the other input to determine motor direction.

Finally, to get the same resolution as a motor controller IC, you could add an external PLD that generates interrupts on any input transition. This would also let you filter the signals to eliminate spurious edges if necessary.

Another way to get higher resolution in a microprocessor-based controller is to use a microcontroller that can generate interrupts on either clock edge. The Microchip PIC16C series has an interrupt-on-change feature that can generate an interrupt when selected pins change state.

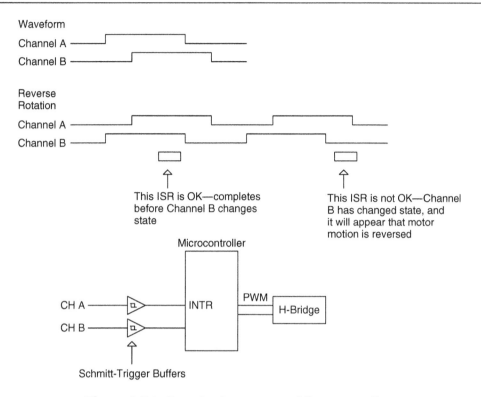

Figure 1-34: Encoder Interrupt to Microcontroller.

1.4 Tradeoffs Between Motors

The tradeoffs between DC motors, brushless DC motors, and steppers are as follows:

- Stepper motors require no encoder and no feedback system to determine motor position. The position of the shaft is determined by the controller, which produces step pulses to the motor. However, this can also be a disadvantage. If the load is too high, the stepper may stall and there is no feedback to report that condition to the controller. A system using a DC motor with an encoder can tell when this condition occurs.

- Steppers have no brushes, so they produce less EMI.

- A stepper can produce full torque at standstill, if the windings are energized. This can provide the ability to hold the rotor in a specific position.

- A stepper can produce very low rotation speed with no loss of torque. A DC motor loses torque at very low speeds because of lower current.

- DC motors deliver more torque at higher speeds than equivalent steppers.

- Because there is no feedback, a stepper-based system has no means to compensate for mechanical backlash.

- Brushless DC motors require electronic commutation, which is more complex than the drive required for brushed DC motors. However, the availability of driver ICs for brushless DC motors makes this less of a concern.

Without feedback, there is no way to know if a stepper is really doing what it is told to do. Consequently, stepper motors are typically used where the load is constant or at least is always known. An example would be stepping the read/write head mechanism in a floppy disk drive. The load in this application is fairly constant. If the load varies greatly during operation, a stepper may stall out or it may overshoot the desired position when trying to stop.

If the load varies but is known, a stepper may be useable by reducing the drive current when the load is low and increasing the current when the drive is high. An example of a known load would be a system that has to move something, but sometimes just has to position the motor into the correct position when there is no load. On the other hand, if the "something" that is being moved varies greatly in mass, friction, and so on, then the load isn't really known and a stepper may not be the best choice. When the load varies a lot, and especially if the load isn't known to the controller, a DC motor with an encoder is usually a better choice than a stepper. The encoder allows the controller to increase the current if the speed and/or position are not correct.

One way to achieve the benefits of the stepper and the encoder/feedback DC motor is to add an encoder to a stepper. This provides most of the advantages of both systems, but at higher cost. The maximum speed of such a system will still be slower than an equivalent DC motor, however.

1.4.1   Power-Up Issues

One problem with DC motors is what happens when power is applied. We've already looked at the issues surrounding the power-up state of microcontroller outputs. There are similar issues surrounding any DC motor design, including designs that use packaged controllers.

Typically, the logic that controls the motor H-bridge or analog amplifier operates from 5 V or 3.3 V. The motor power supply may be 12 V, 24 V, or even 50 V. If the motor power supply comes up first, the inputs to the H-bridge or amplifier may be in an invalid state and the motor may jerk momentarily. In a system with a limited range of motion, such as a robotic arm, the motor may slam up against whatever limits the travel. This can be hard on the mechanical components and gears connected to the motor shaft. A DC motor can apply considerable torque in this condition—it is equivalent to a 100% PWM duty cycle.

The best way to eliminate this problem is to ensure that the motor power supply comes on after the logic supply is on and everything is stable. Some multiple-output power supplies have an inhibit input for the high-voltage output that can be used for this purpose. But how do you control the inhibit signal if the power supplies come up together? The logic supply is not available to power the logic that inhibits the motor supply. Some supplies have a low voltage (5 V or 12 V) output that comes up before all the other supplies and is intended for precisely this purpose. This auxiliary output is usually designed to supply minimal current (<100 ma). In some cases, you can just connect the inhibit input on the motor supply to a pull-up resistor from the auxiliary supply (to inhibit the motor supply) and then pull the inhibit input to ground when the logic electronics is stable. Figure 1-35 illustrates a one-transistor approach to this.

If the motor power supply cannot be controlled in this way, it may be necessary to inhibit the H-bridge in some manner, possibly by using a gate between the PWM output of the controller and the PWM input to the H-bridge. Of course, the gate logic has to operate from the motor supply or another supply that is stable when the motor voltage is.

Figure 1-36 shows a method I used in such a situation. The system used a National LMD18200 H-bridge. The LMD18200 has a brake input that is normally used for braking the motor. In this application we weren't using braking, so the brake input pin was available. When the 24 V motor supply is turned on and the 5 V supply is not yet on, the MOSFET is turned off (because the gate is low). A resistor pulls the MOSFET drain up to +24 V, but the voltage is clamped to 4.7 V by a zener diode. This voltage is recognized as a logic HIGH by the LMD18200, which brakes the motor and prevents motion. Some time after +5 V comes up (delay determined by R/C values at gate of MOSFET), the MOSFET gate goes high, the MOSFET turns off, and the motors can operate normally.

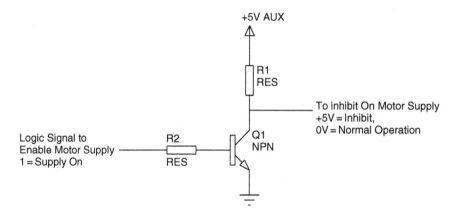

Figure 1-35: Motor Inhibit Using Auxiliary Power Supply and Power Supply Inhibit.

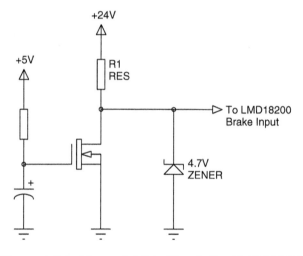

Figure 1-36: Motor Inhibit Circuit for LMD18200.

1.5 *Motor Torque*

How do you know if the motor you have chosen is powerful enough for the application? How do you know if you've picked a motor that is too big, adding unnecessary cost to the system? Motors are specified with a particular torque, the amount of force they can exert. The Pacific Scientific 4 N series of brushed DC motors is specified with torque ranging from 55 to 163 oz-in (0.39–1.15 N-m), depending on the model. These values are at some specific rated current—6.8 to 14.1 amps in this case. There is also a maximum current that the motor can withstand momentarily. The torque determines how much force the motor can exert and therefore how fast it can accelerate a load to a given speed.

1.5.1 Stall Torque

The stall torque is the torque that the motor will generate if the rotor is locked so that it can't turn.

1.5.2 Back EMF

When you spin a coil of wire in a magnetic field, you generate electricity—this is how the generator in a car works. A DC motor is a coil of wire spinning in a magnetic field. When operating the motor generates a DC voltage, a back EMF, that voltage opposes the voltage applied to make it move. The faster the motor spins, the more back EMF is generated.

1.5.3 Torque versus Speed

The torque of a DC motor falls off with speed. This is due to several factors, including the back EMF. This limits the maximum speed of a DC motor in a practical application and the maximum torque it can generate at a given speed.

1.5.4 A Real-World Stepper Application

A final example will serve to illustrate certain real-time concepts and bring together some of the concepts described in this chapter. Figure 1-37 shows a microcontroller controlling a five-phase stepper motor. This circuit is a simplified diagram of an actual application that I designed. In this circuit, the microcontroller directly controls the high- and low-side driver transistors for the five stepper phases. This motor controlled an agitator that mixes the contents of bottles for a medical application. The gate drive logic allowed the microcontroller to turn either transistor in the pair on or off, allowing the phase to be driven high, driven low, or allowed to float.

The PWM output is wrapped back around to another timer, which generates an interrupt. This causes an interrupt every T states, where T is the value in the second timer. This interrupt rate is the step rate of the motor; larger values of T result in a slower step rate and higher values of T result in a faster step rate. By clocking the step rate timer from the PWM timer, the circuit ensures that changes to the output state occur while the drive transistors are turned off.

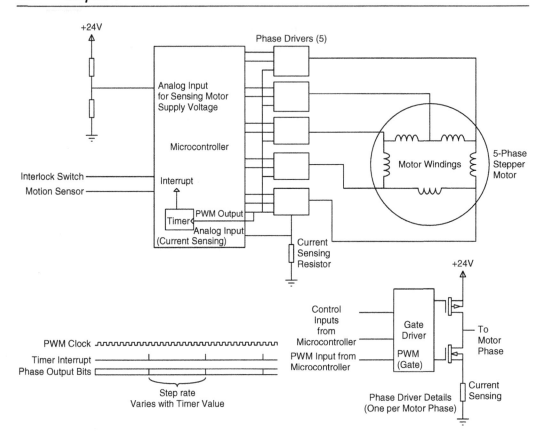

Figure 1-37: Five-Phase Stepper Motor System.

This circuit has several requirements:

- When the door is closed, the microcontroller will ramp the stepper up to a predetermined speed. If the user opens the door, the interlock switch opens and the stepper ramps down.

- Motor current is controlled by the duty cycle of the PWM output, which chops the current in the low-side MOSFETs. Motor current is increased as speed increases to ensure that sufficient torque is maintained.

- When the motor stops, it must stop in a specific position to allow the operator to add and remove bottles.

- The microcontroller has to monitor motor current, shutting down the motor and generating a fault output if excessive current is drawn. An internal ADC is used for this purpose.

- The position sensor generates a pulse once per revolution; the microcontroller has to count steps from the position sensor pulse to the stopping position. A fault output is generated if a full revolution is made without a pulse from the position sensor. The motor continues to operate in that case, but the stopping position will be undefined and the external system displays a message for the operator.

- The motor has to be ramped up when starting because the stepper motor will stall if the full step rate is applied while the motor is standing still.

- The motor must not start until the motor supply voltage is present. If the motor supply voltage fails, the motor must stop until the voltage is restored. This is handled by providing the motor supply voltage to one of the microcontroller analog inputs via a voltage divider. The microcontroller will not start motor operation until the motor voltage is present.

- Finally, another timer generates a timeout every five milliseconds for debouncing the interlock. This timer does not generate an interrupt, but is polled by the main loop. This is to ensure that the step rate interrupt is serviced immediately, while the PWM output still has the transistors off. If the debounce timer were to generate an interrupt, the code would sometimes be executing the debounce interrupt routine when the step rate interrupt occurred, and this would delay servicing the step rate interrupt.

The system state is based on what the motor is doing:

- Stopped

- Ramping up to speed

- Ramping down to stop

- Running constant speed

- Seeking stop position

- Overcurrent (fault condition, requires power cycle to clear)

- Bad voltage (motor drive voltage not present)

1.5.5 Firmware

The firmware executes a continuous loop, servicing various functions as needed. A simplified description of the various routines follows. The logic for the main loop looks like this:

If interrupt flag is set, update speed and current.

If 5-millisecond timer times out, debounce interlocks.

If position sensor pulse occurs, reset position count.

Update motor state.

If PWM active, start ADC conversion.

If ADC conversion complete, process motor current.

If position count passes one revolution, set position fault.

Speed and current update logic is as follows:

If motor Ramping Up,

    Decrease speed timer value (increases speed).

    Increase PWM duty cycle (increases motor current).

    If speed timer value = terminal value, change motor mode to Running Constant Speed.

If motor Ramping Down,

    Increase speed timer value (decreases speed).

    Decrease PWM duty cycle (decreases motor current).

    If speed timer value = minimum, change motor mode to Stopped and turn off PWM output.

Motor state update logic is:

If motor drive voltage not present and motor mode not Overcurrent, set motor mode to Bad Voltage.

If motor drive voltage present and motor state = Bad Voltage, set motor mode to Stopped.

If door open and motor Running Constant Speed and no position sensor fault, change motor state to Seeking Stop Position.

If door open and motor Running Constant Speed and position sensor fault, change motor state to Ramping Down.

If door open and motor Seeking Stop Position and position count is at rampdown point, change motor state to Ramping Down.

If door closed and motor stopped, start PWM and change motor state to Ramping Up.

Switch debounce logic is:

If interlock indicates door closed and if door state indicates door open, increment debounce counter 1 and clear debounce counter 2.

If debounce counter 1 = debounce value, change door state to closed.

If interlock indicates door open and if door state indicates door closed, increment debounce counter 2 and clear debounce counter 1.

If debounce counter 2 = debounce value, change door state to open.

Motor current uses the following logic:

If motor current exceeds overcurrent threshold, stop PWM, set motor to Overcurrent, set fault output.

Interrupt routine (step rate interrupt) logic is as follows:

Generate the next step in the sequence to the drive transistors (uses a lookup table).

Set a flag to indicate that the interrupt occurred.

Increment the position count.

The interrupt routine uses a lookup table. The table contains bits for each high-side and low-side transistor; a 1 turns the transistor on and a 0 turns the transistor off. Each entry in the table is the value needed to advance the motor to the next step in rotation. The key real-time requirement for this design is fast, repeatable update of the motor phases. To accomplish this, some tradeoffs were made, such as making the switch debounce timer polled instead of letting it generate an interrupt.

Debugging outputs from the microcontroller included a port bit that was set at the start of the main routine and cleared at the end. Another port bit was set at the entry to the ISR and cleared just before exiting the ISR. Because most port bits were used in the design, these two bits were shared with bits used for in-circuit programming.

CHAPTER 2

Testing

Arnold S. Berger

Test is the last step in traditional software development. We gather requirements, do high level design, detailed design, create code, do some unit testing, then integrate and start—finally— final test.

Since most projects run late, what do you think gets cut? Test, of course. The implication is that we deliver bug-ridden products that infuriate our customers and drive them to competitive products.

Best practice development includes code inspections. Yet inspections typically find only 70% of a system's bugs, so a fabulous test regime is absolutely essential. Test is like a double-entry bookkeeping system that insures mistakes don't leak into the deployed product.

In every other kind of engineering testing is considered fundamental. In the USA, every Federally funded bridge must undergo extensive wind tunnel tests, for instance. Mechanical engineers subject spacecraft to an almost bizarre series of evaluations. It's quite a sight to see a 15-foot-high prototype being nearly torn to pieces on a shaker, which vibrates at a rate that puts a thousand-Hertz tone into the air. The bridge prototype, as well as that of the shaken spacecraft, are discarded at great expense, but in both cases that cost is recognized as a key ingredient of proper engineering practices.

Yet in the software world test is the ugly stepchild. No one likes to do it. Time spent writing tests feels wasted, despite the fact that test is a critical part of all engineering disciplines. The Agilent community has thankfully embraced test as a core part of their processes, and they advocate creating tests synchronously with writing the code, realizing that leaving such a critical step till the end of the project is folly.

In this chapter Arnie Berger writes extensively about testing strategies, focusing on the peculiar issues that arise from embedded systems. He mentions a short program with only 5 decisions that leads to 10^{14} different execution paths—think of the difficulty of creating proper tests for that! So Arnie also addresses the critical question of when to stop testing. For shipping is ultimately the most important part of the project.

He talks extensively about code coverage tests. In the USA, safety-critical avionics must conform to DO-178B level A (for the most critical components), which mandates that every statement, branch and decision be tested ... and that the developers prove they ran the tests. Expensive? You bet. But such extensive testing is a lot cheaper than the aftermath of a downed airliner. Even if you're not doing safety-critical software, a wise developer steals best practices and does as much of this sort of testing as is possible. For it's clear that for every untested branch at least one bug might be lurking.

Do check out the references listed, many of which are available on the web, as those will give even more insight into this crucial subject.

—Jack Ganssle

Embedded systems software testing shares much in common with application software testing. Thus, much of this chapter is a summary of basic testing concepts and terminology. However, some important differences exist between application testing and embedded systems testing. Embedded developers often have access to hardware-based test tools that are generally not used in application development. Also, embedded systems often have unique characteristics that should be reflected in the test plan. These differences tend to give embedded systems testing its own distinctive flavor. This chapter covers the basics of testing and test case development and points out details unique to embedded systems work along the way.

2.1 Why Test?

Before you begin designing tests, it's important to have a clear understanding of why you are testing. This understanding influences which tests you stress and (more importantly) how early you begin testing. In general, you test for four reasons:

- To find bugs in software (testing is the only way to do this)

- To reduce risk to both users and the company

- To reduce development and maintenance costs

- To improve performance

2.1.1 To Find the Bugs

One of the earliest important results from theoretical computer science is a proof (known as the Halting Theorem) that it's impossible to prove that an arbitrary program is correct.

Given the right test, however, you can prove that a program is incorrect (that is, it has a bug). It's important to remember that testing isn't about proving the "correctness" of a program but about finding bugs. Experienced programmers understand that every program has bugs. The only way to know how many bugs are left in a program is to test it with a carefully designed and measured test plan.

2.1.2 To Reduce Risk

Testing minimizes risk to yourself, your company, and your customers. The objectives in testing are to demonstrate to yourself (and regulatory agencies, if appropriate) that the system and software works correctly and as designed. You want to be assured that the product is as safe as it can be. In short, you want to discover every conceivable fault or weakness in the system and software before it's deployed in the field.

2.1.3 Developing Mission-Critical Software Systems

Incidents such as the Therac-25 radiation machine malfunction—in which several patients died due to a failure in the software monitoring the patients—should serve as a sobering reminder that the lives of real people might depend on the quality of the code that you write. I'm not an expert on writing safety-critical code, but I've identified some interesting articles on mission-critical software development:

- Brown, Doug. "Solving the Software Safety Paradox." *Embedded Systems Programming*, December 1998, 44.

- Cole, Bernard. "Reliability Becomes an All-Consuming Goal." *Electronic Engineering Times*, 13 December 1999, 90.

- Douglass, Bruce Powel. "Safety-Critical Embedded Systems." *Embedded Systems Programming*, October 1999, 76.

- Knutson, Charles, and Sam Carmichael. "Safety First: Avoiding Software Mishaps." *Embedded Systems Programming*, November 2000, 28.

- Murphy, Niall. "Safe Systems Through Better User Interfaces." *Embedded Systems Programming*, August 1998, 32.

- Tindell, Ken. "Real-Time Systems Raise Reliability Issues." *Electronic Engineering Times*, 17 April 2000, 86.

2.1.4 To Reduce Costs

The classic argument for testing comes from *Quality Wars* by Jeremy Main.

In 1990, HP sampled the cost of errors in software development during the year. The answer, $400 million, shocked HP into a completely new effort to eliminate mistakes in writing software. The $400M waste, half of it spent in the labs on rework and half in the field to fix the mistakes that escaped from the labs, amounted to one-third of the company's total R&D budget ... and could have increased earnings by almost 67%.[5]

The earlier a bug is found, the less expensive it is to fix. The cost of finding errors and bugs in a released product is significantly higher than during unit testing, for example (see Figure 2-1).

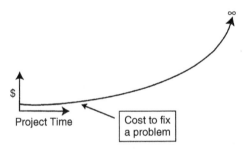

Figure 2-1: The Cost to Fix a Problem. Simplified graph showing the cost to fix a problem as a function of the time in the product life cycle when the defect is found. The costs associated with finding and fixing the Y2K problem in embedded systems is a close approximation to an infinite cost model.

2.1.5 To Improve Performance

Testing maximizes the performance of the system. Finding and eliminating dead code and inefficient code can help ensure that the software uses the full potential of the hardware and thus avoids the dreaded "hardware re-spin."

2.2 *When to Test?*

It should be clear from Figure 2-1 that testing should begin as soon as feasible. Usually, the earliest tests are module or unit tests conducted by the original developer. Unfortunately, few developers know enough about testing to build a thorough set of test cases. Because carefully developed test cases are usually not employed until integration testing, many bugs that could be found during unit testing are not discovered until integration testing. For example, a major network equipment manufacturer in Silicon Valley did a study to figure out the key sources of its software integration problems. The manufacturer discovered that 70 percent of the bugs found during the integration phase of the project were generated by code that had never been exercised before that phase of the project.

2.2.1 Unit Testing

Individual developers test at the module level by writing stub code to substitute for the rest of the system hardware and software. At this point in the development cycle, the tests focus on the logical performance of the code. Typically, developers test with some average values, some high or low values, and some out-of-range values (to exercise the code's exception processing functionality). Unfortunately, these "black-box" derived test cases are seldom adequate to exercise more than a fraction of the total code in the module.

2.2.2 Regression Testing

It isn't enough to pass a test once. Every time the program is modified, it should be retested to assure that the changes didn't unintentionally "break" some unrelated behavior. Called *regression testing*, these tests are usually automated through a test script. For example, if you design a set of 100 input/output (I/O) tests, the regression test script would automatically execute the 100 tests and compare the output against a "gold standard" output suite. Every time a change is made to any part of the code, the full regression suite runs on the modified code base to insure that something else wasn't broken in the process.

From the Trenches

I try to convince my students to apply regression testing to their course projects; however, because they are students, they never listen to me. I've had more than a few projects turned in that didn't work because the student made a minor change at 4:00AM on the day it was due, and the project suddenly unraveled. But, hey, what do I know?

2.3   Which Tests?

Because no practical set of tests can prove a program correct, the key issue becomes what subset of tests has the highest probability of detecting the most errors, as noted in *The Art of Software Testing* by Glen Ford Myers[6]. The problem of selecting appropriate test cases is known as *test case design*.

Although dozens of strategies exist for generating test cases, they tend to fall into two fundamentally different approaches: *functional testing* and *coverage testing*. Functional testing (also known as *black-box testing*) selects tests that assess how well the implementation meets the requirements specification. Coverage testing (also known as *white-box testing*) selects cases that cause certain portions of the code to be executed. (These two strategies are discussed in more detail later.) Both kinds of testing are necessary to test rigorously your embedded design. Of the two, coverage testing implies that your code is stable, so it is reserved for testing a completed or nearly completed product. Functional tests, on the other hand, can be written in parallel with the requirements documents. In fact, by starting with the functional tests, you can minimize any duplication of efforts and rewriting of tests. Thus, in my opinion, functional tests come first. Everyone agrees that functional tests can be written first, but Ross[7], for example, clearly believes they are most useful during system integration . . . not unit testing.

The following is a simple process algorithm for integrating your functional and coverage testing strategies:

1. Identify which of the functions have NOT been fully *covered* by the functional tests.

2. Identify which sections of each function have not been executed.

3. Identify which additional coverage tests are required.

4. Run new additional tests.

5. Repeat.

2.3.1   Infamous Software Bugs

The first known computer bug came about in 1946 when a primitive computer used by the Navy to calculate the trajectories of artillery shells shut down when a moth got stuck in one of its computing elements, a mechanical relay. Hence, the name *bug* for a computer error.[1]

In 1962, the Mariner 1 mission to Venus failed because the rocket went off course after launch and had to be destroyed at a project cost of $80 million.[2] The problem was traced to a typographical error in the FORTRAN guidance code. The FORTRAN statement written by the programmer was

```
DO 10 I = 1.5
```

This was interpreted as an assignment statement, DO10I = 1.5.

The statement should have been

```
DO 10 I = 1,5.
```

This statement is a DO LOOP. Do line number 10 for the values of I from one to five.

Perhaps the most sobering embedded systems software defect was the deadly Therac-25 disaster in 1987. Four cancer patients receiving radiation therapy died from radiation overdoses. The problem was traced to a failure in the software responsible for monitoring the patients' safety.[4]

2.4 When to Stop?

The algorithm from the previous section has a lot in common with the instructions on the back of every shampoo bottle. Taken literally, you would be testing (and shampooing) forever. Obviously, you'll need to have some predetermined criteria for when to stop testing and to release the product.

If you are designing your system for mission-critical applications, such as the navigational software in a commercial jetliner, the degree to which you must test your code is painstakingly spelled out in documents, such as the FAA's DO-178B specification. Unless you can certify and demonstrate that your code has met the requirements set forth in this document, you cannot deploy your product. For most others, the criteria are less fixed.

The most commonly used stop criteria (in order of reliability) are:

- When the boss says

- When a new iteration of the test cycle finds fewer than X new bugs

- When a certain coverage threshold has been met without uncovering any new bugs

Regardless of how thoroughly you test your program, you can never be certain you have found all the bugs. This brings up another interesting question: How many bugs can you tolerate? Suppose that during extreme software stress testing you find that the system locks up about every 20 hours of testing. You examine the code but are unable to find the root cause of the error. Should you ship the product?

How much testing is "good enough"? I can't tell you. It would be nice to have some time-tested rule: "if method Z estimates there are fewer than X bugs in Y lines of code, then your program is safe to release." Perhaps some day such standards will exist. The programming industry is still relatively young and hasn't yet reached the level of sophistication, for example, of the building industry. Many thick volumes of building handbooks and codes have evolved over the years that provide the architect, civil engineer, and structural engineer with all the information they need to build a safe building on schedule and within budget. Occasionally, buildings still collapse, but that's pretty rare. Until programming produces a comparable set of standards, it's a judgment call.

2.5   Choosing Test Cases

In the ideal case, you want to test every possible behavior in your program. This implies testing every possible combination of inputs or every possible decision path at least once. This is a noble, but utterly impractical, goal. For example, in *The Art of Software Testing*, Glen Ford Myers[6] describes a small program with only five decisions that has 10^{14} unique execution paths. He points out that if you could write, execute, and verify one test case every five minutes, it would take one billion years to test exhaustively this program. Obviously, the ideal situation is beyond reach, so you must use approximations to this ideal. As you'll see, a combination of functional testing and coverage testing provides a reasonable second-best alternative. The basic approach is to select the tests (some functional, some coverage) that have the highest probability of exposing an error.

2.5.1   Functional Tests

Functional testing is often called black-box testing because the test cases for functional tests are devised without reference to the actual code—that is, without looking "inside the box." An embedded system has inputs and outputs and implements some algorithm between them. Black-box tests are based on what is known about which inputs should be acceptable and how they should relate to the outputs. Black-box tests know nothing about how the algorithm in between is implemented. Example black-box tests include:

- **Stress tests**: Tests that intentionally overload input channels, memory buffers, disk controllers, memory management systems, and so on.

- **Boundary value tests**: Inputs that represent "boundaries" within a particular range (for example, largest and smallest integers together with −1, 0, +1, for an integer input) and input values that should cause the output to transition across a similar boundary in the output range.

- **Exception tests**: Tests that should trigger a failure mode or exception mode.

- **Error guessing**: Tests based on prior experience with testing software or from testing similar programs.

- **Random tests**: Generally, the least productive form of testing but still widely used to evaluate the robustness of user-interface code.

- **Performance tests**: Because performance expectations are part of the product requirement, performance analysis falls within the sphere of functional testing.

Because black-box tests depend only on the program requirements and its I/O behavior, they can be developed as soon as the requirements are complete. This allows black-box test cases to be developed in parallel with the rest of the system design.

Like all testing, functional tests should be designed to be *destructive*, that is, to prove the program doesn't work. This means overloading input channels, beating on the keyboard in random ways, purposely doing all the things that you, as a programmer, know will hurt your baby. As an R&D product manager, this was one of my primary test methodologies. If 40 hours of abuse testing could be logged with no serious or critical defects logged against the product, the product could be released. If a significant defect was found, the clock started over again after the defect was fixed.

2.5.2 Coverage Tests

The weakness of functional testing is that it rarely exercises all the code. Coverage tests attempt to avoid this weakness by (ideally) ensuring that each code statement, decision point, or decision path is exercised at least once. (Coverage testing also can show how much of your data space has been accessed.) Also known as white-box tests or glass-box tests, coverage tests are devised with full knowledge of how the software is implemented, that is, with permission to "look inside the box." White-box tests are designed with the source code handy. They exploit the programmer's knowledge of the program's APIs, internal control structures, and exception handling capabilities. Because white-box tests depend on specific implementation decisions, they can't be designed until after the code is written.

From an embedded systems point of view, coverage testing is the most important type of testing because the degree to which you can show how much of your code has been exercised is an excellent predictor of the risk of undetected bugs you'll be facing later.

Example white-box tests include:

- **Statement coverage**: Test cases selected because they execute every statement in the program at least once.

- **Decision or branch coverage**: Test cases chosen because they cause every branch (both the true and false path) to be executed at least once.

- **Condition coverage**: Test cases chosen to force each condition (term) in a decision to take on all possible logic values.

Theoretically, a white-box test can exploit or manipulate whatever it needs to conduct its test. Thus, a white-box test might use the JTAG interface to force a particular memory value as part of a test. More practically, white-box testing might analyze the execution path reported by a logic analyzer.

Gray-Box Testing

Because white-box tests can be intimately connected to the internals of the code, they can be more expensive to maintain than black-box tests. Whereas black-box tests remain valid as long as the requirements and the I/O relationships remain stable, white-box tests might need to be re-engineered every time the code is changed. Thus, the most cost-effective white-box tests generally are those that exploit knowledge of the implementation without being intimately tied to the coding details.

Tests that only know a little about the internals are sometimes called gray-box tests. Gray-box tests can be very effective when coupled with "error guessing." If you know, or at least suspect, where the weak points are in the code, you can design tests that stress those weak points. These tests are gray box because they cover specific portions of the code; they are error guessing because they are chosen based on a guess about what errors are likely. This testing strategy is useful when you're integrating new functionality with a stable base of legacy code. Because the code base is already well tested, it makes sense to focus your test efforts in the area where the new code and the old code come together.

2.6 Testing Embedded Software

Generally the traits that separate embedded software from applications software are:

- Embedded software must run reliably without crashing for long periods of time.

- Embedded software is often used in applications in which human lives are at stake.

- Embedded systems are often so cost-sensitive that the software has little or no margin for inefficiencies of any kind.

- Embedded software must often compensate for problems with the embedded hardware.

- Real-world events are usually asynchronous and nondeterministic, making simulation tests difficult and unreliable.

- Your company can be sued if your code fails.

Because of these differences, testing for embedded software differs from application testing in four major ways. First, because real-time and concurrency are hard to get right, a lot of testing focuses on real-time behavior. Second, because most embedded systems are resource-constrained real-time systems, more performance and capacity testing are required. Third, you can use some real-time trace tools to measure how well the tests are covering the code. Fourth, you'll probably test to a higher level of reliability than if you were testing application software.

2.6.1 Dimensions of Integration

Most of our discussion of system integration has centered on hardware and software integration. However, the integration phase really has three dimensions to it: hardware, software, and real-time. To the best of my knowledge, it's not common to consider real time to be a dimension of the hardware/software integration phase, but it should be. The hardware can operate as designed, the software can run as written and debugged, but the product as a whole can still fail because of real-time issues.

Some designers have argued that integrating a real-time operating system (RTOS) with the hardware and application software is a distinct phase of the development cycle. If we accept their point of view, then we may further subdivide the integration phase to account for the non-trivial task of creating a board support package (BSP) for the hardware. Without a BSP, the RTOS cannot run on the target platform. However, if you are using a standard hardware platform in your system, such as one of the many commercially available single-board computers (SBC), your BSP is likely to have already been developed for you. Even with a well-designed BSP, there are many subtle issues to be dealt with when running under an RTOS.

Simon[8] does an excellent job of covering many of the issues related to running an application when an interrupt may occur at any instant. I won't attempt to cover the same ground as Simon, and I recommend his book as an essential volume in any embedded system developer's professional library.

Suffice to say that the integration of the RTOS, the hardware, the software and the real-time environment represent the four most common dimensions of the integration phase of an embedded product. Since the RTOS is such a central element of an embedded product, any discussion about tools demands that we discuss them in the context of the RTOS itself. A simple example will help to illustrate this point.

Suppose you are debugging a C program on your PC or UNIX workstation. For simplicity's sake, let's assume that you are using the GNU compiler and debugger, GCC and GDB, respectively. When you stop your application to examine the value of a variable, your computer does not stop. Only the application being debugged has stopped running; the rest of the machine is running along just fine. If your program crashes on a UNIX platform, you may get a core dump, but the computer itself keeps on going.

Now, let's contrast this with our embedded system. Without an RTOS, when a program dies, the embedded system stops functioning—time to cycle power or press RESET. If an RTOS is running in the system and the debugging tools are considered to be "RTOS aware," then it is very likely that you can halt one of the running processes and follow the same debugging procedure as on the host computer. The RTOS will keep the rest of the embedded system functioning "mostly normally" even though you are operating one of the processes under the control of the debugger. Since this is a difficult task to do and do well, the RTOS vendor is uniquely positioned to supply its customers with finely tuned tools that support debugging in an RTOS environment. We can argue whether or not this is beneficial for the developer; certainly the other tool vendors may cry "foul," but that's life in the embedded world.

Thus, we can summarize this discussion by recognizing that the decision to use an RTOS will likely have a ripple effect through the entire design process and will manifest itself most visibly when the RTOS, the application software, and the hardware are brought together. If the tools are well designed, the process can be minimally complex. If the tools are not up to the task, the product may never see the light of day.

2.6.2 Real-Time Failure Modes

What you know about how software typically fails should influence how you select your tests. Because embedded systems deal with a lot of asynchronous events, the test suite should focus on typical real-time failure modes.

At a minimum, the test suite should generate both typical and worst case real-time situations. If the device is a controller for an automotive application, does it lock up after a certain sequence of unforeseen events, such as when the radio, windshield wipers, and headlights are all turned on simultaneously? Does it lock up when those items are turned on rapidly in a certain order? What if the radio is turned on and off rapidly 100 times in a row?

In every real-time system, certain combinations of events (call them *critical sequences*) cause the greatest delay from an event trigger to the event response. The embedded test suite should be capable of generating all critical sequences and measuring the associated response time.

For some real-time tasks, the notion of deadline is more important than latency. Perhaps it's essential that your system perform a certain task at exactly 5:00P.M. each day. What will happen if a critical event sequence happens right at 5:00P.M.? Will the deadline task be delayed beyond its deadline?

Embedded systems failures due to failing to meet important timing deadlines are called hard real-time or time-critical failures. Likewise, poor performance can be attributed to soft real-time or time-sensitive failures.

Another category of failures is created when the system is forced to run at, or near, full capacity for extended periods. Thus, you might never see a mal loc( ) error when the system is running at one-half load, but when it runs at three-fourths load, mal loc( ) may fail once a day.

Many RTOSs use fixed size queues to track waiting tasks and buffer I/O. It's important to test what happens if the system receives an unusually high number of asynchronous events while it is heavily loaded. Do the queues fill up? Is the system still able to meet deadlines?

Thorough testing of real-time behavior often requires that the embedded system be attached to a custom hardware/simulation environment. The simulation environment presents a realistic, but virtual, model of the hardware and real world. Sometimes the hardware simulator can be as simple as a parallel I/O interface that simulates a user pressing switches. Some projects might require a full flight simulator. At any rate, regression testing of real-time behavior won't be possible unless the real-time events can be precisely replicated.

Unfortunately, budget constraints often prohibit building a simulator. For some projects, it could take as much time to construct a meaningful model as it would to fix all the bugs in all the embedded products your company has ever produced. Designers do not spend a lot of time developing "throw-away" test software because this test code won't add value to the product. It will likely be used once or twice and then deleted, so why waste time on it?

A VHDL simulator could be linked to a software driver through a bus functional model of the processor. Conceptually, this could be a good test environment if your hardware team is already using VHDL- or Verilog-based design tools to create custom ASICs for your product. Because a virtual model of the hardware already exists and a simulator is available to exercise this model, why not take advantage of it to provide a test scaffold for the software team? This was one of the great promises of co-verification, but many practical problems have limited its adoption as a general-purpose tool. Still, from a conceptual basis, co-verification is the type of tool that could enable you to build a software-test environment without having to deploy actual hardware in a real-world environment.

2.6.3 Measuring Test Coverage

Even if you use both white-box and black-box methodologies to generate test cases, it's unlikely that the first draft of the test suite will test all the code. The interactions between the components of any nontrivial piece of software are just too complex to analyze fully. As the earlier "shampoo" algorithm hinted, we need some way to measure how well our tests are covering the code and to identify the sections of code that are not yet being exercised.

The following sections describe several techniques for measuring test coverage. Some are software-based, and some exploit the emulators and integrated device electronics (IDE) that are often available to embedded systems engineers.

Because they involve the least hardware, I'll begin with the software-based methods. Later I'll discuss some less intrusive, but sometimes less reliable, hardware-based methods. Despite the fact that the hardware-based methods are completely nonintrusive, their use is in the minority.

Software Instrumentation

Software-only measurement methods are all based on some form of execution logging. Statement coverage can be measured by inserting trace calls at the beginning of each "basic block" of sequential statements. In this context, a basic block is a set of statements with a single entry point at the top and one or more exits at the bottom. Each control structure, such as a goto, return, or decision, marks the end of a basic block. The implication is that after the block is entered every statement in the block is executed. By placing a simple trace statement, such as a print f ( ), at the beginning of every basic block, you can track when the block—and by implication all the statements in the block—are executed. This kind of software-based logging can be an extremely efficient way to measure statement coverage.

Of course, print f ( ) statements slow the system down considerably, which is not exactly a low-intrusion test methodology. Moreover, small, deeply embedded systems might not have any convenient means to display the output (many embedded environments don't include print f ( ) in the standard library).

If the application code is running under an RTOS, the RTOS might supply a low-intrusion logging service. If so, the trace code can call the RTOS at the entry point to each basic block. The RTOS can log the call in a memory buffer in the target system or report it to the host.

An even less-intrusive form of execution logging might be called *low-intrusion* print f ( ). A simple memory write is used in place of the print f ( ). At each basic block entry point, the logging function "marks" a unique spot in excess data memory. After the tests are complete, external software correlates these marks to the appropriate sections of code.

Alternatively, the same kind of logging call can write to a single memory cell, and a logic analyzer (or other hardware interface) can capture the data. If, upon entry to the basic block, the logging writes the current value of the program counter to a fixed location in memory, then a logic analyzer set to trigger only on a write to that address can capture the address of every logging call as it is executed. After the test suite is completed, the logic analyzer trace buffer can be uploaded to a host computer for analysis.

Although conceptually simple to implement, software logging has the disadvantage of being highly intrusive. Not only does the logging slow the system, the extra calls substantially change the size and layout of the code. In some cases, the instrumentation intrusion could cause a failure to occur in the function testing—or worse, mask a real bug that would otherwise be discovered.

Instrumentation intrusion isn't the only downside to software-based coverage measurements. If the system being tested is ROM-based and the ROM capacity is close to the limit, the instrumented code image might not fit in the existing ROM. You are also faced with the additional chore of placing this instrumentation in your code, either with a special parser or through conditional compilation.

Coverage tools based on code instrumentation methods cause some degree of code intrusion, but they have the advantage of being independent of on-chip caches. The *tags* or markers emitted by the instrumentation can be coded as noncachable writes so that they are always written to memory as they occur in the code stream. However, it's important to consider the impact of these code markers on the system's behavior.

All these methods of measuring test coverage sacrifice fine-grained tracing for simplicity by assuming that all statements in the basic block will be covered. A function call, for example, might not be considered an exit from a basic block. If a function call within a basic block doesn't return to the calling function, all the remaining statements within the basic block are erroneously marked as having been executed. Perhaps an even more serious shortcoming of measuring statement coverage is that the measurement demonstrates that the actions of an application have been tested but not the reasons for those actions.

You can improve your statement coverage by using two more rigorous coverage techniques: Decision Coverage (DC) and Modified Condition Decision Coverage (MCDC). Both of these techniques require rather extensive instrumentation of the decision points at the source code level and thus might present increasingly objectionable levels of intrusion. Also, implementing these coverage test methods is best left to commercially available tools.

2.6.4 Measuring More than Statement Execution

DC takes statement coverage one step further. In addition to capturing the entry into the basic blocks, DC also measures the results of decision points in the code, such as looking for the result of binary (true/false) decision points. In C or C++, these would be the if, for, while, and do/while constructs. DC has the advantage over statement coverage of being able to catch more logical errors. For example, suppose you have an if statement without an else part:

```
if (condition is true)
{
< then do these statements >;
}
< code following elseless if >
```

You would know whether the TRUE condition is tested because you would see that the then statements were executed. However, you would never know whether the FALSE condition ever occurred. DC would allow you to track the number of times the condition evaluates to TRUE and the number of times it evaluates to FALSE.

MCDC goes one step further than DC. Where DC measures the number of times the decision point evaluates to TRUE or to FALSE, MCDC evaluates the terms that make up the decision criteria. Thus, if the decision statement is:

```
if (A || B)
{
< then do these statements >;
}
```

DC would tell you how many times it evaluates to TRUE and how many times it evaluates to FALSE. MCDC would also show you the logical conditions that lead to the decision outcome. Because you know that the if statement decision condition would evaluate to TRUE if A is TRUE and B is also TRUE, MCDC would also tell you the states of A and B each time the decision was evaluated. Thus, you would know why the decision evaluated to TRUE or FALSE not just that it was TRUE or FALSE.

Hardware Instrumentation

Emulation memories, logic analyzers, and IDEs are potentially useful for test-coverage measurements. Usually, the hardware functions as a trace/capture interface, and the captured data is analyzed offline on a separate computer. In addition to these three general-purpose tools, special-purpose tools are used just for performance and test coverage measurements.

Emulation Memory: Some vendors include a *coverage bit* among the attribute bits in their emulation memory. When a memory location is accessed, its coverage bit is set. Later, you can look at the fraction of emulation memory "hits" and derive a percent of coverage for the particular test. By successively "mapping" emulation memory over system memory, you can gather test-coverage statistics.

One problem with this technique is that it can be fooled by microprocessors with on-chip instruction or data caches. If a memory section, called a *refill line*, is read into the cache but only a fraction of it is actually accessed by the program, the coverage bit test will be overly optimistic in the coverage values it reports. Even so, this is a good upper-limit test and is relatively easy to implement, assuming you have an ICE at your disposal.

Logic Analyzer: Because a logic analyzer also can record memory access activity in real time, it's a potential tool for measuring test coverage. However, because a logic analyzer is designed to be used in "trigger and capture" mode, it's difficult to convert its trace data into coverage data. Usually, to use a logic analyzer for coverage measurements, you must resort to statistical sampling.

For this type of measurement, the logic analyzer is slaved to a host computer. The host computer sends trigger commands to the logic analyzer at random intervals. The logic analyzer then fills its trace buffer without waiting for any other trigger conditions. The trace buffer is uploaded to the computer where the memory addresses, accessed by the processor while the trace buffer was capturing data, are added to a database. For coverage measurements, you only need to know whether each memory location was accessed; you don't care how many times an address was accessed. Thus, the host computer needs to process a lot of redundant data. For example, when the processor is running in a tight loop, the logic analyzer collects a lot of redundant accesses. If access behavior is sampled over long test runs (the test suite can be repeated to improve sampling accuracy), the sampled coverage begins to converge to the actual coverage.

Of course, memory caches also can distort the data collected by the logic analyzer. On-chip caches can mask coverage holes by fetching refill lines that were only partly executed. However, many logic analyzers record additional information provided by the processor. With these systems, it's sometimes possible to obtain an accurate picture of the true execution coverage by post-processing the raw trace. Still, the problem remains that the data capture and analysis process is statistical and might need to run for hours or days to produce a meaningful result.

In particular, it's difficult for sampling methods to give a good picture of ISR test coverage. A good ISR is fast. If an ISR is infrequent, the probability of capturing it during any particular trace event is correspondingly low. On the other hand, it's easy to set the logic analyzer to trigger on ISR accesses. Thus, coverage of ISR and other low-frequency code can be measured by making a separate run through the test suite with the logic analyzer set to trigger and trace just that code.

Software Performance Analyzers: Finally, a hardware-collection tool is commercially available that facilitates the low-intrusion collection method of hardware assist without the disadvantage of intermittent collection of a logic analyzer. Many ICE vendors manufacture hardware-based tools specifically designed for analyzing test coverage and software performance. These are the "Cadillac™" tools because they are specifically designed for gathering coverage test data and then displaying it in a meaningful way. By using the information from the linker's load map, these tools can display coverage information on a function or module basis, rather than raw memory addresses. Also, they are designed to collect data continuously, so no gaps appear in the data capture, as with a logic analyzer. Sometimes these tools come already bundled into an ICE, others can be purchased as hardware or software add-ons for the basic ICE. These tools are described in more detail later in the following section: "Performance Testing."

2.7 Performance Testing

The last type of testing to discuss in this chapter is performance testing. This is the last to be discussed because performance testing, and, consequently, performance tuning, are not only important as part of your functional testing but also as important tools for the maintenance and upgrade phase of the embedded life cycle. Performance testing is crucial for embedded system design and, unfortunately, is usually the one type of software characterization test that is most often ignored. Dave Stewart, in "The Twenty-Five Most Common Mistakes with Real-Time Software Development" [9], considers the failure to measure the execution time of code modules the number one mistake made by embedded system designers.

Measuring performance is one of the most crucial tests you need to make on your embedded system. The typical response is that the code is "good enough" because the product works to specification. For products that are incredibly cost sensitive, however, this is an example of engineering at its worst. Why overdesign a system with a faster processor and more and faster RAM and ROM, which adds to the manufacturing costs, lowers the profit margins, and makes the product less competitive, when the solution is as simple as finding and eliminating the hot spots in the code?

On any cost-sensitive embedded system design, one of the most dramatic events is the decision to redesign the hardware because you believe you are at the limit of performance gains from software redesign. Mostly, this is a gut decision rather than a decision made on hard data. On many occasions, intuition fails. Modern software, especially in the presence of an RTOS, is extremely difficult to fully unravel and understand. Just because you can't see an obvious way to improve the system throughput by software-tuning does not imply that the next step is a hardware redesign. Performance measurements made with real tools and with sufficient resources can have tremendous payback and prevent large R&D outlays for needless redesigns.

2.7.1   How to Test Performance

In performance testing, you are interested in the amount of time that a function takes to execute. Many factors come into play here. In general, it's a nondeterministic process, so you must measure it from a statistical perspective. Some factors that can change the execution time each time the function is executed are:

- Contents of the instruction and data caches at the time the function is entered

- RTOS task loading

- Interrupts and other exceptions

- Data-processing requirements in the function

Thus, the best you can hope for is some statistical measure of the minimum, maximum, average, and cumulative execution times for each function that is of interest.

Figure 2-3 shows the Code TEST performance analysis test tool, which uses software instrumentation to provide the stimulus for the entry-point and exit-point measurements. These tags can be collected via hardware tools or RTOS services.

2.7.2 Dynamic Memory Use

Dynamic memory use is another valuable test provided by many of the commercial tools. As with coverage, it's possible to instrument the dynamic memory allocation operators malloc() and free() in C and new and delete in C++ so that the instrumentation tags will help uncover memory leakages and fragmentation problems while they are occurring. This is infinitely preferable to dealing with a nonreproducible system lock-up once every two or three weeks. Figure 2-2 shows one such memory management test tool.

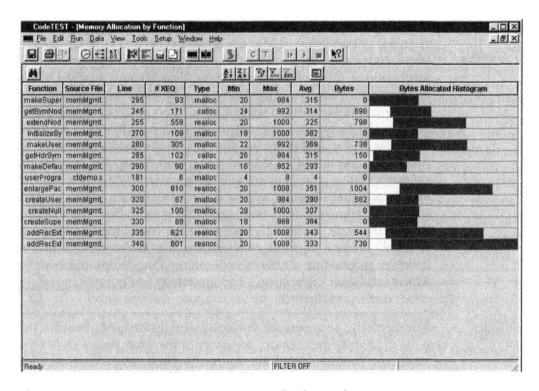

Figure 2-2: Memory Management Test Tool. The Code TEST memory management test program (courtesy of Applied Microsystems Corporation).

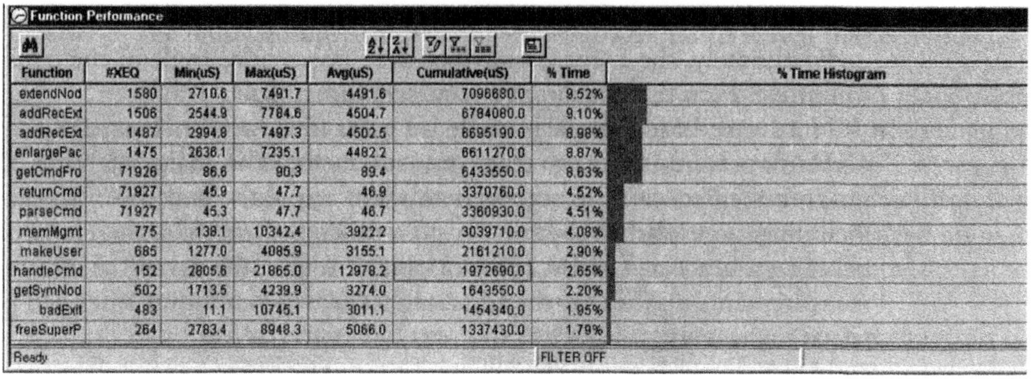

Function Performance

Function	#XEQ	Min(uS)	Max(uS)	Avg(uS)	Cumulative(uS)	% Time	% Time Histogram
extendNod	1580	2710.6	7491.7	4491.6	7096680.0	9.52%	
addRecExt	1506	2544.9	7784.6	4504.7	6784080.0	9.10%	
addRecExt	1487	2994.8	7497.3	4502.5	6695190.0	8.98%	
enlargePac	1475	2636.1	7235.1	4482.2	6611270.0	8.87%	
getCmdFro	71926	86.6	90.3	89.4	6433550.0	8.63%	
returnCmd	71927	45.9	47.7	46.9	3370760.0	4.52%	
parseCmd	71927	45.3	47.7	46.7	3360930.0	4.51%	
memMgmt	775	138.1	10342.4	3922.2	3039710.0	4.08%	
makeUser	685	1277.0	4085.9	3155.1	2161210.0	2.90%	
handleCmd	152	2805.6	21865.0	12978.2	1972690.0	2.65%	
getSymNod	502	1713.5	4239.9	3274.0	1643550.0	2.20%	
badExit	483	11.1	10745.1	3011.1	1454340.0	1.95%	
freeSuperP	264	2783.4	8948.3	5066.0	1337430.0	1.79%	

Ready FILTER OFF

Figure 2-3: Code TEST Test Tool. CodeTEST performance analysis tool display showing the minimum, maximum, average, and cumulative execution times for the functions shown in the leftmost column (courtesy of Applied Microsystems Corporation).

From the Trenches

Performance testing and coverage testing are not entirely separate activities. Coverage testing not only uncovers the amount of code your test is exercising, it also shows you code that is never exercised (dead code) that could easily be eliminated from the product. I'm aware of one situation in which several design teams adapted a linker command file that had originally been written for an earlier product. The command file worked well enough, so no one bothered to remove some of the extraneous libraries that it pulled in. It wasn't a problem until they had to add more functionality to the product but were limited to the amount of ROM space they had. Thus, you can see how coverage testing can provide you with clues about where you can excise code that does not appear to be participating in the program. Although removing dead code probably won't affect the execution time of the code, it certainly will make the code image smaller. I say probably because on some architectures, the dead code can force the compiler to generate more time-consuming long jumps and branches. Moreover, larger code images and more frequent jumps can certainly affect cache performance.

Conceptually, performance testing is straightforward. You use the link map file to identify the memory addresses of the entry points and exit points of functions. You then watch the address bus and record the time whenever you have address matches at these points. Finally, you match the entry points with the exit points, calculate the time difference between them, and that's your elapsed time in the function. However, suppose your function calls other functions, which call more functions. What is the elapsed time for the function you are trying to measure? Also, if interrupts come in when you are in a function, how do you factor that information into your equation?

Fortunately, the commercial tool developers have built in the capability to unravel even the gnarliest of recursive functions.

Hardware-based tools provide an attractive way to measure software performance. As with coverage measurements, the logic analyzer can be programmed to capture traces at random intervals, and the trace data—including time stamps—can be post-processed to yield the elapsed time between a function's entry and exit points. Again, the caveat of intermittent measurements applies, so the tests might have to run for an extended period to gather meaningful statistics.

Hardware-only tools are designed to monitor simultaneously a spectrum of function entry points and exit points and then collect time interval data as various functions are entered and exited. In any case, tools such as these provide unambiguous information about the current state of your software as it executes in real time.

Hardware-assisted performance analysis, like other forms of hardware-assisted measurements based on observing the processor's bus activity, can be rendered less accurate by on-chip address and data caches. This occurs because the appearance of an address on the bus does not necessarily mean that the instruction at that address will be executed at that point in time, or any other point in time. It only means that the address was transferred from memory to the instruction cache.

Tools based on the instrumentation of code are immune to cache-induced errors but do introduce some level of intrusion because of the need to add extra code to produce an observable tag at the function's entry points and exit points. Tags can be emitted sequentially in time from functions, ISRs, and the RTOS kernel itself. With proper measurement software, designers can get a real picture of how their system software is behaving under various system-loading conditions. This is exactly the type of information needed to understand why, for example, a functional test might be failing.

From the Trenches

From personal experience, the information that these tools provide a design team, can cause much disbelief among the engineers. During one customer evaluation, the tool being tested showed that a significant amount of time was being spent in a segment of code that none of the engineers on the project could identify as their software.

Upon further investigation, the team realized that in the build process the team had inadvertently left the compiler switch on that included all the debug information in the compiled code. Again, this was released code. The tool was able to show that they were taking a 15-percent performance hit due to the debug code being present in the released software. I'm relatively certain that some heads were put on the block because of this, but I wasn't around to watch the festivities.

Interestingly, semiconductor manufacturers are beginning to place additional resources on-chip for performance monitoring, as well as debugging purposes. Desktop processors, such as the Pentium and AMD's K series, are equipped with performance-monitoring counters; such architectural features are finding their way into embedded devices as well. These on-chip counters can count elapsed time or other performance parameters, such as the number of cache hits and cache misses.

Another advantage of on-chip performance resources is that they can be used in conjunction with your debugging tools to generate interrupts when error conditions occur. For example, suppose you set one of the counters to count down to zero when a certain address is fetched. This could be the start of a function. The counter counts down; if it underflows before it's stopped, it generates an interrupt or exception, and processing could stop because the function took too much time. The obvious advantages of on-chip resources are that they won't be fooled by the presence of on-chip caches and that they don't add any overhead to the code execution time. The downside is that you are limited in what you can measure by the functionality of the on-chip resources.

2.8 Maintenance and Testing

Some of the most serious testers of embedded software are not the original designers, the Software Quality Assurance (SWQA) department, or the end users. The heavy-duty testers are the engineers who are tasked with the last phases of the embedded life cycle: maintenance and upgrade. Numerous studies (studies by Dataquest and *EE Times* produced similar conclusions) have shown that more than half of the engineers who identify themselves as embedded software and firmware engineers spend the majority of their time working on embedded systems that have already been deployed to customers. These engineers were not the original designers who did a rotten job the first time around and are busy fixing residual bugs; instead, these engineers take existing products, refine them, and maintain them until it no longer makes economic sense to do so.

One of the most important tasks these engineers must do is understand the system with which they're working. In fact, they must often understand it far more intimately than the original designers did because they must keep improving it without the luxury of starting over again.

From the Trenches

I'm often amused by the expression, "We started with a clean sheet of paper," because the subtitle could be, "And we didn't know how to fix what we already had." When I was an R&D Project Manager, I visited a large telecomm vendor who made small office telephone exchanges (PBX). The team I visited was charged with maintaining and upgrading one of the company's core products. Given the income exposure riding on this product, you would think the team would have the best tools available. Unfortunately, the team had about five engineers and an old, tired PBX box in the middle of the room. In the corner was a dolly with a four-foot high stack of source code listings. The lead engineer said someone wheeled that dolly in the previous week and told the team to "make it 25 percent better." The team's challenge was to first understand what they had and, more importantly, what the margins were, and then they could undertake the task of improving it 25 percent, whatever that meant. Thus, for over half of the embedded systems engineers doing embedded design today, testing and understanding the behavior of existing code is their most important task.

It is an unfortunate truth of embedded systems design that few, if any, tools have been created specifically to help engineers in the maintenance and upgrade phases of the embedded life cycle. Everyone focuses on new product development. Go to any Embedded Systems Conference™ and every booth is demonstrating something to help you improve your time to market. What if you're already in the market? I've been to a lot of Embedded System Conferences™ and I've yet to have anyone tell me his product will help me figure out what I'm already shipping to customers. Today, I'm aware of only one product idea that might come to market for a tool specifically focusing on understanding and categorizing existing embedded software in a deployed product.

Additional Reading

Barrett, Tom. "Dancing with Devils: Or Facing the Music on Software Quality." *Supplement to Electronic Design*, 9 March 1998, 40.

Beatty, Sean. "Sensible Software Testing." *Embedded Systems Programming*, August 2000, 98.

Myers, Glenford J. *The Art of Software Testing*. New York: Wiley, 1978.

Simon, David. *An Embedded Software Primer*. Reading, MA: Addison-Wesley, 1999.

Summary

The end of the product development cycle is where testing usually occurs. It would be better to test in a progressive manner, rather than waiting until the end, but, for practical reasons, some testing must wait. The principal reason is that you have to bring the hardware and software together before you can do any kind of meaningful testing, and then you still need to have the real-world events drive the system to test it properly.

Although some parts of testing must necessarily be delayed until the end of the development cycle, the key decisions about what to test and how to test must not be delayed. Testability should be a key requirement in every project. With modern SoC designs, testability is becoming a primary criterion in the processor-selection process.

Finally, testing isn't enough. You must have some means to measure the effectiveness of your tests. As Tom DeMarco[3], once said, "You can't control what you can't measure."

If you want to control the quality of your software, you must measure the quality of your testing. Measuring test coverage and performance are important components but for safety critical projects, even these aren't enough.

References

1. Hopper, Grace Murray. "The First Bug." *Annals of the History of Computing*, July 1981, 285.

2. Horning, Jim. *ACM Software Engineering Notes*. October 1979, 6.

3. DeMarco, Tom. *Controlling Software Projects*. New York: Yourdon, 1982.

4. Leveson, Nancy, and Clark S. Turner. "An Investigation of the Therac-25 Accidents." *IEEE Computer*, July 1993, 18–41.

5. Main, Jeremy. *Quality Wars: The Triumphs and Defeats of American Business*. New York: Free Press, 1994.

6. Myers, Glen Ford J. *The Art of Software Testing*. New York: Wiley, 1978.

7. Ross, K.J. & Associates. http://www.cit.gu.edu.au/teaching/CIT2162/991005.pdf, p. 43.

8. Simon, David. *An Embedded Software Primer*. Reading, MA: Addison-Wesley, 1999.

9. Stewart, Dave. "The Twenty-Five Most Common Mistakes with Real-Time Software Development." A paper presented at the Embedded Systems Conference, San Jose, 26 September 2000.

CHAPTER 3

System-Level Design

Keith E. Curtis

How do you learn to design embedded systems?

It's relatively easy to translate a good design into working code. But creating that design is the true art of engineering. It's something that's not taught in college. Till I read Keith Curtis's book I told people, "Design five systems, then you'll know how," feeling that this is an experiential process, one that comes from doing, from making mistakes, and from recovering from those errors.

But Keith, in a clear, concise and detailed way, lays out a step-by-step process for turning a requirements document into a design. For that reason this is the most important chapter in this book.

Many studies have shown that most bugs in released products—and especially the most expensive-to-correct defects—stem not from coding errors, but from some bit of unimplemented functionality. The developers miss a requirement. But in this chapter you'll learn to "dissect" (Keith's morbid but absolutely spot-on description) a list of requirements into detailed specifications that you can use to generate code.

Note that the words "requirements" and "specifications" are distinct. In the software community we tend to use them interchangeably even though they are completely different concepts. Requirements define how a system interacts with the real world; they are the clearly itemized benefits realized by the customer. Specifications define the way the system operates. To quote "Software for Dependable Systems" by Jackson, Thomas and Millett (ISBN 0-309-10857-8): "Indeed, many failures of software systems can be attributed exactly to a failure to recognize this distinction, in which undue emphasis was placed on the specification at the expense of the requirements. The properties that matter to the users' of a system are the requirements; the properties the software developer can enforce are represented by the specification; and the gap between the two should be filled by the properties of the environment itself."

For more on eliciting requirements see Karl Wieger's excellent "Software Requirements" (ISBN 978-0735618794). Then use this chapter to create both a spec and a detailed design.

Keith's description is probably the only one extant that deals with the real-time issues faced by embedded developers. Follow Keith's process and you'll have separate specs for timing parameters, tasks, task priorities, and more. He also writes about tolerances in timing figures. Software people hardly ever think about tolerances, even though they are an essential part of the nature of the real world. Every EE worries about a resistor's +/− 5% rating; similarly, external events have some timing error band whose nature must be understood to have a correct design.

—Jack Ganssle

In this chapter, we will start the actual software design process. Because we are using a top-down approach to the design, it follows that this chapter will deal primarily with the top level of the design. This level of design is referred to as the *system level*. At this level, the general organization of the software will be developed, including definition of the tasks, layout of the communications, determination of the overall system timing, and the high-level definition of the priority structure.

These four areas—tasks, communications, timing, and priorities—are the four basic requirements for multitasking. The development of the system tasks includes context switching, but for our purposes, it is expanded to include the creation of the tasks; the development of a communications plan to handle all the communications between tasks; a timing control system to insure that each task is active at the right time to accomplish its function; and, finally, a priority manager to shift execution time to those tasks that are important to the system at any given moment.

To begin the system-level design, the designer needs a clear understanding of what the final software design must accomplish. The source of this information is the *system requirements document*, or simply the requirements document. The requirements document should contain the functions required, their timing, their communications needs, and their priorities.

If the requirements document does not contain all of these answers, and it typically doesn't, then it is up to the designer to obtain this information. The answer may come through asking questions of the department that generated the document, such as Marketing. Some of the information may be implied through a reference to another document, such as an industry standard on RS-232 serial communications. And, in some cases, the designer may simply have to choose.

Wherever the answers come from, they should end up in the requirements document. As part of the design, this document will be a living entity throughout the design process. As the requirements change, either through external requests from other departments or through compromises that surface in the design, the changes must be documented and must include an explanation of the reason for the change. In this way, the requirements document not only defines what the system should be, but also shows how it evolved during the development.

Some may ask, "Why go to all this trouble? Isn't commenting in the listing sufficient?" Well, yes, the commenting is sufficient to explain how the software works, but it does not explain why the software was designed in a certain way. It can't explain that the allocation of the tasks had to be a certain way to meet the system's priorities. It can't explain that halfway through the design additional functions were added to meet a new market need. And it can't explain why other design options were passed over because of conflicts in the design. Commenting the listing conveys the how and what, while the requirements document conveys the why.

An effective shorthand technique is to also list the information in a *design notes* file. This file should be kept simple; a text file is typically best. In this file, all of the notes, decisions, questions, and answers should be noted.

Personally, I keep a text file open in the background to hold my design notes when I dissect a requirements document. That way, I can note important information as I come across it. Another good reason to keep a design notes text file is that it is an excellent source of documentation for commenting. Whether generating a header comment for a software function or source information for a user's guide, all a designer has to do is copy and paste applicable information out of the design notes file. This saves time and eliminates errors in typing and memory. It also tends to produce more verbose header comments.

3.1   Dissecting the Requirements Document

While this may sound a little gruesome, it is accurate. The designer must carve up the document and wring out every scrap of information to feed the design process. In the following sections, we will categorize the information, document it in a couple of useful shorthand notations, and check the result for any vague areas or gaps. Only when the designer is sure that all the information is present and accounted for, should the design continue on. If not, then the designer runs the risk of having to start over. The five most frustrating words a designer ever hears are "What I really meant was."

So what is needed in a requirements document? Taking a note from the previous section, the four basic requirements are:

- **Tasks**: This includes a list of all the functions the software will be required to perform and any information concerning algorithms.

- **Communications**: This includes all information about data size, input, output, or temporary storage and also any information about events that must be recognized, and how.

- **Timing**: This includes not only the timing requirements for the individual tasks, but also the overall system timing.

- **Priorities**: This includes the priorities for the system, priorities in different system modes, and the priorities within each task.

Together, these four basic requirements for the system define the development process from the system level, through the component level, down to the actual implementation. Therefore, they are the four areas of information that are needed in a requirements document.

So, where to start? As the saying goes, "Start at the beginning." We start with the system tasks, which means all the functions that are to be performed by the tasks. And that means building a *function list*.

To aid in the understanding of the design process, and to provide a consistent set of examples, we will use the design of a simple alarm clock as an example. The following is a short description of the design and the initial requirement document:

Requirements Document

The final product is to be a 6-digit alarm clock with the following features:

1. 6-digit LED display, showing hours : minutes : seconds. The hours can be in either a 12 hour or 24 hour format. In the 12 hour format a single LED indicator specifying AM / PM is included.

2. 6 controls, FAST_SET, SLOW_SET, TIME_SET, ALARM_SET, ALARM_ON, SNOOZE.

3. The alarm shall both flash the display, and emit a AM modulated audio tone.

3.1.1   *Function List*

The first piece of documentation to build from the requirements document is a comprehensive *function list*. The function list should include all of the software functions described in the requirements document, any algorithms that may be specified or implied, and the general flow of the functions operation.

Reviewing the requirements document above, the following preliminary list of functions was compiled.

Preliminary Function List

1. Display functions to output data onto the displays

    a. 12-hour display function for time

    b. 24-hour display function for time

    c. 12-hour display function for alarm

    d. 24-hour display function for alarm

    e. Display flashing routine for the alarm

2. An input function to monitor and debounce the controls

    a. Control input monitoring function

    b. Debounce routine

3. A Command decoder function to decode the commands entered by the controls

4. An alarm function to check the current time and generate the alarm when needed.

    a. Turn alarm on / off

    b. Snooze

    c. Generate alarm tone

    d. Set alarm

5. Real-time clock

    a. Increment time at 1 Hz

    b. Set Time

3.1.2   Function List Questions

1.   Display function questions

   1.1.   Are displays scanned or driven in parallel?

   1.2.   How is 12 / 24 hour operation selected?

2.   Input function questions

   2.1.   How do the control inputs work?

3.   A command decoder questions

   3.1.   What are the commands?

   3.2.   How do the commands work?

4.   An alarm function questions

   4.1.   How does the user turn the alarm on and off?

   4.2.   How does the user know the alarm is on or off?

   4.3.   How does the snooze function work?

   4.4.   How is the alarm set?

   4.5.   What frequency is the alarm tone?

5.   Real-time clock questions

   5.1.   What is the time reference for 1 Hz?

   5.2.   How does the time reference operate?

   5.3.   What happens if the power fails?

   5.4.   How is the time set?

How can something as simple as an alarm clock generate so many functions and so many questions? I know how an alarm clock works, so why can't I just start writing code? While the designer may have a very good idea of how an alarm clock works, the purpose of this exercise is to get a very good idea of how marketing thinks the alarm clock should work, so we can design the alarm clock they want. Remember those five terrifying words, "what I really meant was."

Note: The designer should not be concerned if some of the functions appear to be repeated, such as the functions for time set, alarm set, and the function to flash the display, for example. Duplicates will be removed when the functions are combined into the various system tasks. In addition, duplicate listings indicate that the functionality may be split across a couple functions, so they also serve to indicate some of the function design choices that are yet to be made. Don't delete them until after the design decision is made.

The questions raised are also important:

- How will the LED display system be implemented in hardware? How are the controls implemented? How does the time reference operate and what will the software have to do?

The group designing the hardware will have the answer to these questions.

- How is the time and alarm time set? How is snooze initiated? How is 12/24 hour operation selected?

The answer to these questions will have to be answered by the group envisioning the product's look and feel.

As part of the function list, the designer should also include information about any algorithms used by a function. For example, the algorithm for converting data into a 7-segment format, any math routines for the 60 second/minute roll over, and even the algorithm for calculating the new alarm time when a snooze is activated. All of these will be a factor in the development of the different tasks in the system and should be recorded.

One final piece of information to note is the flow of the functions. Flow deals with the order in which things happen in a function. It can be simple and linear. For example: Increment seconds, if seconds = 60 then seconds = 0 and increment minutes. Or, it can be complex and require a graphical flow chart to accurately depict its functionality (see Figure 3-1).

Either way, it needs to be clearly defined so the designer has a clear idea of how the function works, with a list of any exceptions.

Note that there is nothing wrong with drawing pictures, and flow charts are very useful for graphically depicting the flow of a function. The use of pseudocode is another useful tool for describing how a function operates. Designers should not feel reluctant to drag out a large piece of paper and start drawing. If electronic copies of the documentation are required, the drawings can always be scanned and stored in a digital form.

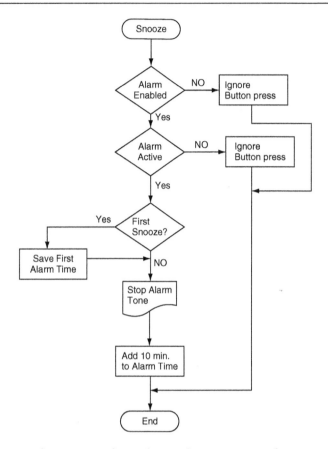

Figure 3-1: Flow Chart of Snooze Function.

Finally, when this section on the requirements document started, it was stated that any answers to questions should be included in a revision of the requirements document. So, including answers from all the groups, the document is rewritten with the new information:

Requirements Document

The final product is to be a 6-digit alarm clock with the following features:

1. A scanned 6-digit numeric LED display.

    a. Time display is in either 24-hour or 12-hour AM/PM format with hours, minutes, and seconds displayed.

b.   Single LED enunciators are included for both ALARM ON and PM time.

c.   No indicator is used for AM or 24-hour operation.

d.   No indication of snooze operation is required.

e.   The alarm function can flash the display.

f.   Battery operation can blank the display.

2.   6 controls, FAST_SET, SLOW_SET, TIME_SET, ALARM_SET, ALARM_ON, SNOOZE.

a.   All controls, except ALARM_ON are push buttons. Combinations of button presses initiate the various commands. ALARM_ON is a slide switch.

b.   See below for command function information.

3.   A Command decoder function to decode the commands entered by the controls.

a.   See below for detailed command operation.

4.   An alarm function.

a.   Alarm time shall be displayed in hours and minutes with the seconds display blank when in the alarm set mode. The format shall match the current time display.

b.   The maximum number of snooze commands is not limited.

c.   The display shall flash in time to the tone.

d.   Turning the alarm on and off, setting the alarm time, and initiating snooze is described in the Command function section of the document.

e.   The alarm tone shall be 1 kHz, modulated at a 1 Hz rate (50% duty cycle).

5.   The clock shall use the 60-Hz power cycle as a time-keeping reference for the real-time clock function.

a.   If 5 consecutive 60-Hz cycles are missed, the clock shall revert to the microcontroller clock.

b. A battery back-up system shall be included that requires no action from the microcontroller to operate.

c. While on battery operation, the display and alarm functions shall be disabled. If the alarm time passes during battery operation, then the alarm shall sound when 60-Hz power is restored.

d. When the microcontroller detects 5 consecutive 60-Hz cycles, it shall revert to the power line time base.

e. See below for setting the time and selecting 12/24-hour operation.

The new document, while verbose, is also much less ambiguous concerning the functionality of the system. Most of the questions have been answered and a significant amount of information has been added. The edits to the document are by no means complete, since there is information concerning communications, timing, and priorities yet to be examined. If you look carefully at the revised document, none of the questions concerning the operation of the commands have been answered. However, at this point most of the functionality of the various software functions has been clarified.

It is now time to answer the questions concerning the user interface, or command structure, of the system. In the previous section, questions concerning this information were asked but not answered. The reason is that the user interface, while contributing to the list of functions, is a sufficiently unique subject that it warrants special attention. Therefore, it is the next section to be covered.

3.1.3 The User Interface

A good user interface can make a product useful and a joy to use, while a bad user interface can be a source of frustration and pain. Although the science of developing a good user interface is sufficiently complex to fill several books this size, a fairly simple analysis of the proposed system can typically weed out most of the more common problems and inefficiencies. Additionally, the technique described in this section clearly documents the command structure and clearly shows any missing information. Even if the interface has been used extensively in older systems, it never hurts to revisit the evaluation, if only to get a clear picture of the command flow.

The first step is to storyboard, or flow chart, the command structure. This is accomplished by graphically showing the step-by-step sequence required to perform a command entry. For instance, see Figure 3-2 for setting the time on our alarm clock.

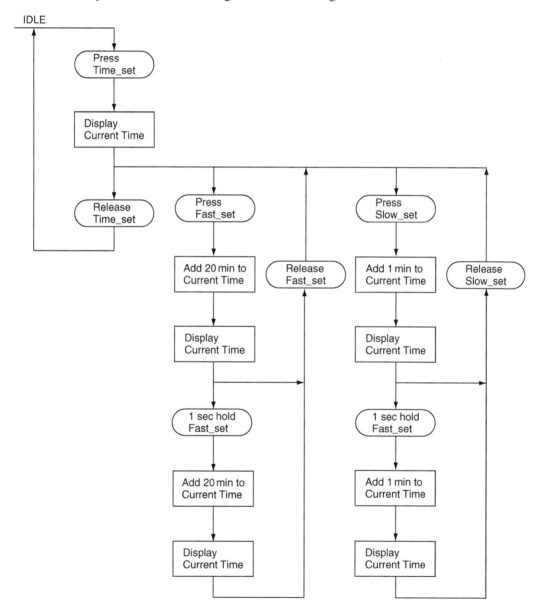

Figure 3-2: Command Structure Flow Chart of time_set.

In the example, round bubbles are used to indicate inputs from the users and rectangular boxes indicate responses from the system. Arrows then indicate the flow of the process, with the point of the arrow indicating the next event in the sequence. Some arrows have two or more points, indicating that two or more different directions are possible. For example, after the current time has been displayed by the system, the user has the option to release the TIME_SET button and terminate the command, or press either the FAST_SET or SLOW_SET buttons to change the current time.

At the top of the diagram is a line labeled IDLE and this is where the time set command sequence begins and ends. IDLE has been defined for this diagram to be the normal state of the system with the alarm disabled. Other system modes with mode-specific command sequences could include ALARM_ON, SNOOZE, and ALARM_ACTIVE. By using a specific system mode as a starting point, the diagram is indicating that the command is only available or recognized in that specific mode. If the label was ALL, then the command would be available in all system modes. Combinations of modes, such as ALARM_ON and ALARM_ACTIVE, can also be specified to indicate that a command is only available in the listed modes. However, most commands are typically available in all modes of the system, with only special-purpose commands restricted to a specific mode. For example, the ALARM_SET command would be available whether the alarm is enabled or disabled, while the SNOOZE command is only useful when the alarm is active, so it makes sense to only allow it for that specific mode.

Each command diagram should be complete, in that it shows all legitimate actions available for the command. It can also be useful to diagram sequences that generate an error, as this clarifies the error-handling functions in the user interface. In our example of an alarm clock, the system's response to an improper input is simply to ignore it. More complex systems may not have this luxury and may need a specific response to the unwanted input. To separate legitimate actions from errors, it is typically sufficient to draw the arrows for error conditions in red and the legitimate course of action in black. For diagrams that will be copied in black and white, a bold line to indicate improper input can also be used.

In more complex systems, the storyboards for a command structure can become large and cumbersome. To avoid this problem, the designer can replace sections of the diagram with a substitution box indicating additional information is available in a subdiagram. This is particularly useful if a commonly used edit sequence, used in multiple places in the diagram, can be replaced with a single subdiagram. The only prudent limitation on the practice is that the substituted section should only have one entrance and one exit. Some systems may in fact be so complex that an overall command storyboard may be required, with the individual commands listed as subdiagrams.

When all the storyboards are complete, they should be shown to the group that designed the system so they can clarify any misunderstandings. This is best done at the beginning, before several hundred lines of software are written and debugged.

Once all of the storyboards are complete, take each storyboard and note down how many key presses are required to set each function, worst case. For the clock time set example, the worst-case number of key presses is 83, 1 for the initial press and hold of the TIME_SET button, 23 presses of the FAST_SET to set hours, and 59 presses of the SLOW_SET to set minutes. Next, calculate the time required to perform that number of button presses. Assume that a key can be pressed repeatedly at a rate of 2–3 presses per second. For the clock this means that the worst-case time required to set the time is 42 seconds if each key press is made individually, and as much as 83 seconds if the autorepeat feature is used.

Now, for the complete command structure, list the commands based on the frequency that each command is likely to be used, with most often used at the top of the list, and least often used at the bottom. Next to each command sequence name, list the worst-case number of key presses required to perform the command and the estimated time required. See the following list for the commands used in the alarm clock:

Table 3-1

Frequency of Use	Function Name	Button Presses	Time
Most infrequent	Set Time	83	42/83 sec
Infrequent	Set Alarm	83	42/83 sec
Frequent	Enable Alarm	Slide Switch	1 sec
Frequent	Disable Alarm	Slide Switch	1 sec
Very frequent	Snooze	1	½ sec

The times and number of key presses should be the inverse of the frequency of use. Specifically, the most common commands should have the least number of key presses and the fastest time to perform, and the least-often used commands should have the largest number of key presses and the longest time to set. If any command is out of sequence, then the flow of that command should be reconsidered, so that it falls in line with the other commands. From the example, Set Time and Set Alarm time are the longest to set and the least frequently used. The Snooze command is the most frequently used and the fastest to activate.

Another criterion for menu-based command structures is the depth of the menu that holds a command. Commonly used commands should be at the top level, or at the most, one level deep in the command structure. Commands deeper in the menu structure should have progressively less frequent use. See the following example menu structure:

Structure 3.1

```
ROOT Menu
   Delete
   Edit  →  Copy
            Paste
            Search  →  Find
            Replace
   File  →  Open
            Save
            Close
            New  →  Blank Template
                    Select Template
```

In this example, the most-often used command is Delete and it is at the top of the menu. Edit commands and File commands come next, with the New file commands buried the deepest in the menu. Typically, a user can remember one or two levels of a menu structure, provided that each level has only three or four functions. Any deeper, and they will typically have to consult a manual (unlikely), or dig through the menus to find the function they want. While designers might wish that users used the manuals more often, making this a requirement by burying commonly used commands at the bottom of a complex menu structure will only drive customers to your competitors.

Another obvious, but nonetheless often overlooked, requirement is that related commands should be in a common subdirectory, and the relationship of the commands should be viewed from the user's point of view, not the designers. Just because Paste and Replace have similar functions does not mean that the user will look for them in the same submenu. The correct choice is to group the commands as shown, by their use by the user, rather than their inner workings.

One hallmark of a good user interface is reusing buttons for similar functions in different commands. For instance, in the clock example, there was a FAST_SET and SLOW_SET button. They are used to set the current time, so it makes sense that the same buttons would also be used to set the Alarm time. Keeping common functions with the same buttons allows the user to stereotype the button's function in their minds and aids in their understanding of the command structure. With this in mind, it would be a major failing in a user interface to change the function of the control, unless *and only unless*, the second function is an extension of the control's original function. For instance, changing from 12 hour to 24 hour by pressing FAST_SET and SLOW_SET buttons together is acceptable because it is an extension of the buttons' original functions. Using the SNOOZE button in combination with the ALARM_SET button is just confusing for the user.

Once the user interface has been clearly defined, the requirements document should be updated to include the storyboards and any changes that may have come out of the analysis. Any changes or additions to the function list, necessitated by the user interface, should also be made at this time.

User Interface Options

So far, we have discussed interfaces based on just displays and buttons. Another method for entry is to use a rotary encoder as an input device. Designers today tend to forget that the original controls on tube radios were all knobs and dials. For all their simplicity, they did provide good resolution and responsive control, plus most users readily identify with the concept of turning a knob. Because they use only a two-bit Grey code to encode their movement, their interface is simple and the output is not tied to the absolute position of the rotary encoder, making them ideal for setting multiple values.

Imagine the simplicity of setting the alarm clock in the previous example using a rotary encoder. Simply hold down the set button and turn the dial until the right time appears on the display. Because the knob can move in two directions and at a rate determined by the user, it gives them additional control that a simple two-button interface does not.

Another trick with a rotary encoder is to tie the increment and decrement stop size to the rate of rotation, giving the control an exponential control resolution. Several quick flips of the knob can run the value up quickly by incrementing the value using a large increment. Then slower, more precise, rotations adjust the value with a smaller increment, allowing the user to fine-tune the value.

Another handy rotary input device is the simple potentiometer combined with an analog-to-digital converter input. This is a single input with many of the same features as the rotary encoder, plus the potentiometer is also nonvolatile, meaning it will not lose its setting when the power is removed. It does present a problem in that it cannot turn through multiples of 360 degrees indefinitely, but depending on the control function, this may not be a problem.

At the end of this phase of the dissection, the designer should have a revised function list. Any missing information in the requirements document should have been identified and answers found. A clear description of the user interface and command structure should have been generated, with storyboards. Any cumbersome or complicated sequences in the command structure should have been identified and rewritten to simplify the interface. And, finally, the requirements document should have been updated to include the new information. As always, any general notes on the system, with any applicable algorithms or specific information concerning the design, should also have been compiled.

The revised documents should look like the following:

Revised Function List

1. Display functions to output data onto the displays

    a. 12-hour display function for time

    b. 24-hour display function for time

    c. 12-hour display function for alarm

    d. 24-hour display function for alarm

    e. Display flashing routine for the alarm

    f. PM indicator display function

    g. Alarm on indicator display function

    h. Function to scan LED displays

2. An input function to monitor and debounce the controls

    a. input function to monitor buttons

    b. Debounce routine

    c. Auto repeat routine

    d. 60-Hz monitoring routine

    e. 60-Hz Fail / Recovery monitoring routine

3. A Command decoder function to decode the commands entered by the controls

   a. An alarm function to check the current time and generate the alarm when needed.

   b. Snooze function to silence alarm for 10 minutes.

   c. Alarm on / off toggling routine

   d. Initiate Snooze

   e. Generate alarm tone routine

   f. Set alarm function

      i. Routine to increment alarm by 1 min

      ii. Routine to increment alarm by 20 min

   g. Set Time function

      i. Routine to increment Time by 1 min

      ii. Routine to increment Time by 20 min

   h. Toggle 12/24 hour mode

4. Real-time clock routine

   a. Time increment routine based on 60-Hz power line time base

   b. Time increment routine based on internal clock time base

   c. Display blanking routine for operation from internal clock time base

   DESCRIPTION OF THE USER INTERFACE
     Display
       6-digit scanned LED display
       1 indicator for PM operation in 12-hour mode
       1 indicator to show alarm is active

     Controls (inputs)
       1 slide switch to enable / disable the alarm
       1 push button for ALARM_SET
       1 push button for TIME_SET
       1 push button for FAST_SET
       1 push button for SLOW_SET
       1 push button for SNOOZE

```
Time base inputs
    60-Hz line time base
    System clock

DESCRIPTION OF THE COMMAND STRUCTURE
  To set Time
    Hold the TIME_SET button
    (display will show current time with seconds blank)
        Press SLOW_SET to increment time by 1 min
        Hold SLOW_SET to auto-increment time by
          1 min at 1-HZ rate
        Press FAST_SET to increment time by 20 min
        Hold FAST_SET to auto-increment time by
          20 min at 1-HZ rate
        (in 12-hour mode, time will roll over at 12:59)
        (in 24-hour mode, time will roll over at 23:59)
        Release the TIME_SET button to return to normal
          operation
        (Seconds will appear and start incrementing from 0)


  To set alarm time
    Hold the ALARM_SET button
    (display will show current alarm time with seconds
        blank)
        Press SLOW_SET to increment alarm time by 1 min
        Hold SLOW_SET to auto-increment alarm time by
          1 min at 1-HZ rate
        Press FAST_SET to increment alarm time by 20 min
        Hold FAST_SET to auto-increment alarm time by
          20 min at 1-HZ rate
        (in 12-hour mode, time will roll over at 12:59)
        (in 24-hour mode, time will roll over at 23:59)
        Release the ALARM_SET button to return to
          normal operation
        (display will show current time)


  To turn alarm on
    Slide alarm control switch to on
    (alarm indicator will light)


  To turn alarm off
    Slide alarm control switch to off
    (alarm indicator will go blank)
```

```
To activate snooze mode, alarm must be active
   Press the SNOOZE button
   (alarm will be remain enabled)
   (tone will stop for for 10 min and then sound again)

To toggle 12 hour / 24 hour mode
   Release ALARM_SET and TIME_SET buttons
   Hold the FAST_SET button
   Press the SLOW_SET button
   (12/24 hour mode will toggle)
   (if result is 24-hr mode, time is displayed in
   24-hr format on press)
   (if result is 12-hr mode, time is displayed in
   12-hr format on press)
```

As no major changes have been made to the requirements document since the last section, the document will not be repeated here.

3.2 Communications

The next area of information to extract from the requirements document relates to communication pathways, both within the system and between the system and any external systems—specifically, information concerning the volume and type of data that will have to be handled by each pathway. This gives the designer a basis to plan out the communications system and to estimate the necessary data memory space required. Some of this information will be specified in the form of communications protocols between the system and external entities such as terminals, remote systems, or autonomous storage. Some of the information will be dictated by the operation of the peripherals in the system, such as timers, A-to-D converters, and the system's displays. And, some of the requirements will be dictated by the operations of the tasks themselves. As with the function list, we will have to play detective and determine what information is present, what is missing, and what is implied.

What kind of information are we looking for? We will have two forms of storage: dynamic and static. Dynamic storage handles a flow of information—for example, a serial peripheral that receives messages from another system. The task managing the peripheral will require storage for the message until it can be passed to a control task for processing. Because the peripheral may continue to receive new information while it is processing the old message, the storage will typically be larger to hold both the current message and the new one being received. This storage is therefore considered dynamic because the amount of data stored changes with time. The data storage is also not constant. While messages are being received, then the storage holds data. If all the messages received by the peripheral task have been processed, the storage is empty. Static storage, on the other hand, has a fixed storage requirement because the information is continuous, regardless of the current activity of its controlling task—for example, the variable structures that hold the current time and alarm time information in our clock example. The data may be regularly updated, but it doesn't change in size, and there is always valid data in the variables, so static storage is constant in size and continuously holds data.

All data pathways within a system will fall into one of these two categories. What we as designers need to do at this point in the design is find the various pathways, determine if the storage is static or dynamic, and make a reasonable estimate concerning the amount of storage required.

A good place to start is the peripherals that introduce information to the system. These include serial communications ports, button inputs, A-to-D converters (ADCs), even timers. These peripherals constitute sources of data for the system as their data is new to the system and not derived from other existing data. To determine whether their requirements are static or dynamic, we will have to determine what the information is and how the system will ultimately use it. Let's take a couple of examples, and determine which are static or dynamic:

Example 3.1: An A-to-D that captures sensor data from several sources. In this example the A-to-D continuously samples multiple sources, voltage, current, temperature, and pressure. It then scales the resulting value and stores the information in a collection of status variables. This peripheral is collecting a continuous stream of data, but it is not storing the information as a stream of data. Rather, it is updating the appropriate status variable each time a new sample is converted. This is an example of static storage. The memory requirements are simply the collection of status variables, multiplied by their width. The number of variables does not change, and they all contain valid data continuously.

Example 3.2: An A-to-D that captures a continuous stream of samples from a single source for digital signal processing. The data is stored in a large array, with the most current at the top and the oldest at the bottom. While this certainly sounds dynamic, it is actually static. As in the previous example, the amount of data does not change, but simply flows through the array of values. Each time a new value is added, the old value falls off the other end. The amount of storage required is the size of the array holding the collection of values, multiplied by their width. The number of variables does not change, and they all contain valid data continuously.

Example 3.3: A command decoder that converts individual button presses into control commands for the system. This certainly sounds static: a button is pressed and a command comes out. However, the storage requirement is actually dynamic. In the definition of static and dynamic storage, it was stated that the amount of valid information in a static system must be constant. Here the output of the system can be a valid command, or the system may be idle with no valid date output. The amount of data changes, even if only from one command to zero commands, so the system is dynamic.

Example 3.4: A system that reads data from a disc drive. This system is definitely dynamic, since the data is read from the disc as the information passes under the heads in the drive, so the timing of the data's arrival is dictated by the physics of the spinning disc. The system that uses the information is very probably not synchronized to the spin of the disc, so the system reading the disc will have to buffer up the information to handle the timing discrepancy between the disc and the receiving system. Because the timing is asynchronous, there is no way to predict the amount of time between the reception of the data and its subsequent transmission to the receiving system. So, the amount of data stored at any given moment is variable, ranging from zero to the maximum size of the disc file, and that makes this storage requirement dynamic.

OK, so some data is static, and we can readily determine the storage requirements for these functions, but how do we determine the maximum size of dynamic storage? The answer lies in the rate at which the information enters the system. In a typical system, such as a serial port, there will be three potential data rates.

1. **The maximum rate**: Typically this is determined by the electrical characteristics of the peripheral, the maximum conversion rate of the A-to-D, the baud rate of a serial port, or the roll-over time of a timer. It represents the theoretical maximum possible rate at which data can be sent, and it should be used to set the timing requirements of the task that will manage the peripheral.

2. **The average rate**: Typically this is an indicator of the average data load on the system. For a serial port, this will be the number of packets sent in a typical second,

multiplied by the average size of the packets. It is not the fastest rate at which the peripheral will have to operate, but it does indicate how much data the system will have to handle on a regular basis.

3. **The peak rate**: This rate is the worst-case scenario, short of the maximum rate defined for the peripheral. It indicates the maximum *amount* of data that will be transmitted in a given second. The word amount is the important distinction between the peak rate and the maximum rate. The maximum rate assumes a continuous flow of data forever. The peak rate indicates the amount of data sent, minus all the delays between packets, and characters in the flow of data. So, the peak rate, by definition, must be less than the maximum rate, and it represents the maximum data load on the system.

So, the maximum rate determines the speed at which the task managing the peripheral must operate, and the average and peak rates determine the average and worst-case data load on the system. How does this determine the amount of storage required? To answer the question, consider the example of a system that must receive serial data from another system.

Data from an external system is transmitted in the following format: 9600 baud, with 8-bit data, no parity, and 1 stop bit. Further, the data will be received in packets of 10 bytes, at an average rate of two packets every second.

So, to store the data received from the serial port, it is pretty obvious the temporary data storage structure will be an 8-bit CHAR. And, given a baud rate of 9600, with 8-bit data, 1 start bit, and 1 stop bit, the maximum rate at which 8-bit CHARs will be generated is 960 characters per second. That means that the receiving task will have to be called at least 960 times a second to keep up with the data. So far, so good, the maximum data rate is 960 characters a second.

$$960 = (9600\text{baud}/(8 \text{ bit data} + 1 \text{ start bit} + 1 \text{ stop bit})) \qquad \textbf{Equation 3-1}$$

However, how big a buffer will be needed to handle the data? Well, the packet size is 10 bytes, so a packet requires 10 bytes of storage. Given that the average rate at which a packet can be received is 2 per second, then the system will have to process 20 characters a second. And the minimum storage would have to be 20 CHARs, 10 for the current packet, plus 10 more to hold the accumulating data in the second packet.

OK, the system needs a minimum of 20 CHARs to buffer the incoming data. However, what happens if the peak rate is five packets per second? Now we need more storage; a minimal 20 CHAR buffer will be overrun. How much more storage should actually be allocated? At the moment, we don't have sufficient information to determine the exact storage needs, either average or peak. This is because we don't know the rate at which the system will actually process the packets. However, a good guess can be made using the average and peak rate numbers. If the average rate is two packets per second, then the maximum time the system will have to process a packet is limited to ½ second. If the peak rate is five packets per second, and the system can process packets at a rate of two per second, then the buffer storage will have to be at least 41 CHARs. Five incoming packets each second, less one processed packet during the first half of the second, gives four packets of storage. At 10 CHARs per packet, plus one extra for the pointers, that's 41 CHARs. So, a good maximum size guess is 41 bytes for the storage.

One side note to consider, before we leave the discussion on buffer size and packet rates, if the control task is fast enough to process the data as it is received, why even user a buffer? Why not just process the data as it is received? Using this method would seem to be very appealing because it is both faster, and less wasteful of data memory. Unfortunately, there is an opportunity cost that is not readily apparent. If the control task is preoccupied with processing the data as it is received, it will not be able to handle other important conditions that may arise while the packet is in process. The response to other system conditions will quite literally by blocked by the reception of the data packet until it is complete. Using the buffer to queue up the complete packet allows the control task to handle the packet all at once, freeing it up to handle other important events as they occur. So, the buffer system in effect trades off data storage for more efficient use of the control task's execution time.

Another point to consider: if the control task does not use a buffer system and processes the data on a CHAR by CHAR basis, it can potentially be hung up if the data stream from the communications peripheral is interrupted in mid-packet. In fact, if the control task does not include some kind of time out timer, the control task may not notice even notice the interruption and hang the entire system waiting for a character that will never arrive.

At this point, the information that should be collected is:

1. What the data is, and its probable variable size.

2. Whether the storage requirement is static or dynamic.

3. Where the data comes from, and goes to.

4. The approximate amount of storage required for the storage.

5. And all information concerning the rate at which the data will appear.

Decisions concerning the actual format of the data storage and the final amount of data memory allocated will be left until later in the design, when more information concerning processing time is available. Until then, just note the information for each pathway in the system.

Having retrieved the specifications for data entering the system, the next step is to gather requirements for data leaving the system. And, again, the exits, like the entrances, will be through the peripherals and can be either static or dynamic.

In the previous section, we determined that static variables were fixed in length and continuously held valid data. The same is true for output peripherals—for example, the LED displays in our clock example. The task that scans the information onto the displays will get its information from one of two static variables that hold the current time and alarm time for the system. The data feed for the peripheral task has a constant size and continuously holds valid data, so the storage for the display is static.

However, if the peripheral is a serial output port, then the storage is no longer static because the amount of data is probably variable in length, and once transmitted, it probably no longer be valid either. Therefore the output queue for a serial output task is probably dynamic. But be careful, it could be that the serial output task simply grabs data from fixed variables in the system, converts them into ASCII characters, and sends them out. In this case, storage for the serial port task may be static because it is constant in length and always holds valid data. Careful examination of the requirements document is required to make a proper determination.

As in the previous section, a determination of the amount of data memory needed to hold any dynamic storage will also have to be made. Unfortunately, there may not be any explicit peak and average data rates to base the calculation on. Instead, we will have to examine the requirements placed on the peripheral and make a best guess as to what the average and peak rates are for the peripheral.

For example, consider a serial port that will be used to return information in response to queries from another system. Like the previous section, we will assume a 9600 baud rate, with 8-bit data, no parity, and one stop bit. This fixes the maximum rate of data transmission to 960 characters a second. The trick is now to determine what the average and peak data rates will be.

Well, if the data is sent in response to queries, then we can estimate the worst-case needs using a little common sense and some math. For example, assume the largest response packet is 15 characters long. If the maximum rate that new packets can be generated is limited by the peak rate at which packets can be received, then the peak rate for new outgoing packets is 5 per second (from the previous section). Given 15 CHARs per query, then the outgoing rate is 5 packets per second, or 75 characters per second. That means that a reasonable guess for data storage is 75 CHARS.

The final section of communications-related data to retrieve from the requirements document is any significant data storage requirements not covered in the previous sections. It can include on-chip copies of information stored in a nonvolatile memory; scratchpad memory for translating, compressing, or de-compressing files of data; or temporary storage of data to be moved to a secondary memory. Specifically, large blocks of data that hasn't been accounted for in the input or output peripheral data pathways.

As in previous sections, the data here can be static or dynamic as well. Static presents little challenge, as it is a permanent allocation. However, dynamic storage will again depend on the timing of the tasks sending and receiving the data, so we will again need to know the maximum, average, and peak rates at which the data will be transmitted. And, like the dynamic storage for the output peripherals, we will typically have to infer the rates from other specifications.

Let's take a simple example: temporary storage for nonvolatile values stored in an external EEPROM memory. Having nonvolatile storage for calibration constants, identification data, even a serial number, is often a requirement of an embedded design. However, the time to retrieve the information from the external memory can unnecessarily slow the response of the system. Typically, nonvolatile memory requires additional overhead to access. This may involve the manipulation of address and data registers within an on-chip nonvolatile storage peripheral, or even communications with the memory through a serial bus. In either case, retrieving the data each time it is needed by the system would be inefficient and time consuming. The faster method is to copy the data into faster internal data memory on power-up and use the internal copies for all calculations.

And that is where the amount of internal memory becomes an issue, because:

1. It means that internal data memory must be allocated for the redundant storage of the information.

2. It means that the data will have to be copied from the external memory, and possibly decompressed, before the system can start up.

3. It means that all updates to the constants must also be copied out to the external memory, after being compressed, when the change is made.

This adds up to several blocks of data: data memory to hold the on-chip copies of the calibration constants; more data memory will be needed for any compression/decompression of the data during retrieval, or storage of updates; and, finally, data memory to buffer up the communications strings passed back and forth to the external memory.

OK, so a few shadow variables will be needed for efficiency. And, certainly some buffer space for communications with the external memory is reasonable, but who builds a compression/decompression algorithm into a small embedded system? Well, it may be a requirement that data tables are compressed to maximize data storage in an external nonvolatile memory, such as a data logger counting tagged fish migrating in a stream. If the data logger is a 10-mile hike from the nearest road, and compression extends the time between downloads, then it makes sense to compress the data. If on-chip storage is limited, then packing bits from several variables into each byte saves the cost (in both dollars and time) required to augment the storage with external memory.

Decompression may also be required for communications with an external peripheral. Take the example of an RTC, or real-time clock, peripheral. Its design is based on a silicon state machine, and the interface is a simple serial transfer. Given the chip is completely hardware in nature, it follows that the data will typically use a format that is convenient for the state machine and the interface, and not necessarily a format that is convenient for the microcontroller talking to it. So, to retrieve the current data and time from the peripheral, it is certainly possible that the microcontroller will have to parse the required data from a long string of bits before they can be stored in the internal variables. It may also be necessary to translate the data from binary integers into BCD values for display.

All of these functions require data storage, some of it dynamic with an as yet undetermined length, and some of it static with a predictable length. Our purpose here is to gather as much information concerning the communications needs of the system and determine the likely storage requirements.

If we examine our clock project in light of these requirements, we come up with the following notes for our design file:

INPUT PERIPHERAL

Buttons: These inputs generate dynamic values a single bit in length. There are 6 inputs, with a maximum rate of 3 presses per second, an average of 1 press per second, and a peak rate of 3 per second. That means a storage requirement of 18 bits for a worst case.

60 Hz: This input is the 60-Hz line clock for the system. Its rate does not change under normal operating conditions, so the maximum, average, and peak rates are the same. That leaves us with 1 bit of storage.

OUTPUT PERIPHERAL

Display: The display always has the same number of bytes, 7. One for each digit of the display, plus 1 to keep track of the display currently being driven. So, the storage requirement is static. An additional bit is needed for blanking the display during the Alarm_active time.

Audio alarm: The alarm control is a single bit, with a maximum, average, and peak rate of 2 kHz, so a single static bit of storage. Note: The rate is determined by doubling the frequency of the tone, a 1-kHz tone requires a bit rate of 2-kHz. Also, the rate was not in the requirements document, so the question was asked and marketing determined a 1-kHz tone was appropriately annoying to wake some one.

OTHER SIGNIFICANT STORAGE

Storage for the current time is needed, so six static 4-bit variables to hold hours, minutes, and seconds.

Storage for the current alarm time is needed, so four static 4-bit variables to hold hours and minutes.

Storage for the snooze offset alarm time is needed, so another four static 4-bit variables to hold the offset hours and minutes.

```
Storage for the following system set commands;

SLOW_SET_TIME,    FAST_SET_TIME,    SLOW_SET_ALARM_TIME,    and
FAST_SET_ALARM_TIME

These four dynamic variables have the same timing as the FAST_SET
and SLOW_SET inputs, so 3 bits per variable or 12 bits total.

Storage for the following static system variables;

ALARM_ENABLED, ALARM_SET_ACTIVE, ALARM_ACTIVE, SNOOZE_ACTIVE

It is assumed that the button routine will directly set these
status variables based on the inputs.
```

It should be noted that these requirements are just estimates at this point in the design, and they are subject to change as the design evolves.

3.2.1 Timing Requirements

While the topic of timing has already been raised in the previous section, in this section the discussion will be expanded to include the execution and response time of the software functions.

When discussing timing in embedded software, there are typically two types of timing requirements, *rate of execution* and *response time*. Rate of execution deals with the event-to-event timing within a software function. It can be the timing between changes in an output, time between samples of an input, or some combination of both. The important thing is that the timing specification relates to the execution timing of the function only—for example, a software serial input routine that simulates a serial port. The rate of execution is related to the baud rate of the data being received. If the baud rate is 9600 baud, then the routine must be called 9600 times a second to accurately capture each bit as it is received.

Response time, on the other hand, is the time between when a trigger event occurs and the time of the first response to the event within the function. The trigger is, by definition, an event external to the function, so the response-timing requirement is a constraint on the software system that manages and calls the software functions. Specifically, it determines how quickly the main program must recognize an event and begin executing the appropriate software routine to handle it. Using the same software serial port routine as an example, the initial trigger for the routine is the falling edge of the start bit. To accurately capture the subsequent flow of data bits, the routine will have to sample near the center of each bit. So, at a maximum, the response time must be less than 1/4 bit time; this will place the sample for the first bit within 1/4 bit time of 50%. If the sample placement must be more accurate, then the response time must be correspondingly faster.

Both the rate of execution and response timing requirements should be specified in the requirements document, even if they are not critical. Listing the requirement at least indicates what timing the designer has chosen to meet in the design. It will also become important later in this chapter when we determine the system timing.

Note, that for some software functions, the specifications may be missing. It could be an omission in the document or the specification may be hidden within the specification of another function. Either way, it once again falls to the designer to play detective and determine the timing requirements. As an example, consider the control function from our clock example. In the requirements document, there may not be a specific requirement for response time and rate of execution listed for the command decoder function. However, there should be timing specification for the maximum response time to a button command entered by the user. So, if the timing requirement states that the system response to a button press must be less than 200 msecs from the start of the button press, then 200 milliseconds is the maximum time allotted for:

- The response time, plus execution time for the keyboard debounce function responsible for scanning the keyboard, and determining when a valid button press has occurred.

- Plus, the response time allotted to the control task, for the detection of a command.

- Plus, the execution time allotted for processing of the command and making the appropriate change in the system.

- Plus, the maximum time required to display the change of status on the system display.

If we know the button may take as much as 100 ms to stop bouncing, and the debounce routine will require a minimum of 50 ms to detect the stable button, and the display task scans through all the displays 60 times a second, then we can determine that the command function has a maximum of 34 msec to detect and process the command:

$$34\,\text{msec} = 200\,\text{msec} - 100\,\text{msec} - 50\,\text{msec} - (1/60\,\text{Hz}) \qquad \textbf{Equation 3-2}$$

So, even though there is no specification for the individual functions in the system, there may be an overall timing specification for the execution of the combination of functions. In fact, this will typically be the case with timing specifications. Timing requirements are most often for a combination of functions rather than the individual functions determined by the designer. This makes sense, as the writers of the requirements document can only specify the performance for the system as a whole, because they will not know what the specific implementation chosen by the designer will look like in the product definition phase. So, designers should take care in their review of the requirements document; sometimes the important information may not be in the most convenient format, and it may in fact be buried within other specifications.

Both timing parameters should also have tolerance requirements listed as well. The response time will typically have a single tolerance value, expressed as a plus percentage / minus percentage. And the execution rate will have at least one and possibly two, depending on the nature of the function.

Because the response time is less complicated, let's start with it first. The response timing tolerance is the amount of uncertainty in the timing of when a functions starts. Typically, it is specified as a plus/minus percentage on the response time, or it can also be specified as just the maximum response time allowed. If it is listed as a $\pm$ value, then the response time has both a minimum (Tresponse $- X\%$) and maximum (Tresponse $+ X\%$) specification, and the response time is expected to fall within these timing limits. If, on the other hand, the only specification is a maximum response time, the more common form, then the minimum is assumed to be 0 and the maximum is the specified maximum response time. Because the minimum and maximum times are the values important to our design, either form works equally well. The designer need only determine the minimum and maximum and note them in the design document for the appropriate software function.

The reason for two potentially different tolerances on execution rate is that first tolerance will typically specify the maximum variation for a single worst-case event-event timing, while the second specifies the total variation in the execution timing over a group of events. If only a single tolerance is specified, then it is assumed that it specifies both event-event and the total variation for a group of events. To clarify, consider a serial port transmit function implemented in software. The routine accepts a byte of data to be sent, and then generates a string of ones and zeros on an output to transmit the start, data, parity, and stop bits. The event-to-event timing tolerance governs the bit-by-bit timing variation in the transitions of the ones and zeros sent. If the port were configured for 9600 baud, then the individual bit timing would be 104 μs. The event-event timing tolerance specifies how much this timing can shift for a single bit period. Some bits may be longer, and others shorter than the optimal 104 μs, but as long as they are within the specification, the receiving system should be able to receive the data.

The overall timing tolerance governs the accumulated average variation in bit timing for the complete byte sent by the routine, basically specifying the maximum variation over the course of the entire transmission. The reason this is important has to do with the idea of stacked tolerances. For example, say each bit time within a serial data transmission is allowed to vary as much as ±10%. This means that the bit transitions may vary from as short as 94 μs, to as much as 114 μs. This is not a large amount, and for a single bit time, it is typically not critical. However, if the transmitted bits were all long by 10%, the timing error will accumulate and shift the position of the data bits. Over the course of 6 bits, the shift would be sufficient to move the fourth to fifth bit transition so far out that the receiving system would incorrectly think it is sampling the sixth data bit of data. If, on the other hand, the overall average error is kept below 4%, then even though the individual bits may vary by 10%, most of the individual bit timing errors will cancel. In this scenario, the accumulated error should be sufficiently small to allow the receiver a marginal probability of receiving the valid data.

If we consider the problem from a practical point of view, it makes sense. There will typically be some variation in the timing of output changes. As long as the variation averages out to zero, or some value sufficiently small to be tolerable, then the overall frequency of the output changes will be relatively unaffected by the individual variation. So, note both values in the design notes for future use by the system in the timing analysis later in this chapter.

> **One other point to note:** Check for any exceptions to the timing requirements, specifically any exception tied to a particular action in the function, such as, "The bit timing shall be 9600 baud ±3%, except for the stop bit, which shall be 9600 baud +100/−3." What this requirement tells the designer is that the individual bits in the data stream must vary less than 3%. The one exception is the stop bit which can be as short as the other bits, but may be as long as two complete bits, before the next start bit in the data stream. This is a valuable piece of information that will help in the design of both the timing and priority control sections of the design and, again, it should be noted in the design notes for the project.

Using our alarm clock design as an example, we will first have to glean all the available timing information from the requirements document, and then match it up with our preliminary function list. For those functions that are not specifically named with timing requirements, we will have to apply some deduction and either derive the information from the specifications that are provided, research the requirements in any reference specifications, or query the writers of the document for additional information.

The following is the resulting modification to the requirements document. Note that timing information specified in other sections of the document have been moved to this new section, and additional information has been added as well.

5. TIMING REQUIREMENTS

    a. Display function timing information

        i. The display shall scan at a rate greater than 60 Hz per digit (+20%/-0).

        ii. All display changes shall update within 1 digit scan time maximum.

    b. Alarm

        i. The alarm function will flash the display at a 1-Hz rate (+/-10% event-event, +/-0% overall). Timing of flash shall be synchronous to real-time clock update (+50 msec/-0).

        ii. The alarm tone shall be a 1-kHz tone +/-10% event-event, and overall. Modulation to be at a 1-Hz rate, 50% duty cycle (+/-10% event-event, +/-2% overall).

        iii. Alarm shall sound within 200 msec of when alarm time equals current time.

        iv. Alarm shall quiet within 200 msec of snooze detection, or 200 msec of alarm disable.

    c. Commands

        i. The minimum acceptable button press must be greater than 300 msec in duration, no maximum.

        ii. All button bounce will have damped out by 100 msec after initial button press.

     iii. All commands shall provide a visual feedback (if applicable) within 200 msec of the initial button press.

     iv. For all two-button commands, the first button shall have stopped bouncing a minimum of 100 msec before second button stops bouncing for second button press to register as a valid command.

     v. Autorepeat function shall have a 1-Hz rate (+/-10% event-event, +/-0% overall) increment shall be synchronous to real-time clock update (+50 msec/-0)

  d. Time base

     i. If 5 consecutive 60-Hz cycles are missed, the clock shall revert to the microcontroller clock within 8 msec of 5th missing rising edge.

     ii. When the microcontroller detects 5 consecutive 60-Hz cycles, it shall revert to the power line time base within 8 msec of 5th rising edge detected.

     iii. The real-time clock function shall have the same accuracy as its timebase (+/-0%). Updates shall be within 16 msec of update event to the real-time clock function.

Applying this new information to the functions listed in our function list should result in the following timing information for the project:

SYSTEM TIMING REQUIREMENTS BY FUNCTION:

1. The LED scanning function rate of execution is 360 Hz +20% / -0% event-event & overall, (6 digits * 60 Hz)

2. Display related functions have a response time of 1 digit scan time maximum (see 1.) Functions affected by this timing specification

    12-hour display function for time

    24-hour display function for time

    12-hour display function for alarm

24-hour display function for alarm

PM indicator display function

Alarm on indicator display function

3. The rate of execution for the alarm display flashing routine is (1 Hz rate +/-10% event-event, +/-0% overall) (synchronous to time update +50 msec/-0).

4. The response time for display blanking due to a switchover to the internal time-base is 8 msec maximum, following detection of 5th missing rising edge.

5. All command functions have a response time of 34 msec maximum 34 msec = 200 msec (spec) - 100 msec (switch bounce) - 50 msec (debounce) - (1/60 Hz).
   Functions affected by this timing specification are Command decoder function plus

   Alarm on/off toggling routine

   Routine to increment alarm by 1 min

   Routine to increment alarm by 20 min

   Routine to increment Time by 1 min

   Routine to increment Time by 20 min

   Toggle 12/24 hour mode

6. No specification for debounce time is given. However, 100 msec is the maximum bounce time, therefore a 50 msec maximum time is chosen for worst-case debounce detection. Both the Control input monitoring function and debounce function must execute in this time.

7. Rate of execution for the Auto repeat function is 1 Hz rate (+/-10% event-event, +/-0% overall) event synchronous to time update (+50 msec/-0).

8. The response time for the alarm control function is 100 msec following new current time value equal to alarm time (includes tone startup time).

9. The response time for a Snooze function is 50 msec maximum
   (includes tone off time)
   50 msec = 200 msec (spec) - 100 msec (switch bounce) - 50 msec
   (debounce).

10. The execution rate of the alarm tone function routine 1-kHz
    tone +/-10% event-event and overall, modulated at a 1-Hz
    rate, 50% duty cycle +/-10% event-event, +/-2% overall).

11. The total response time of the 60-Hz monitoring and 60-Hz
    Fail/Recovery functions must be less than 8 msec of either
    the 5th detected 60-Hz pulse or its absence.

12. The rate of execution for the 60-Hz time base and internal
    time base shall be 1 Hz +/-0% overall relative to the source
    time base. Trigger to event response time of 16 msec maximum.

Once the information is complete, it should be noted in the design notes file for the project. Include any equations used to calculate the timing requirements and any special timing information—for example, the requirement in 3 and 7 requiring synchronous timing to the time update, and the notes in 8 and 9 concerning the inclusion of the startup and off times for the tone generator. At this point all the timing information for the system should be known and documented.

3.3 System Priorities

An important topic, related to timing, is the priority requirements for the system. From our discussion earlier, priority handling is different from timing in that timing determines the rate at which a function must be executed, while priority handling is determining if a function should execute. With this in mind, the designer must extract information from the requirements document concerning the operating modes of the system, the priorities within each mode, and when and why those modes change must be determined.

The logical place to start is to determine what operational modes the system has, specifically:

1. Does the system have both an active and passive mode?

2. Does it have an idle mode in which it waits for an external event?

3. Does it have two or more different active modes in which the system has different priorities?

4. Does it have a shut-down mode in which the system is powered but mostly inactive?

5. Does it have a configuration mode in which operational parameters are entered?

6. Is there a fault mode where system errors are handled?

For example, let's generate a priority list for the alarm clock we are designing. From the requirements document, we know:

- The alarm can be either enabled or disabled.

- If enabled, the alarm can either have gone off, or not. Let's call these pending/active.

- If the alarm is active, then it can be temporarily silenced by a snooze command.

- Both the current time and alarm time can be set by button commands.

- If the power fails, the display is blank, time is kept, and alarm functions are inhibited.

If we assign different modes to the various combinations of system conditions, we get the following preliminary list of system modes:

- **Timekeeping mode**: Current time display, alarm is disabled, no commands are in progress, and normal power.

- **Time set mode**: Current time display, alarm is disabled, normal power, and time set commands are in operation.

- **Alarm pending mode**: Current time display, alarm is enabled, normal power, no commands in progress, and alarm is not active.

- **Alarm set mode**: Alarm time display, normal power, alarm set commands are in operation, and alarm is not active.

- **Alarm active mode**: Flashing display of current time, alarmed is enabled, alarm is active, no commands in progress, and normal power.

- **Snooze mode**: Current display time, alarm is enabled, snooze is in progress, and normal power.

- **Power fail mode**: Display is blank, internal time base in operation, alarm is inhibited, and battery supplied power.

Note that each of the system modes is unique in its operation. Some modes are differentiated by the fact that commands are active, others because of the status of the alarm. In fact three of the modes are different states within the alarm function. It doesn't really matter at this point in the design if we have five system modes, or thirty. What we want to determine is all the factors that affect how the system operates. When we get to the priority handler design section of this chapter, we will expand or contract the system mode list as needed to fit the design. For now we just need to generate a reasonable list of modes to hang some additional information on.

If we compare the preliminary list of modes to the previous criteria, we should notice that there is one mode missing, the error mode. We will need a mode to handle error conditions, such as the initial power up, when the system does not know the current time. If we establish this error mode, and define its behavior, we might have something like the following:

- **Error mode**: Display flashing 12:00, alarm is inhibited, no command is in progress, and normal power.

Once the preliminary list of system modes has been established, the next step is to determine which functions are important in each mode. Each mode will have some central operation, or group of operations, that are important and others that are not so important. This translates into some software functions having a higher priority than other functions. In fact, some functions may have such a low priority that they may not even be active. So, using the description of the modes as a guide, we can take the list of functions and determine if each has a high, medium, or low priority in a given mode. Those that are not needed in a specific mode are left off the list. So, once again using our alarm clock as an example, the following preliminary priority list can be compiled:

Priority List

1. Timekeeping mode

   1.1 High Priority
       60-Hz monitoring function
       Time increment function based on 60-Hz power line time base

   1.2 Medium Priority
       Function to scan LED displays
       12-hour display function for time
       24-hour display function for time
       PM indicator display function

    1.3 Low Priority
          60-Hz Fail/Recovery monitoring function
          Control input monitoring function
          Debounce function
          Toggle 12/24 hour mode
          Alarm on/off toggling function

2. Time set mode

    2.1 High Priority
          Control input monitoring function
          Debounce function
          Auto repeat function
          Set Time function
          Routine to increment Time by 1 min
          Routine to increment Time by 20 min

    2.2 Medium Priority
          Function to scan LED displays
          12-hour display function for time
          24-hour display function for time
          PM indicator display function

    2.3 Low Priority
          60-Hz monitoring function
          60-Hz Fail/Recovery monitoring function

3. Alarm pending mode

    3.1 High Priority
          60-Hz monitoring function
          Time increment function based on 60-Hz power line time base
          Alarm control function

    3.2 Medium Priority
          Function to scan LED displays
          12-hour display function for time
          24-hour display function for time
          PM indicator display function

    3.3 Low Priority
          60-Hz Fail/Recovery monitoring function
          Control input monitoring function

Debounce function
Toggle 12/24 hour mode
Alarm on/off toggling function

4. Alarm set mode

    4.1 High Priority
         Time increment function based on 60-Hz power line time base
         Control input monitoring function
         Debounce function
         Auto repeat function
         Alarm control function
         Set alarm function
         Routine to increment alarm by 1 min
         Routine to increment alarm by 20 min

    4.2 Medium Priority
         Function to scan LED displays
         12-hour display function for alarm
         24-hour display function for alarm
         PM indicator display function

    4.3 Low Priority
         60-Hz monitoring function
         60-Hz Fail/Recovery monitoring function

5. Alarm active mode

    5.1 High Priority
         60-Hz monitoring function
         Time increment function based on 60-Hz power line time base
         Generate alarm tone function
         Alarm control function

    5.2 Medium Priority
         Function to scan LED displays
         Display flashing function for the alarm
         12-hour display function for time
         24-hour display function for time
         PM indicator display function

    5.3 Low Priority
         60-Hz Fail/Recovery monitoring function
         Control input monitoring function

Debounce function
Toggle 12/24 hour mode
Alarm on/off toggling function
Snooze function

6. Snooze mode

    6.1 High Priority
         60-Hz monitoring function
         Time increment function based on 60-Hz power line time base
         Snooze function
         Alarm control function

    6.2 Medium Priority
         Function to scan LED displays
         12-hour display function for time
         24-hour display function for time
         PM indicator display function

    6.3 Low Priority
         60-Hz Fail/Recovery monitoring function
         Control input monitoring function
         Debounce function
         Toggle 12/24 hour mode
         Alarm on/off toggling function

7. Power fail mode

    7.1 High Priority
         Time increment function based on 60Hz power line time base
         60-Hz monitoring function

    7.2 Medium Priority
         Function to scan LED displays
         Display blanking function for operation from internal
         clock time base

    7.3 Low Priority
         60-Hz Fail/Recovery monitoring function
         Time increment function based on internal clock time base

8. Error mode

    8.1 High Priority
         60-Hz monitoring function

```
8.2   Medium Priority
      Function to scan LED displays
      12-hour display function for time

8.3   Low Priority
      60-Hz Fail/Recovery monitoring function
      Control input monitoring function
      Debounce function
```

The eight modes are listed with the functions that are important in each mode. The priorities of each function, in each mode, are also established and those functions that are not required are left off the list indicating that they are not used in that particular mode. The result is a clear list of system modes and priorities. The only thing missing are the specific conditions that change the mode. These transitions are generally due to external conditions, such as a command entry or power failure. Transitions can also be due to internal events, such as the alarm time. Whatever the reason, the transition and the event triggering the transition need to be determined and noted. The following are the events triggering a transition in the alarm clock design:

Table 3-2

Original Mode	Next Mode	Trigger Event
Powered down	Error	Initial power up
Error	Time set	Press of the TIME SET button
Error	Alarm set	Press of the ALARM SET button
Timekeeping	Time set	Press of the TIME SET button
Timekeeping	Alarm set	Press of the ALARM SET button
Time set	Timekeeping	Release of the TIME SET button
Alarm set	Timekeeping	Release of the ALARM SET button
Timekeeping	Alarm pending	Alarm control switch to enabled
Alarm pending	Timekeeping	Alarm control switch to disabled
Alarm active	Timekeeping	Alarm control switch to disabled
Alarm pending	Alarm active	Alarm time = current time
Alarm active	Snooze	Snooze command
Snooze	Alarm active	Alarm time + snooze time = current time
{all modes}	Power fail	Fifth consecutive missing 60-Hz pulse
Power fail	Timekeeping	Fifth consecutive 60-Hz pulse
{all modes}	Error	Error condition

With these additions, the system modes and priorities are sufficiently defined for the design.

The only functions that haven't been specified are those functions that fall into the category of *housekeeping functions*. These functions have no specific timing or priority; rather, they are just executed when execution time is available. This could be because their typical timing is infrequent compared to other higher priority functions, or it could be that they are run as a sort of preventive maintenance for the system. Typical examples of this kind of function can include the following:

1. Periodic checks of the voltage of the battery used for battery backup.

2. Periodic checks of the ambient temperature.

3. Periodic verification of a data memory checksum.

4. Functions so low in priority that any other functions are run before they are.

5. Functions that may have a higher priority in other modes, but do not in the current mode.

Any function that is not in the system list of priorities could be included in the list of housekeeping functions, so it can be included in the priority control system. Note that it is perfectly acceptable to have no housekeeping functions. And it is also acceptable to have functions in the list that are only present in some system modes. The only purpose of the list is to guarantee that all functions get execution time, some time during the operation of the system. For our example with the alarm clock, there are no housekeeping functions beyond those with low priority in the various system modes.

3.4   Error Handling

The final section of information to glean from the requirements document is error handling—specifically, what set of errors is the system designed to recognize and how will the system handle the errors. Some errors may be recoverable, such as syntax error in an input, out of paper in a printer, or a mechanical jam. Other errors are more serious and may not be recoverable, such as low battery voltage, failed memory data check sum, or an illegal combination of inputs from the sensors indicating a faulty connection. Whatever the illegal condition, the system should be able to recognize the error, indicate the error to the operator, and take the appropriate action.

The first step is to compile a list of errors and classify them as *soft errors, recoverable*, or nonrecoverable *hard errors*. Soft errors include faults that can safely be ignored, or can be handled by clearing the fault and continuing operations. Typically soft faults are user input faults which can be safely either ignored, or handled by reporting a simple error condition. These include minor user input faults, incorrect syntax, or even the entry of out-of-bound values. Recoverable errors are errors in the system due to transitory system faults that, once cleared, will allow the system to continue operation. These include corrupted data memory, control faults that require user intervention to clear, or a lost program counter. Finally, hard errors are those errors classified as a failure in the system hardware requiring diagnostics and repair to clear. These include the detection of an impossible combination of inputs, failure of the program memory checksum, or failure in the operation of a system peripheral.

After the list of errors has been compiled and classified, the criteria for detecting the error should be specified and all acceptable options for responding to the error. As an example, consider a simple lawn sprinkler controller. It is designed to accept data in the form of water time and duration. When the time corresponding to a watering time is equal to the current time, it turns on the sprinkler for the specified duration.

However, what happens if a specified watering time of 25:20 is entered? Or the current time is 24:59? Or the checksum on the time and duration data memory fails a routine check? These are examples of potential faults for a system. Compiling them into a list and classifying them, we get:

Soft Fault

Fault: User enters a start time > 23:59.

Test: Determined at input by comparison to 23:59.

Response: Display "bad time" for 5 seconds and clear input.

Recoverable Fault

Fault: Data checksum fails.

Test: Determined by checksum housekeeping function.

Response: Display "MEMORY FAULT" and turn off all sprinklers, clear data memory, and wait for user to reset time and duration values.

Hard Fault

Fault: Clock peripheral reports > 24:59.

Test: Determined at time check by comparison to 23:59.

Response: Display "system failure" and turn off all sprinklers and shut down system.

In each of these possible problems, the system has both a means of detecting the fault, and a way to respond to the fault. If the fault, its detection, or recovery are not listed in the requirements document, then it is up to the designer to find answers to these questions and add them to the document.

Note that some faults should be included as a matter of good programming practice, such as watchdog timer (WDT) fault, brownout reset (BOR) fault, and program/data corruption faults. In most microcontrollers, there will typically be flags to indicate that the last reset was the result of a BOR or WDT. Handling these forms of reset will depend on the specific requirements of the system.

Program and data corruption faults are a little different because they rely on software functions to check the CRC or checksum of the data in data memory. While this can be, and typically is, relegated to a housekeeping function for a spot check, it should also be included in any routine that makes changes to the affected data. If it is not included in the modifying functions, the function could make it change, recalculate the checksum and never know that it just covered up a corrupted data value. So it is important to take data corruption seriously and make an effort to provide adequate checking in the design.

For our alarm clock example, the range of faults is fairly limited, but they must still be documented for the next phase of the design.

Soft Fault

Fault: Button pressed is not valid for current mode or command. Press of SLOWSET without FASTSET, ALARMSET, or TIMESET held.

Press of SNOOZE when not in alarm active mode. Press of any key in power fail mode.

Test: Comparison of decoded button command with legal commands, by mode.

Response: Ignore button press.

Soft Fault

Fault: Button combination is invalid.

Press of SNOOZE with FASTSET, SLOWSET, ALARMSET, TIMESET.

Press of ALARMSET with TIMESET.

Test: Checked against acceptable combinations in command function.

Response: Ignore button press.

Recoverable Fault

Fault: Alarm time is out of range (Alarm time $> 23:59$).

Test: Alarm control function test of value before current time comparison.

Response: If alarm is enabled, sound alarm until ALARMSET button press. If in any other mode, ignore (fault will be identified when alarm is enabled).

Recoverable Fault

Fault: Power failure.

Test: 5th missing 60-Hz time base pulse.

Response: Goto power fail mode until 5th detected 60-Hz pulse.

Hard Fault

Fault: Watchdog timer timeout, brownout reset.

Test: Hardware supervisor circuits.

Response: System is reset. If BOR, then system held in reset until power is restored.

System will power up in error mode.

With the compilation of the error condition list, this completes the dissection of the requirements document, and all the relevant information required for the design should now be in the design notes file. In addition, all updates to the requirements document should be complete at this point in the design. If it is not, then the designs should make those updates now, before embarking on the system design. This is not just good coding practice—it will also save confusion and disagreement at a later date when the group responsible for testing the design begins comparing the operation of the design against the requirements document. So, fix it now while the change is simple and still fresh in the designer's mind, rather than later when the reasons for the change may have been forgotten.

3.5 System-Level Design

At this point, the system level of the design is generated. All the information has been retrieved from the requirements document, and the designer should have a clear picture of how the design must operate. What happens now is the top, or system, level definition of the system.

Tasks will be created and the various functions will be assigned to them. A communications plan will be developed to handle data transfers between the tasks. A system timing analysis will be performed to determine the system timing tick. The system modes and priorities will be analyzed, and a system-level error detection and handling system will be defined. Basically, a complete high-level blueprint for the system will be generated, with module specifications for each of the tasks and major systems in the design.

3.5.1   Task Definition

The first step in the system-level design is *task definition*. Task definition is the process of gathering the various software functions from the requirements document dissection together and grouping them into a minimal number of tasks. Each task will be a separate execution module, with its own specific timing, priority, and communications pathways. Because of this, the functions within the module must be compatible, or at least capable of operating without interfering with one another. Now a typical question at this point is "Why a minimal number of tasks—why not create a task for every function?" That would eliminate the need to determine whether or not the various functions are compatible. However, there are two main problems: overhead and synchronization. Overhead is the amount of additional code required to manage a function, the switch statement, the timing handler, and any input/output routines required for communications. Synchronization is the need for some of the software functions to coordinate their function with other functions in the system. Placing compatible functions into a single task accomplishes both goals, the overhead for a group of functions is combined into a single task, and because the functions share a common task, they can coordinate activities without complex handshaking. An example would be combining a cursor function and a display-scanning function into a common task. Putting the two functions together reduces the additional code by half, and it allows the designers to coordinate their activity by combining them into a single execution string. So, there are valid reasons why some of the functions should be combined into a common task.

This is not to say that all software functions should be combined into common tasks. After all, the whole purpose of this design methodology is to generate software that can execute more than one task simultaneously. And there are very good reasons why some software functions are so incompatible that they can't or shouldn't be combined into a common task. Part of task definition is to analyze the various software functions and determine which, if any, functions should be combined.

So, how does a designer decide which functions are compatible and which are not? The simplest method is to start combining similar functions into tasks, and then determine if the combination is compatible. To do this, start by writing the name of each function on a piece of tape. Drafting tape works best because it is designed to be stuck down and taken up repeatedly without much trouble. Next, take a large piece of paper and draw 10–15 large circles on it, each about 5–8 inches in diameter. The placement of the circles is not critical; just distribute them evenly on the paper. Then take the strips of tape with the function names, and place them within the circle on the sheet of paper. Try to group like functions together, and try to limit the number of circles used. Don't worry at this point if some circles have more names inside than others do. We are just trying to generate a preliminary distribution of the functions.

Once all the functions have been distributed into the circles on the paper, take a pencil (not a pen) and name the circles that have pieces of tape in them. Use a name that is generally descriptive of the collection of functions within the circle. For example, if a circle contains several functions associated with interpreting and executing user commands, then COMMAND would be a good label. Try not to be too specific, as the exact mix of functions will most likely change over the course of the analysis for compatibility. And don't be concerned if all the functions are moved out of a specific circle. The names are just for convenience at this point. The final naming and grouping of functions will be decided at the end of the process.

Now that a preliminary grouping is complete, we can begin evaluating the compatibility of the various software functions within each circle. The first step in the process is to place the strips of tape equidistant around the circumference of the circle. If there is not enough room for the tape to lay along the circle, place it on the circle, extending out radially like a star. Next, draw a line from each function to all of the other functions, and then repeat the process for any functions that are not connected to all the other functions. This web of lines defines all the possible relationships between all the functions in the circle, one line for each relationship.

Now that we know all the different combinations to examine, we need a set of basic criteria on which to base our decisions. The criteria will be based on timing, priorities, and functionality. However, the designer should remember that the criteria are just guidelines, not hard and fast rules. The final choice will come down to a judgment call on the part of the designers as to which functions should be combined. For some functions there will be one criterion that states that two functions should be combined, and another that states they should not. This should not come as a surprise; no single set of rules will apply to 100% of all designs. When this happens, the designer should review the reasons given for compatibility and incompatibility and decide which is more important. For example, two functions could have completely different timing and priorities, which would demand that they couldn't be combined. However, if they are also mutually exclusive in execution (they never execute at the same time), then they could be combined into a common task without conflict. The task will simply have to adjust its timing and priority level based on which function is currently active. It would then be up to the designer to decide whether the combination is worth the trouble, or if one or both of the functions should be shifted to another task.

> **Note:** If two functions are combined against the recommendation of one or more criteria, the designer should note the reason in the design notes and make sure that the verbiage is included in the header comments for the resulting task. This will save any later engineer the trouble of determining why one of the compatibility criteria was disregarded.

If the designer finds a function that is incompatible with most or all of the other functions in a circle, it should be moved to another circle with similar functions, and evaluated there. The new circle should be an existing named task, but if it cannot be placed in an existing circle, it can be placed in a new empty circle as a last resort. Remember, we are trying to minimize the total number of tasks, but if the function is completely incompatible, it needs to have its own task.

There will also be cases in which a function should be separated into its own task for priority reasons, specifically if the task is intermittent in operation. In the next chapter, we will examine a priority handler that can make use of this lone function characteristic to reduce the processing load on the system. Against that possibility, the intermittent task should be labeled and set within its own circle for later evaluation.

3.5.2   Criteria for Compatible Software Functions

The criteria in this section should be used to determine if a pair of software functions should or must be combined into a single task. Any criterion that states two functions *should* be combined is making a recommendation. Any criterion that states two functions *must* be combined is stating that the combination should be required and only overruled in the event of a serious incompatibility. Note that this list should be considered a good starting point for developing a designer's own personal list; it is by no means all-inclusive. Over a designer's career, a personal list of criteria should be compiled and fine-tuned as the complexity of their designs increase. Like a good library of custom functions, the design methodology of a designer should grow and improve with time.

Therefore designers should feel free to add or modify these criteria to fit their level of experience and programming style.

- **Software functions that execute sequentially**: This one is pretty obvious: if two software functions always execute one after the other, then it makes sense to put them in a common task. The state machine that implements the task will just execute the states required for the first function, and then continue on, executing the states of the second function. The only restriction to this criterion is that software functions that have to execute simultaneously may need to be separated into different tasks. For more, see the next criterion.

- **Software functions that execute synchronously**: This criterion has a number of restrictions on it. The functions must always execute at the same time, never separately. The functions must also be linear. This means no branches, computed GOTOs, loops, or conditional statements—just a straight linear sequence for both functions. This type of task can also be difficult to implement because the two functions must be interleaved together into a single set of states. As a result, it is only recommended for functions that meet the restrictions exactly. If not, then they must be combined.

- **Software functions that control a common peripheral**: This criterion has to do with managing control over a peripheral. If two tasks exercise control over a common peripheral, then there is the possibility that they may come into contention. This happens when one task is using the peripheral with a specific configuration, and then the other task inadvertently takes control and changes that configuration without the first task's knowledge. If both functions are placed in a common task, it removes the question of control arbitration entirely because the state machine can typically only execute one function at a time. However, if the two functions are incompatible for other reasons, a good alternative is to generate a third task specifically designed to handle the arbitration between the original functions. This kind of task takes on the role of gatekeeper for the peripheral, granting control to one task and holding the other until the first task completes its operation. The second task is then granted control until its operation is complete. Because the separate peripheral task is the only software in direct control of the peripheral, and all data transfers must go through the peripheral task, contention is avoided and both controlling tasks eventually obtain undisturbed use of the peripheral.

- **Software functions that arbitrate control of common data**: This criterion is very similar to the last criterion concerning peripheral control, with the exception that it deals with control over a commonly controlled data variable. Just as two functions may come into contention over the control of a peripheral, two functions may also come into contention over control of a variable. So, this criterion is designed to simplify the arbitration of control, by recommending the combination of the software functions into a common task. However, as with the peripheral criterion, if the two functions are incompatible for other reasons, then a third arbitrating function may need to be created to handle the actual updates to the variable.

- **Software functions that are mutually exclusive in operation**: Often in a design it may be necessary to force two functions to be mutually exclusive in their operations. The two functions may have opposite functions, such as heating and cooling, or they may control a common resource. In any event, mutually exclusive functions are defined as functions that never execute at the same time, or with any degree of overlap. So, functions that meet this requirement must be combined into a single task. This criterion may sound unimportant; after all, the reduction in overhead from combining functions is not so great that it would warrant the arbitrary combination of functions. However, what combining the functions into a single task will do is guarantee their mutually exclusive operation. This is because the state machine can typically only execute a single function at one time. By combining the two functions into a single task, the two functions are accessed by the same state variable, and it will require a specific transition event to move from one function to the other, guaranteeing the mutually exclusive nature of the functions.

- **Software functions that are extensions of other functions**: This criterion is fairly obvious: if two or more functions are related in function, then they should reside in a common task. A good example of this relationship is the display function in our alarm clock example. The functions for scanning the LED display and flashing the display in the case of an alarm are related, and the flashing function is really an extension of the scanning function. Both functions deal with the LED display, and the flashing function is really just a timed blanking of the displays, so combining them together into a single function makes sense. They affect a common resource, their operation is related, and their control of the common display peripheral may require arbitration between the functions. So, combining the functions is a must, it will reduce overhead, simplify the arbitration, and places both display related functions into a single object.

- **Software functions with common functionality**: This criterion has to do with functions that share common aspects with one another—for example, two functions that require a common multistep math sequence, such as a running average. If the functions are placed in a common task, then the math functions can be coded into a common set of states within the task. If the functions are not combined, then the steps for the math function may have to be repeated in both tasks, at the cost of additional program memory. Combining the functions into a common task does save program memory by eliminating the repeated states, but there is a restriction. By placing the two functions into a common task, the two functions are forced to be mutually exclusive in operation. So, if the two functions do not operate in a mutually exclusive fashion, then this criterion does not apply. See the incompatibility criterion following concerning subfunctions.

3.5.3 Criteria for Incompatible Software Functions

The criteria in this section should be used to determine if a pair of software functions should not or must not be combined into a single task. Any criterion that states two functions *shouldn't* be combined is making a recommendation. Any criterion that states two functions *must not* be combined is stating the combination should never be attempted. Note, as previously, that this list should be considered a good starting point for developing a designer's own personal list and is by no means all-inclusive.

- **Software functions that have asynchronous timing**: This criterion is pretty obvious. If two functions can execute at any time and with any degree of overlap in execution, then they must not be combined into a single task. Separating the functions gives them the freedom to execute at any time appropriate to their operation without any interference from the other function. And, this is, after all, the reason for designing a multitasking system, so different functions can execute independent of each other's timing.

- **Software functions that execute at different rates**: This criterion is another obvious restriction, in that it excludes functions that have to operate at different rates. As an example, consider a software serial port operating at 1200 baud and a sound generator operating at 3 kHz. Due to its timing, the software serial port will be required to execute 1200 times a second, and the tone generator function will be required to execute at 6000 a second. While a common state machine could be created to handle the two different functions, the overhead and timing problems make separate tasks a simpler solution. So, separating the two functions is a more efficient solution. However, if the two functions are mutually exclusive, then the complexity in the timing functions is alleviated, and the two functions could be combined. The timing for the task would then depend upon which function is currently operating, with the task itself switching the timing as needed for the two functions.

- **Software functions with different priorities**: Just as with the previous criterion concerning timing, functions with different priorities should also be separated into different tasks. If two functions with differing priorities were to be combined into a single task, the decision of whether to execute the task or not would have to take into account the current function being performed by the task state machine. It would also require that some of the state transitions within the state machine might have to include additional input from the priority handler. This would unnecessarily complicate both the priority handler and the state machine, and any savings in program memory due to the combined overhead could be consumed in the more complicated coding of state machine and the priority handler. So, while it is recommended that the functions should reside in separate tasks, it is up to the designer to weigh any potential savings against the increased complexity.

- **Software functions that operate as subfunctions to other tasks**: Just as common routines in a linear program can be written as a single subroutine and called from two or more places in the program, a subroutine task can be used in a similar fashion by other tasks. While the optimal solution for minimal program memory would have been to combine the subfunction and both calling functions into a common task, incompatibilities between the calling functions may not allow that option. Breaking the subroutine function out into a separate task, which can then be called by the calling tasks, may be preferable to duplicating the function in both controlling tasks, even with the added overhead of a separate task. Separating the subfunction into a separate task will also alleviate any problems with arbitrating control of the subfunction.

- **Software functions that operate intermittently**: Some tasks only need to be active intermittently. If a function is not needed full time, then from the standpoint of efficient use of processing time, it makes sense to only call the function when it is needed. So part-time functions are good candidates for this type of priority control, provided the function is separated into its own task. Note, this does not preclude the combination of two or more intermittent functions into a common task, provided the functions are either synchronous or mutually exclusive in operation.

One or more additional tasks may also be required to handle error conditions within the system. These tasks typically monitor the error condition of the various other tasks in the system and coordinate the recovery from all errors. For example, if a serial input task detects an error in an incoming packet, an error-handler task may have to perform several different functions to clear the error:

1. Reset the serial input task to clear the error.

2. Notify the sender of the current packet of data that an error has occurred.

3. Reset any tasks that might be in the process of operating on the serial data.

4. Reset any data buffer between the tasks.

In addition, the order of the sequence used to clear the error may be critical as well, so building this functionality into a separate error-handling task gives the system the flexibility to handle the error outside the normal operation of the other tasks, especially if the response to the error requires the cooperation of more than one task. Complex systems may even require multiple error-handling tasks if the potential exists for more than one type of error to occur asynchronously. The designer should review the list of potential errors and list all the required recovery mechanisms. Then group them like the software functions in the previous section and apply the criteria for compatible and incompatible functions. Don't be surprised if the list of tasks grows by two or more tasks by the time the evaluation is complete.

Once all the software functions and error recovery functions have been placed in a circle of compatible functions, a final descriptive name for each task/circle can be decided, and a Task list can be compiled. The list should include the name and descriptions of the individual functions in each task, plus any special reasons for including the functions in the task, or excluding it from another task.

Once the list is complete, it should be included in the design notes for the project. Again, be complete in documenting the task list, and be verbose. When the documentation is complete, it should look something like the following:

TASK LIST FOR THE ALARM CLOCK PROJECT

Task1   Display
a)   Function to scan LED displays
b)   12 hour display function for time
c)   24 hour display function for time
d)   12 hour display function for alarm
e)   24 hour display function for alarm
f)   PM indicator display function
g)   Alarm on indicator display function
h)   Display flashing function for the alarm
i)   Display blanking function for operation
        from internal clock time base

Task2   TimeBase
a)   Time increment function based on 60 Hz power line time base
b)   Time increment function based on internal clock time base
c)   60-Hz monitoring function
d)   60-Hz Fail/Recovery monitoring function

Task3   Buttons
a)   Control input monitoring function
b)   Debounce function
c)   Auto repeat function
d)   Command Decode function (combined SetAlarm
        and SetTime functions)
e)   Routine to increment alarm by 1 min
f)   Routine to increment alarm by 20 min
g)   Routine to increment Time by 1 min
h)   Routine to increment Time by 20 min
i)   Toggle 12/24 hour mode
j)   Alarm on/off toggling function
k)   Activate Snooze

Task4   AlarmControl
a)   An alarm control function
b)   Snooze function

Task5   AlarmTone
a)   Generate alarm tone function

Task6   Error Task

The decisions that lead to this combination of functions and tasks are listed below:

Task1 Display

1. The function which scans the LED displays seems to be the primary function of this task.

2. All of the display functions use a common peripheral with the LED display scanning function.

3. The 12/24 hour display functions for the alarm and current time drive a common aspect of a peripheral, the numeric LED display.

4. The 12/24 hour display functions for the alarm and current time are mutually exclusive in operation.

Task2 Timebase

1. The 60-Hz monitoring function seems to be the driving function of this task.

2. Both time base increment functions and the failure/recover monitoring function are extensions of the 60-Hz monitoring function.

3. The 60-Hz time increment function executes sequentially following the 60-Hz monitoring function.

4. The internal clock increment function is mutually exclusive in operation to the 60-Hz increment function, and the control of both functions is via the failure/recover monitoring function.

5. The failure/recover monitoring function is executed sequentially after the 60-Hz monitoring function.

6. Both the 60-Hz time increment function and the internal time base increment function control a common variable, the current time.

Task3 Buttons

1. The control input monitoring function is seen as the overall function of this task.

2. The debounce function is executed under the control of the control input monitoring function.

3. The auto-repeat function is an extension of the debounce function.

4. The command decode function, a combination of the set alarm and set timer functions, is executed sequentially after the debounce and auto-repeat functions.

5. The four alarm and time increment function perform nearly identical functions on the alarm and current time variables, denoting common functionality.

6. The four alarm and time increment functions are mutually exclusive in operation.

7. The four alarm and time increment functions, plus the 12/24 hour toggle function, and the alarm on/off function are executed sequentially following the command decode function.

> **Note:** In this example, it proved to be more efficient not only to combine the alarm and time set functions in a common task, but to also combine the SetTime, and SetAlarm functions into a common function within the task.

Task4 Alarm Control

1. Both the alarm control and snooze functions control two common peripheral functions, the display and the tone generator function.

2. Both the alarm control and snooze functions have common functionality in the form of the alarm / current time comparison function.

Task5 Alarmtone

1. Looking toward the priority control section of the design, the tone generation function is isolated into a separate task due to its intermittent operation.

2. Two functions within the alarm control task control this function, so a separate task will allow arbitration, if needed.

Task6 Error

This task is separate for control of other tasks.

So we have five separate tasks, with one additional task for error handling. All the tasks were generated using the same criteria listed previously, for compatible and incompatible functions. With the compilation of the final task list, this completes the task definition at the system-level design. The final task list, with the rationale behind the decisions, should be copied into the system design notes, and any changes or addendum to the requirements list should be made at this time.

3.5.4 Communications

The next step in the system level of the design is to map out the communications between the various tasks and peripherals in the system. This accomplishes a couple of things for the design: one, it helps provide the designer with a rough estimate on the amount of data memory that the system will require and, two, it defines all of the variables in the system, which is not specific to a task so they can be defined in a common header file. And, three, it provides a quick check for a very troublesome systemic communications problem called *state lock*.

The method employed to generate the communications plan is graphical, just like the method used in the task definition phase of the system-level design. The type of diagram used to map out the communications pathways is called a *data flow diagram*. It consists of a collection of 1- to 2-inch circles, each circle representing a peripheral or task within the system. The circles will be the sources and destinations for information moving around the system. Between the circle are arrows that represent the data pathways along which the information will flow. The direction of the arrow indicates the direction of the data flow. The resulting diagram should show graphically all the communications between the various tasks within the system. Any significant data storage associated with the various tasks are also noted on the diagram. A variable list and dictionary is then compiled, based on the information in the data flow diagram. The resulting documentation will then form the basis of all system-level variable definitions. So, designers are encouraged to be as accurate as possible in both the diagram and the resulting variable documentation.

> **Note:** The usefulness of the data flow diagram does not end once the variable list and dictionary is completed. It also a graphical representation of all system-level data storage that is a convenient reference diagram during the component and implementation phases of the design.

To start the diagram, take large piece of paper and draw a 2- to 3-inch circle for each of the tasks and peripherals in the system. Try to evenly space the circles on the entire sheet, with as much space between the circles as possible. Then, label each circle with the name of its associated task or peripheral.

> **Note:** Don't try to optimize the placement of the circle at this point in the design, as the diagram will be revised at least once during the course of this exercise. Just make sure that there is a circle for each source and destination for data in the system.

For systems that link two or more subsystems by communications pathways, place circles in the diagram for the tasks in both systems. Separate them on the diagram, with a boundary line to show the separation of the two systems, and label the tasks charged with communications between the systems. A short heavy line is used to indicate the system-to-system communications pathway.

Once all the circles have been placed on the diagram, use the communications information from requirements document dissection and the function listing in the task list, to draw arrows between the circles to represent information passed between the various tasks and peripherals. The arrows denote the various task-to-task and task-to-peripheral communication pathways. Start the arrow at the circle representing the task, which contains the sending function, and place the head of the arrow on the circle representing the task, which contains the receiving function. Each of the arrows should then be labeled with a name descriptive of the information being passed along the pathway. See Figure 3-3 for an example of a data flow diagram for our alarm clock project.

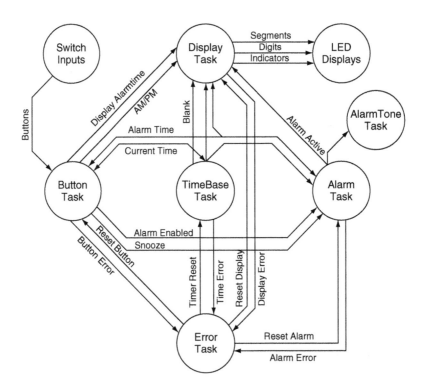

Figure 3-3: Alarm Clock Data Flow Diagram.

Note: The direction of the arrow should indicate the direction of the information flow. Some data pathways may have handshaking flags, which will pass in both directions as part of their communication. However, the direction of the arrow in this diagram is to indicate the direction of the actual communications, so even though handshaking may return, the direction of interest is the direction in which information is actually moving.

For pathways that transfer information from one sending task to multiple receiving tasks, start each pathway arrow at the same point on the sending task's circle to indicate that the same information is being sent to multiple destinations. Then, place the head of each arrow on the circle of each receiving task. Figure 3-4a shows this form of data pathway. It is also acceptable to branch an arrow off from an existing arrow, partway down its length. In fact, a very handy method of showing the distribution of data from one task to multiple other tasks is to create pseudo distribution bus in the diagram, originating at the sending task, with arrows branching off to the receiving tasks as it passes near. Our only purpose here is to clearly indicate that multiple receivers are listening to a common sending task. There are no hard and fast rules to the diagram, and the designer is encouraged to generate whatever form of shorthand is convenient.

In the very likely event that the diagram starts to become cluttered and confusing, try overwriting the pathways with different color pens to distinguish one pathway from

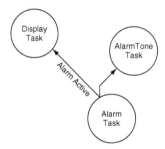

Figure 3-4(a): One Source, Multiple Destinations.

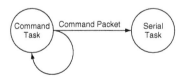

Figure 3-4(b): Storage Loop.

another in the diagram. Be careful not to overwrite two crossing pathways with the same color as this will only add to the confusion. Also, make sure that pathway arrows only cross at right angles, to further reduce confusion.

If the diagram becomes too cluttered or confusing, it is perfectly acceptable to redraw it on a larger piece of paper and relocate the circles that are causing the problem. Remember, I did say that we would be redrawing this diagram at least once, and probably more than once. Plus, after making a few of the pathway connections, the designer will have a better feel for where the task and peripheral circles should be located to simplify the connections. Just remember to follow same procedure and verify that no pathways are inadvertently left off the diagram.

The next step is to label each data pathway with a name and a general description of the data moving along the pathway. If the diagram is sufficiently large, this information can be noted along the length of the data pathway arrow. If, on the other hand, the diagram is too convoluted or cramped, it is also acceptable to legend the arrow with a number and then build a list with the information. Particularly for larger systems, this method is typically easier to manage, and it also lends itself to electronic documentation better than trying to place the text along the arrow in the diagram.

Once all the data pathways between tasks are documented, it is time to add the information related to significant data storage. This information was gathered at the end of the communications section of the requirements document dissection. To show the storage required by the individual tasks, draw an arrow from the task associated with the storage, wrap it around 180 degrees, and place the head on the same task. Then label the loop with a name indicating the nature of the storage. Use the same form of notation used in the last section when describing task-to-task pathways.

In the event that the information is also passed to another task, start the tail of the arrow at the same point on the circle as the arrow leading to the other task, and then loop the arrow around just like the other storage arrows. Label both the loop and the arrow with the same name to show that the information is local storage and a data pathway. Figure 3-4b demonstrates an example of a storage loop.

When the diagram is complete, the designer should go back through the information from the requirements document dissection to verify that all task inputs and outputs have connections to other tasks. The designer should also review the function and task lists to verify that new connections have also been made. Often in the process of task definition, new communications pathways may be created, but through an oversight, the information was not back-annotated to the requirements document. Checking the function and task lists should catch these missed pathways.

> **Note:** The designer is strongly discouraged from skipping over this step as it is a valuable check on the design of the tasks as well as the communications layout of the system.

Unconnected pathways can indicate any one of a number of system design problems:

- The inadvertent generation of redundant data.

- Missing data that must be generated.

- An omission in the task list documentation.

- Or, even a failure in the designer's understanding of the operation of the system.

In any event, the problem should be identified and corrected before continuing on with the design and the affected documentation should also be revised to include the corrections. And yes, the function and task lists, *as well as* the requirements document should be updated.

Once all questions have been resolved and the documentation updated, the diagram should be redrawn one last time in a single color of ink with related peripherals and tasks grouped together so that the pathway arrows are reasonably straight and easy to follow. The diagram should also leave plenty of room for the addition of new pathways. And there will be additional data pathways generated as the design evolves. No design methodology, regardless of how methodical, can accurately predict every possible need in advance. A good methodology though, should be sufficiently adaptable to handle new requirements as the project progresses.

Next, make a list of all the data pathways, prioritizing the list by name of the pathway and the name of the sending task. For pathways with multiple destinations or sources, make a single entry in the list, but list all sources and destinations for the pathway. For each pathway, note the type of data to be transferred, whether the storage is static or dynamic, plus the estimated width and storage requirements. This information should have come from the dissection of the requirements document earlier in this chapter. The designers should take their time in the generation of this list, making it as comprehensive as possible, as the list will be the basis for the final variable dictionary and the header file that will declare the variables used for communications. For dynamic variables, make a note of any estimates made concerning input and output data rates as well.

Once the preliminary list is complete, it is time to assign an appropriate data transfer protocol to each pathway. The protocol used, either broadcast, semaphore, or buffer, will depend on the needs of the pathway and the speeds of the sending and receiving tasks.

How do we determine which protocol is the right one for a given data path? Each protocol has specific advantages and limitations. The buffer protocol has the ability to cache data between fast and slow senders and receivers, but has difficulty with more than one receiving task. The semaphore protocol transfers not only information but also event timing information. However, it can introduce state lock problems if a circular link of pathways in generated. And the broadcast protocol is useful for sending data from one or more senders, to one or more receivers, but it does not transfer event timing. The secret is to match the needs of the pathway to the correct protocol.

The best place to start is with the pathways that were identified as needing dynamic. Because this type of storage is variable, it is best implemented with a buffer type of protocol. The buffer handles variable-length storage well, and the circular storage format allows the sending task to start a second message, prior to the receiving task completing the first message. The only exception to this recommendation is for those pathways that use dynamic for the transmission of a single variable, such as a command flag. Leave these pathways unassigned for now.

Once the pathways using dynamic storage are identified, overwrite them with a green pencil or marker to identify them as buffer protocol pathways.

The next group of pathways to identify are those that need to include event-timing information as part of their data transmission. These pathways will typically fall into a couple of categories:

- **Commands**: Data that initiate an activity by the system; this is typically a user-initiated command or request from external to the system.

- **Events**: An event within the system requiring a response or action be taken in response to the event. This could be a flag indicating that a critical temperature or time has been reached.

- **Changes**: A notification to the system that some important parameter has changed and the system must respond in some fashion. For example, a notification from one task to another that it has completed a its task and a common resource is now available for use.

The semaphore protocol is typically used for these pathways due to its ability to transmit both data and event timing information. The very nature of handshaking requires that both the sending and receiving tasks must temporarily synchronize their operation to complete the transfer. So, it makes the protocol invaluable not only for making sure the receiving task has current information, but also for making the receiving task aware that the current data has changed. Data pathways using the semaphore protocol should be overwritten using a red pencil or marker in the data flow diagram to identify them as semaphore protocols.

The remaining data pathways can be assigned the broadcast protocol. These pathways should be static, and should not require event timing information as part of the transfer. Pathways with multiple destinations should also use the broadcast protocol, due to the complexity involved in handshaking between multiple senders and multiple receiving tasks. These will typically be system or task-specific status information within the system. For example, the current time in our alarm clock design should use a broadcast protocol. This is because the various tasks within the system will not need to know each and every change in the current time. Or the receiving tasks can poll the current time with sufficient speed to see any changes with out the need for an event timing information. Finally, overwrite all the broadcast protocol pathways in the data flow diagram with a blue pencil or marker to identify them.

Once protocols have been assigned and identified by color on the data flow diagram, the diagram should be examined to determine if a potential state lock condition is possible. To find this systemic problem, follow each semaphore pathway, head to tail, from task to task, to determine whether any combination of pathways will produce a complete loop. If they do, then the system is potentially susceptible to a state lock condition. Be sure to check not only pathways within the design, but also pathways that may travel over a communications link into another system. This is the reason that the data flow diagram of multiple linked systems must be drawn on a common diagram.

In a state lock condition, two cross-coupled tasks have both initiated a semaphore data transfer to the other before recognizing the each other's transfer request. This can be between two adjacent tasks, or it can happen between two tasks that have several intermediate tasks between them. The only requirement is that all pathways that form the circle must be semaphore, as it is the handshaking nature of the semaphore that causes the problem.

Because both tasks in a state lock condition have sent data and are now waiting for the other to acknowledge the transfer, they have become locked, perpetually waiting for the other to respond. But, because they themselves are waiting, the condition cannot be resolved. Once in the state lock condition, the only remedy is to break the protocol for one of the transfers.

There are several methods to recover from state lock; however, the best solution is simply to avoid the condition in the first place. The first step is to recognize the possibility. Graphically representing the communications in the system makes this very easy; any complete loop formed exclusively by semaphore communications has the potential to exhibit state lock. The next step is to simply break the circle by replacing one of the semaphore pathways with either a broadcast or a buffer protocol. Even a buffer protocol with only a two-variable storage capability is sufficient to break the cycle. All that has to happen is that one of the two tasks must have the ability to initiate a transfer and then continue on executing within the task. Eventually, the task will notice the other transfer and complete it, breaking the lock.

If all of the pathways in a circular link must be semaphore due to the nature of the software functions in the tasks, then the designer should back up one step and determine if the specific combination of functions is actually necessary. Often, by simply moving a function from one task to another, one or more of the semaphore pathways will shift to a different task and the circle will be broken. Designers should remember that a design need not be fixed at the end of each step; sometimes a decision early in the design leads to a configuration that simply won't work. When this happens, take the design back a step or two in the methodology and try something different. Because the design notes for the design detail every decision in the process, it is a simple process to back up and take the design in a different direction to avoid a problem.

If the problem can't be avoided, the designer need not despair, there are other solutions for avoiding, recognizing, and recovering from state lock conditions. Unfortunately, they are not as simple as just changing a protocol, and they will require some additional overhead in the design, so the discussion on their operation will be tabled until the component phase of the design. For now, the designer should note the problem on the data flow diagram, so it can be addressed in a later phase of the design.

Once all of the potential state lock conditions have been addressed, the variable list should be updated with the selection of communications protocol. Any pathways that still have the potential for state lock should be identified and highlighted with a note concerning corrective action later in the design. The variable list for our alarm clock example is included in Table 3-3, with its associated data flow diagram (Figure 3-3).

Table 3-3 Preliminary Communications Variable List

Variable	Source	Destination	Number & Size	Type	Protocol
• Current_Time	TimeBase Buttons	Display Alarm	6 BCD nibbles	static	Broadcast
• Alarm_time	Alarm	Display Buttons	4 BCD nibbles	static	Broadcast
• Blank	TimeBase	Display	flag	static	Broadcast
• Alarm_enabled	Buttons	Alarm Display	flag	static	Broadcast
• Alarm_active	Alarm	Display	flag Alarm_tone	static	Broadcast
• Snooze	Button	Alarm	flag	static	Semaphore
• AMPM_mode	Button	Display	flag	static	Broadcast
• Display_alarm	Button	Display	flag	static	Broadcast
• Segments	Display	LEDs	7 bit word	static	Broadcast
• Digits	Display	LEDs	6 bit word	static	Broadcast
• Indicators	Display	LEDs	2 flags	static	Broadcast
• Command buttons	Switches	Button	6 flags	static	Broadcast
• Time_error	Timebase	Error	flag	static	Broadcast
• Alarm_error	Alarm	Error	flag	static	Broadcast
• Display_error	Display	Error	flag	static	Broadcast
• Button_error	Button	Error	flag	static	Broadcast
• Reset_time	Error	Timebase	flag	static	Semaphore
• Reset_alarm	Error	Alarm	flag	static	Semaphore
• Reset_button	Error	Button	flag	static	Semaphore
• Reset_display	Error	Display	flag	static	Semaphore

There are several interesting things to note about the variable list compiled for our alarm clock example. One, all of the variables are static, even though several dynamic variables were identified in the requirements document dissection. This is because the dynamic storage was needed for communications between functions that were later combined into a single task. As a result, the sending and receiving functions are now operating at the same speed and no longer need dynamic storage to communicate. Two, there are no pathways using a buffer protocol in the list; this is because the only multibyte data transfers in the system are the time and alarm time values and they are a parallel transfer. And three, there are only five pathways that use a semaphore protocol. This is because the designer chose to put most of the user's commands in the same task with the button test, debounce and command decoder. As a result, the only communications requiring event-timing information are the snooze command and the error reset flags from the error task.

3.5.5   Timing Analysis

One of the key points in this design methodology is that it must generate real-time programming. So, it follows that the analysis of the system's timing requirements should be part of the system's design. In this section, we will examine the timing requirements of each of the software functions in the various tasks, and from this information, determine a timing system that will meet the system's needs.

The first step is to list all the timing specifications from the requirements document. Note, if the functions grouped into a task have different requirements, then the specifications for each function should be included separately. Now is also a good time to review the reasons for combining the function to verify that they should really be in a common task.

In the example shown following, the timing requirements for our alarm clock design example are listed. Entries for both the event-to-event timing and response timing are included in the time domain. If the timing requirement is listed in the form of a frequency, it should be converted to the time domain at this time for easier comparison with the other timing requirements.

```
Task1    Display
         360Hz +20/-0    2.635 - 2.777mS
         Alarm flash     0-50mS following time update (1Hz)
                         50% duty cycle +/-10%
         Blank           909.9mS to 1111.1mS +/-0 overall
         Sync to Time update
         Response to Blank 8mS maximum

Task2    TimeBase
         1sec +/-0 overall relative to internal or 60Hz
         timebase switchover must occur within 8mS of
         presence or absence of 5th pulse

Task3    Buttons
         Button bounce is 100mS
         Debounce is 50mS
         Response to decoded command 34mS maximum
         Auto Repeat 909.9mS to 1111.1mS +/-0 overall
         Sync to time update 0-50mS following time update

Task4    AlarmControl
         Alarm response to time increment, 100mS
         maximum including tone startup
         Snooze response time 50mS including tone shutoff
```

Task5	**AlarmTone**	
	Alarm Tone	.454mS min, .5mS typ, .555mS max
	Modulation	454mS min, 500mS typ, 555mS max event to event
		492mS min, 500mS typ, 510mS max overall

Task6	**Error Task**
	no timing specified.

From this information an overall timing chart for the individual tasks of the system can be compiled. This should list all optimum, minimum, and maximum timing values for both event-to-event and response timing requirements. Any notes concerning exceptions to the timing requirement should also be included.

Table 3-4

	Minimum	Optimum	Maximum
Task1			
scan	2.635	2.777	2.777
flash response	0.000	25.000	50.000
flash offtime	450.000	500.000	550.000
blank	909.900	1000.000	1111.100
blank response	0.000	4.000	8.000
Task2			
timebase	1000.000	1000.000	1000.000
switch response	0.000	4.000	8.000
Task3			
bounce	100.000	100.000	100.000
debounce	0.000	25.000	50.000
command	0.000	17.000	34.000
autorepeat	909.900	1000.000	1111.100
aoutr response	0.000	25.000	50.000
Task4			
time response	0.000	50.000	100.000
snooze response	0.000	25.000	50.000
Task5			
tone	0.454	0.500	0.555
var modulation	454.000	500.000	555.000
modulation	492.000	500.000	510.000
Note: all values in milliseconds			

All the information needed to determine the system *tick* is now present. The system tick is the maximum common time increment, which fits the timing requirements of all the tasks in the system. The tick chosen must be the largest increment of time that will be divided into all of the timing requirements an integer number of times. While this sounds simple, it seldom is in practice. Timing requirements are seldom integer multiples of each other, so the only solution is to choose a tick that fits most of the requirements, and fits within the tolerance of all the rest. When a suitable tick is found, it should be noted in large letters at the bottom of the chart. This number is the heartbeat of the system and will be at the very center of all timing decisions from this point on.

The best tick for our alarm clock is 250 microseconds.

Sometimes even the tolerances on the timing specifications will not allow a reasonable size tick that will fit every requirement. When this happens, the designer is left with a limited number of options:

1. The designer can review the timing requirements for the system, looking for values that can be changed without changing the operation of the system. Timing requirements for display scanning, keyboard scanning, tone generation, and others maybe a matter of esthetics rather than an externally imposed requirement. The only real requirement may only be that they have consistent timing. If the timing for one of these functions is the hard to fit value, experiment with the timing requirements for these functions. Often this will suggest other tick increments that may fit within the requirements of all the functions. For example, the timing for the scanning routine in our example is 2.635 ms to 2.777 ms. However, if it were reduced to 2.5 ms for the minimum, then the system Tick could be increased from 250 μS to 500 μS. This still scans the displays at a greater than 60-Hz rate, so no flicker would be introduced.

2. The second option is to consider moving some of the more difficult tasks to accommodate tasks to a timer-based interrupt. The interrupt can be configured to operate at a faster rate that accommodates the difficult tasks, and frees up the balance of the tasks to operate at a different rate.

    Note: if a task is moved to an interrupt, communications to and from the task will require either a semaphore or buffer protocol. This is because the task will be completely asynchronous to the other tasks, much as the tasks in a preemptive operating system. So, additional handshaking is required to prevent the transmission of partially updated communications variables, in the event that the timer interrupt falls in the middle of a task's update.

3. The third option is to consider using a tick that is smaller than the smallest task timing increment. Sometimes, using a tick that is 1/2 or 1/3 of the smallest task timing increment will create integer multiples for hard to accommodate tasks.

> **Note:** This option will decrease the time available in each pass of the system and increase the scheduling job for the priority handler, so it is not generally recommended. If fact, the original tick value of 250 μS was obtained using this option. However, shifting the display timing would eliminate the need for a smaller tick, so it was chosen instead.

At this point there should also be a quick mention of the system clock. Once the system tick has been determined, a hardware mechanism within the microcontroller will be needed to measure it accurately. Typically, this job falls to one of the system's hardware timers. The timers in small microcontrollers usually have the option to either run from a dedicated crystal oscillator or from the main microcontroller oscillator. If a dedicated oscillator is available, then the oscillator frequency must be set at a 256 multiple of the desired system tick frequency. In our example, that would be 512 kHz, or 256 times 1/.5 ms. If the system clock is employed, a pre- or postscaler will be needed to allow the system clock to operate in the megahertz range. Assuming a prescaler based on powers of two, that means a 1.024 MHz, 2.048 MHz, 4.096 MHz, 8.192 MHz, or 16.384 MHz oscillator. If none of these options are available, then an interrupt routine can be built around the timer, for the purposes of preloading the timer with a countdown value. This value is chosen so that the timer will overflow at the same rate as the desired tick. Note that an interrupt routine is needed for this job because there will very probably be task combinations that will periodically overrun the system tick. An interrupt routine is the only way to guarantee a consistent time delay between the roll-over and the preload of the timer. For our example, we will use a 4.096-MHz main system clock and a divide-by-8 prescaler to generate the appropriate timer roll-over rate for our system tick, and avoid the interrupt option.

Once a suitable timing tick is chosen, the skip rates for all of the system tasks can be calculated. This value will be used by software timers which will hold off execution of the state machine associated with the task, for X number of cycles. This slows the execution of the state machine, so its operation is within its desired timing. Using the timing information from our alarm clock design, and assuming the modified Task1 scan timing, Table 3-5 is constructed.

Note the values in parentheses following the skip rates. These are the skip rates for the maximum times. Assuming that the optimum time is not the maximum, then these values constitute the amount of leeway that is still available in the task's timing. We noted this information for its potential use later in the design, when we define the priority handlers.

Table 3-5

	Optimum	Skip Rate
Task1		
scan	2.500	5
flash response	25.000	50 (100)
flash offtime	500.000	1000 (1100)
blank	1000.000	2000 (2222)
blank response	4.000	8 (16)
Task2		
timebase	1000.000	2000
switch response	4.000	8 (16)
Task3		
bounce	100.000	200
debounce	25.000	50 (100)
command	17.000	34 (68)
autorepeat	1000.000	2000 (2222)
aoutr response	25.000	50 (100)
Task4		
time response	50.000	100 (200)
snooze response	25.000	50 (100)
Task5		
tone	0.500	1
var modulation	500.000	1000 (1110)
modulation	500.000	1000 (1020)

Up to this point in the design, we have assumed that the system would use a rigid timing system that regulates the timing of the software loop holding the task state machines. However, there is another option for systems that are not required to comply with specific timing requirements. The option is to run the system without a timing control. By far, the first option using a rigid timing control is the most common. However, in rare instances, when the timing tolerances are very broad or nonexistent, the second option can be implemented. Now as a designer, you may ask, "What is the advantage to a completely unregulated system and what possible design could possibly operate without some regulation?" The truth is, no system can operate completely without timing regulation, but some systems can operate by only regulating the functions that actually require specific timing. The other tasks in the system are run at the maximum speed of the main system loop.

For example, consider a simple user interface terminal with a display and keyboard. Button presses on the keyboard result in ASCII data being sent to the host system, and data received from the host is scanned onto the display. The only functions in the system that require specific timing are the serial transmit and receive functions interfacing with the host system. The display and keyboard scanning rates only have to comply with a reasonable minimum scanning rate. In this example, the serial input and output tasks are typically regulated by the baud rate of the serial interface. The control, display scanning, and keyboard scanning tasks could then be run at the fastest rate possible given the micro-controller clock frequency. The rate at which these three tasks operate would be variable, based on the execution time of each task on each pass through the system loop. However, as long as the minimum scanning rates are achieved, the system should operate properly.

The advantage to this type of system is that it operates more efficiently and more quickly than regulated systems. There is no dead time at the end of each cycle as the system waits for the next tick; the system just jumps back to the top of the loop and starts into the next task. This saves program memory, complexity, and it means that every available system instruction cycle is used to perform a system function. As a result, the system is very efficient, and will outperform a more rigidly regulated system. The only downside is that the tasks within the loop cannot use the loop timing to regulate their operation. Instead, they must rely on hardware-based timer systems for accurate timing.

The major downside to this system is that it requires a hardware timer for every software-timed function, and only works well for systems with few, if any, routines with strict timing requirements.

3.5.6 Priority Handler

So far in this chapter, we have gathered together the various priority requirements and used them to define the system's modes. This covers the majority of the work at this level of the design. The only additional work is to update the table with the information from the task definition performed earlier in the chapter. Basically, we need to rewrite the priority list and the criteria for mode change list using the task names. We also need to note any functions that should be disabled by a specific system mode.

So, to review the information from the requirements document dissection, we have defined the following list of system modes:

- **Timekeeping mode**: Current time display, alarm is disabled, no commands are in progress, and normal power.

- **Time set mode**: Current time display, alarm is disabled, normal power, and time set commands are in operation.

- **Alarm pending mode**: Current time display, alarm is enabled, normal power, no commands in progress, and alarm is not active.

- **Alarm set mode**: Alarm time display, normal power, alarm set commands are in operation, and alarm is not active.

- **Alarm active mode**: Flashing display of current time, alarmed is enabled, alarm is active, no commands in progress, and normal power.

- **Snooze mode**: Current display time, alarm is enabled, snooze is in progress, and normal power.

- **Power fail mode**: Display is blank, internal time base in operation, alarm is inhibited, and battery supplied power.

Replacing the individual functions with the tasks that now incorporate the functions, we have the following priority list:

Table 3-6

System Mode	High Priority	Med Priority	Low Priority
Timekeeping mode	Time Base Task	Display Task	Button Task Error Task
Time set mode	Button Task	Display Task	Time Base Task Error Task
Alarm pending mode	Time Base Task Alarm Control Task	Display Task	Button Task Error Task
Alarm set mode	Button Task Time Base Task	Display Task	Error Task
Alarm active mode	Time Base Task Alarm Tone Task Alarm Control Task	Display Task	Button Task Error Task
Snooze mode	Time Base Task Alarm Control Task	Display Task	Button Task Error Task
Power fail mode	Time Base Task	Display Task	Error Task
Error mode	Error Task Time Base Task	Display Task	Button Task

There are several interesting things to note about the new priority list. Many of the newly defined tasks include both low- and high-priority functions. This means that some tasks can be classified as either low, mid, or high priority. When compiling the table, always list the task only once, and at its highest priority. When we get to the implementation of the priority handler, we can adjust the task priority based on the value in the state variable, if needed.

Also, note that some of the functions do not change in priority. For example, the display task is always a medium priority. Other tasks do shift in priority based on the system mode; they may appear and disappear, like the alarm tone and alarm control tasks, or they may just move up or down as the button and time base tasks do.

Once the priority list has been updated to reflect the task definition information, we also have to perform a quick sanity check on the criteria for changing the system modes. To be able to change mode, it make sense that the task charged with providing the information that triggers the change must be active before the change can occur. What we want to do at this point is review each criterion, checking that the task providing the trigger for the change is in fact active in the original mode. If not, then the priority list needs to be updated to include the task, typically at a mid or low level of priority. For example, using our alarm clock design example:

Table 3-7

Original mode	Next mode	Trigger event
Powered down	Error	Initial power up
Error	Time set	Press of the TIME SET button
Error	Alarm set	Press of the ALARM SET button
Timekeeping	Time set	Press of the TIME SET button
Timekeeping	Alarm set	Press of the ALARM SET button
Time set	Timekeeping	Release of the TIME SET button
Alarm set	Timekeeping	Release of the ALARM SET button
Timekeeping	Alarm Pending	Alarm Control Switch to enabled
Alarm Pending	Timekeeping	Alarm Control Switch to disabled
Alarm Active	Timekeeping	Alarm Control Switch to disabled
Alarm Pending	Alarm Active	Alarm time = Current time
Alarm Active	Snooze	Snooze Command
Snooze	Alarm Active	Alarm time + snooze time = Current time
{all modes}	Power Fail	5th consecutive missing 60-Hz pulse
Power Fail	Timekeeping	5th consecutive 60-Hz pulse
{all modes}	Error	Error condition

In each of the original modes, the task responsible for providing the trigger, whether it is a button press or missing time base pulses, must be active at some priority level to provide the necessary triggering event. If the task is not active, then the system will hang in the mode with no means to exit. Note that there may be instances in which the response time requirement for a system mode change requires a higher priority for the task providing the mode change trigger. If so, then both system priority and timing requirements may have to shift in order to accommodate a faster response. Make sure to note the reason for the change in priority and timing in the design notes and adjust the priority list accordingly.

Once all the priority information has been cataloged and the necessary task trigger event information verified, copy both the priority list and the list of criteria for making a system mode change into the design notes for the system. Include any information relating the changes made to the design and list any options that were discarded and why they were discarded. Be clear and be verbose; any question you can answer in the text will save you time explaining the choices later when the support group takes over the design.

3.5.7   Error Recovery

So far in our design of the system, we have touched on a few error detection and recovery systems. These include error and default states for the task state machines, a system error task to handle errors that affect more than one task, and a definition of the severity of several system-level failures. In fact, one of the primary software functions in the design of the alarm clock is the automatic switch over to an internal time base if the 60-Hz time base stops; this is also an example of an error detection and recovery system.

What we have to do now is define how these faults will be handled and what tasks will be affected by the recovery systems. In our dissection of the requirements documents, we define soft, recoverable, and hard errors for the system:

Soft Fault

Fault: Button pressed is not valid for current mode or command.

Press of SLOWSET without FASTSET, ALARMSET, or

TIMESET held.

Press of SNOOZE when not in alarm active mode.

Press of any key in power fail mode.

Test: Comparison of decoded button command with legal commands, by mode.

Response: Ignore button press.

Fault: Button combination is invalid.

Press of SNOOZE with FASTSET, SLOWSET, ALARMSET, TIMESET.

Test: Checked against acceptable combinations in command function.

Response: Ignore button press.

Recoverable Fault

Fault: Alarm time is out of range (Alarm time $> 23:59$).

Test: Alarm control function test of value before current time comparison.

Response: If alarm is enabled, sound alarm until ALARMSET button press.

If in any other mode, ignore (fault will be identified when alarm is enabled).

Recoverable Fault

Fault: Power failure.

Test: 5th missing 60-Hz time base pulse.

Response: Goto power fail mode until 5th detected 60-Hz pulse.

Hard Fault

Fault: Watchdog timer timeout, brownout reset.

Test: Hardware supervisor circuits.

Response: System is reset. If BOR, then system held in reset until power is restored.

System will power up in error mode.

We now need to add any new faults that have come to light during the course of the design. These include error conditions within the state machines, or any communications errors between the tasks. We also need to decide on recovery mechanisms, the scope of their control, and whether the recovery system resides in the state machine, or the error task state machine.

Let's start with a few examples. Consider a state variable range fault in the display task state machine. The detection mechanism is a simple range check on the state variables, and the recovery mechanism is to reset the state variable. Because the display task is a control end point, meaning it only accepts control and does not direct action in another task, the scope of control for the recovery mechanism is limited to the task state machine. As a result, it makes sense that the recovery mechanism can be included within the state machine and will not require coordination with recovery mechanisms in other tasks.

A fault in the time base task, however, could have ramifications that extend beyond the task state machine. For example, if the state machine performs a routine check on the current time and determines that the value is out of range, then the recovery mechanism will have to coordinate with other tasks to recover from the fault. If the alarm control task is active, it may need to suspend any currently active alarm condition until after the current time value is reset by the user. The display task will have to display the fact that the current time value is invalid and the user needs to reset the current time. The time base task will have to reset the current time to a default value. And, the system mode will have to change to Error until the user sets a new current time value. All of this activity will require coordination by a central entity in the system, typically a separate error task acting as a watchdog. In fact, the specific value present in the error task state variable can be used as an indicator as to the presence and type of error currently being handled by the system.

To document all this information, we will use the same format as before, classifying the fault as to severity, soft, recoverable, or hard. Name the fault with a label descriptive of the problem and the task generating the fault condition. List the method or methods for detecting the fault, and detail the recovery mechanism used by the system. Remember that each task will have a state machine, and each state machine will have at least one potential error condition, specifically the corruption of its state variable. In addition, there will likely be other potential error conditions, both in the operation of the task and its communications with external and internal data pathways.

Another potential source of errors is from the communications system. Semaphore protocol pathways have the potential to create potential state lock conditions. If the problem cannot be averted by changing one or more of the pathway protocols, then the state lock condition will be an error condition that must be detected and recovered from by the system. Buffers also have the potential to create error conditions, should they fill their buffer space. While these errors are typically considered soft errors because they don't require user intervention, the error-handling system may need to be aware of the problem. Once all the potential system errors have been identified, the severity of the error condition must be determined, a test developed to detect the condition, and a recovery mechanism devised to handle the problem.

This can be particularly problematic for communications errors, specifically potential state lock conditions. This is because both communications in a state lock condition are legitimate data transfers. However, due to the nature of the lock, one of the two pathways will likely have to drop their data, to allow the other communications to continue. So, basically, the error recovery system will have to decide which data pathway to flush and which to allow to continue.

Using our clock design as an example, the following additional error should be added to the system-level design:

Soft Error

Fault: Display task state variable corruption.

Test: Range check on the state variable.

Response: Reset the state variable.

Recoverable Error

Fault: Button task state variable corruption.

Test: Range check on the state variable.

Response: Reset the state variable.

Cancel any current command semaphores.

Reset all debounce and autorepeat counter variables.

Recoverable Error

Fault: Time base task state variable corruption.

Test: Range check on the state variable.

Response: Reset the state variable.

Range check time base timer variables.

If out of range, then reset and notify error task to clear potential alarm fault.

Recoverable Error

Fault: Alarm control task state variable corruption.

Test: Range check on the state variable.

Response: Reset the state variable.

If alarm is active, disable then retest for alarm time.

If alarm enabled or active, range check alarm time.

If alarm time out of range, then notify error task of fault condition.

Soft Error

Fault: Alarm tone task state variable corruption.

Test: Range check on the state variable.

Response: Reset the state variable.

Recoverable Error

Fault: Error task state variable corruption.

Test: Range check on the state variable.

Response: Reset the state variable.

Check status on other system state machines.

If error condition, then set error system mode, set current time to default.

Wait for user control input.

Recoverable Error

Fault: Alarm disabled but also active.

Test: Routine check by error task.

Response: Reset alarm control task state variable.

Recoverable Error

Fault: Snooze active when alarm is disabled.

Test: Routine check by error task.

Response: Reset alarm control task state variable.

Hard Error

Fault: Program memory fails a CRC test.

Test: CRC check on power-up.

Response: System locks, with a blank display.

These additional fault conditions and recovery mechanisms are then added to the design notes. The description of the fault condition should include an appropriate, verbose description of the type of error condition, the error condition itself, the method for detection of the error, and the recovery systems. Include notes on the placement of the new software functions to detect and correct the error condition, plus any options in the design that were discarded and the reasons why.

Notes concerning any additional software functions required to handle the error detection and recovery should also be added to the appropriate task descriptions so they can be included in the state machine design. This includes both errors from the corruption of data variables and the corruption of the state variable for the task state machine.

All notes concerning an Error task or tasks should also be added to the design notes. This includes updates to the task list, the system data flow diagram and variable dictionary, timing calculations, and priority handling information. Remember to review any additions to the communications plan, for potential state lock conditions.

3.5.8 System-Level Design Documentation

At this point, the design should include all of the system-level design information for the design. It may not be final, but it should be as complete as possible. Remember, the next level of the design will use this information as the basis for design, so the information from this level must be as complete as possible.

To recap, the information generated so far includes the following:

- **The requirements document**: Should be updated with all the current system information, including functions required for operation, communications and storage requirements, timing information, and priority information. It should also include detailed information concerning the user interface and finally, all information available on potential system errors, methods used to identify the error conditions, and methods for recovering from the errors.

- **Information retrieved from the requirements document**: Should include information concerning the following:

  Task Information: This includes a list of all the functions the design will be required to perform, any information concerning algorithms used by the functions, and a descriptive write-up detailing the general flow of the functions.

Communication Information: This includes all information about the size and type of data, for internal communications between functions, external communications with off-system resources, and any significant temporary storage. Also any information about event timing that is tied to the variables used, as well as the classification of the data storage as either static or dynamic, plus all rate information for dynamic variables. Both peak and average should also be included.

Timing Information: This includes not only the timing requirements for the individual tasks, but also the overall system timing, including both event-to-event and response-time timing. Should also include all timing tolerance information, as well as any exceptions to the timing requirements based on specific system modes.

Priority Information: This includes a detailed description of all system modes and the trigger events that change the system mode. Should also include the overall priorities for the system, changes in function priorities due to changes in the system mode, and the priorities within each task based on current activities.

- **Documentation on the task definition phase of the system-level design**: This should include descriptive names for the various new tasks in the system, what software functions have been grouped into the functions, and the reasons for combining or excluding the various software functions. In the event that conflicting criteria recommend both combining and excluding a function, the reasoning behind the designer's decision should also be included. The final documentation should also include the preliminary task list, plus any updates due to changes in subsequent areas of the system-level design.

- **Documentation on the communications plan for the design**: This should include all revisions of the system data-flow diagram, the preliminary variable list and all related documentation concerning protocol assignments, memory requirements, and timing information. Special note should be made of any combination of pathways that can result in a state lock condition, and the reasons for not alleviating the problem through the assignment of a different protocol for one of the problem pathways.

- **Documentation on the timing analysis for the system**: This should include all calculations generated to determine the system tick, including both optimum and worst-case timing requirements. Reasons for the choice of system tick should be included, and any functions that are to be handled through an interrupt-based timing system. For systems with unregulated timing, the reasons for the decision to use an unregulated system should be included, along with the plan for any timing critical functions. Finally, the tick itself should be documented along with the skip timer values for all tasks in the system.

- **Documentation on the systems priorities**: Include the updated priority list, using the task name generated in the task definition phase of the design. Note any tasks that combine lower priority and higher priority functions, and the new priority assigned to the task. Note all events that trigger a change in system mode and all information generated in the validation of the trigger event information.

- **Documentation on the error detection and recovery system in the design**: Particularly any new error conditions resulting from the task state machines, potential communications problems, and general data corruption possibilities.

One final note on documentation of the system-level design: in all the design decisions made at this level, some will require back annotation to earlier design notes and even the requirements document for the system. As a designer, please do not leave this to the last moment; there will always be something missed in the rush to release the documentation to the next level of the design. As a general rule, keep a text editor open on the computer desktop to make notes concerning the design. A second instantiation holding the requirements document is also handy. Bookmarks for tagging the main points of the design, such as task definition, communications, priorities, and timing make accessing the documents quick and help to organize the notes. If the notes are made as the information is found, then the information is fresh in the mind of the designer, and the notes will be more complete.

I know this sounds like a broken record, but remember that good documentation allows support designers to more readily take up the design with only minimal explanation for the designer. Good documentation also aids designers if they ever have to pick up the design in the future and rework all or part of the design. And, good documentation will help the technical writers in the development of the manuals and troubleshooting guides for the system. So, there is a wealth of reasons for being accurate and verbose in the documentation of the design, both for the designers themselves and for any other engineers that may have to pick up the design in the future.

At this point in the design, it is also a good idea to go back through the design notes and organize the information into four main areas: task, communications, timing, and priorities. The information in the design notes will be the basis for all of the design work, so spending a few hours at this point to clean it up and organize the data will be time well spent. Do save the original document under a different name in case information is lost in the translation and cleanup.

CHAPTER 4

Some Example Sensor, Actuator, and Control Applications and Circuits (Hard Tasks)

Lewin A. R. W. Edwards

Lewin Edwards' "Open-Source Robotics and Process Control Cookbook: Designing and Building Robust, Dependable Real-time Systems" is a popular book that gives practical advice about creating applications that have to run in the real world. He bridges the huge gap between the theory one learns in college and the gritty realities of making stuff actually work.

This chapter, excerpted from that book, focuses on building the drive and sense electronics needed for a small submarine. He shows how to make the sub go using two different sorts of motors, and how it can sense its position in the undersea environment.

This chapter is an ideal follow-up to Stuart Ball's earlier section about analog electronics. Stuart's discussion is from a "how it all works" standpoint. Here, Lewin goes into the details about actually building the hardware and software. Listings and schematics convert lofty ideas into practical instantiations that one can use to build a working system. Or, for those of us who will never work on a submarine, into ideas we can steal for other, similar projects, ranging from robotic controllers to actuators for industrial control applications.

The stepper motors that Stuart describes so well are here provided with an API—a wrapper, if you will, that application-level developers can invoke to make the sub move. Amateur designers typically write only the simplest interfaces (start motor, stop motor). Lewin provides a much more sophisticated set of resources, like step forward at a particular speed till a limit switch closes. That drastically reduces the workload of application developers, and is the way professional programmers craft their systems.

DC motors, which Stuart also described from a high level, get equal coverage here. Lewin doesn't create a circuit of discrete transistors to give these the bidirectional drive capability this application demands; instead he (as would a professional engineer) picks a standard, off-the-shelf IC that provides the mojo needed to spin up the motors. He then addresses the primary problem all users of these devices face: that of feedback. How far have we gone? Unlike a stepper, whose translation corresponds exactly with the number of pulses generated by the computer, a DC motor's speed varies widely by torque and applied power. Here Lewin uses a simple Hall-effect tachometer to monitor motion and provide feedback. Even better, he uses a hardware counter as a time base, instead of taking the lazy and CPU-cycle-burning approach of a delay loop.

MEMS is not just for researchers! Lewin shows how to add a MEMS accelerometer to the system to measure roll and pitch.

The code all runs on a common AVR processor. It's in assembly. Would you use assembly language in a real-world application? That, of course, depends on a number of factors. But the code is so clear that porting it to C, should that be desired, would be trivial.

—Jack Ganssle

4.1   Introduction

In this chapter, I will present a few useful "cookbook" applications for real-time control circuits designed to perform some specific low-level task and interface with a master controller for instructions and overmonitoring. For the moment, we will deal principally with the design and firmware of the peripherals themselves. The purpose of this chapter is to provide introductory-level information on how to interface with some common robotics-type sensors and actuators, and in particular to show how these can be tied into the type of system we have been discussing. Although the projects are standalone and don't directly develop on each other, you should read at least the description of the stepper motor controller in full, because that section describes how the SPI slave interface is implemented. This information isn't repeated in the descriptions of the other projects.

Note that in this chapter, we will discuss an overall system configuration where all devices are connected directly to the Linux SBC, as illustrated in Figure 4-1.

This configuration is easy to develop and test, and is an excellent basis for many types of projects; in fact, this is how I prototyped all the E-2 hardware. For the sake of completeness, however, I should point out that in the actual E-2 system, all of the peripherals are connected to a single master controller (an Atmel ATmega128, in fact). This controller is connected to the SBC over an RS-232 link as illustrated in Figure 4-2.

The master controller is the real brains of the vehicle. In fact, in E-2 the Linux system can be considered just another peripheral of the master controller. The Linux board

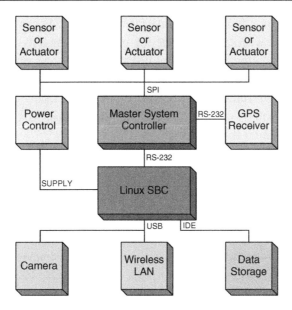

Figure 4-1: Simplified System Layout.

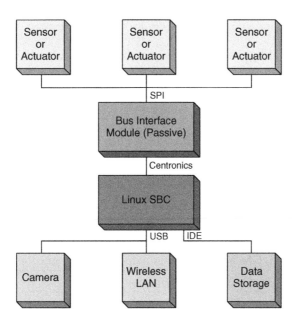

Figure 4-2: Actual E-2 System Layout.

performs strictly high-level functions; it interfaces to two USB cameras and an 802.11b WLAN adapter, besides writing the vehicle log on a high-capacity storage medium and performing some computationally intensive tasks such as image analysis and digital spectrum analysis of audio coming in from the exterior microphones. This design is basically an engineering refinement of the system we'll be talking about in this chapter; discussing it in detail really wouldn't add much to the material you already have. Pay no attention to that man behind the curtain!

4.2 E2BUS PC-Host Interface

Internal control signals in E-2 are carried on a simple SPI-style ("three-wire") interface using a 10-conductor connector referred to as the "E2BUS" connector. The PCB layouts I have provided use JST's PH series 2 mm-pitch disconnectable crimp type connectors. These are commonly used for inter-board connections in applications such as VCRs, printers and CD-ROM drives; they provide fairly good vibration resistance and they hit an excellent price-performance point, as long as you don't mind investing in the appropriate crimp tool. If, however, you are building these circuits on breadboards, you will probably prefer to use standard 5.08 mm (100 mil) headers.

The E2BUS pinout used by the circuits in this chapter is shown in Table 4-1:

Table 4-1

	Name	Function
1	+12 V	+12 VDC regulated supply
2	GND	Ground
3	+5 V	+5 VDC regulated supply
4	GND	Ground
5	MOSI	SPI data input (to peripheral)
6	MISO	SPI data output (from peripheral)
7	SCK	SPI clock
8	_SSEL	Active low slave device select line
9	_RESET	Active low reset input
10	GND	Ground

E2BUS is specified to carry up to 500 mA on each of the 12 V and 5 V lines. Peripherals that expect to draw more than 500 mA on either rail should have separate power input connectors (the main drive motor controller is one example that falls into this category).

> **Note:** 3-wire SPI is in no way related to "three-wire serial" RS-232 interfaces, which are simply a normal serial connection with only RxD, TxD and ground connected. SPI is a synchronous protocol.

There are two useful things to note about the E2BUS connector:

1. It's possible to assemble a cable that will let you connect a PC's parallel port directly to an E2BUS peripheral (in a pinch, you can dispense with buffering and simply run wires direct from the parallel port signals to the E2BUS device). A fairly simple bit-banging piece of software on the PC will allow you to communicate with the peripheral.

2. The E2BUS interface brings out all the signals necessary to perform in-system reprogramming of the flash and EEPROM memory of the AVR microcontrollers we are using, so in theory this port could be used to update the code and nonvolatile parameter data, if any, in an E2BUS module without needing to remove the microcontroller. For various reasons, however, it isn't always possible to achieve this with an AVR-based circuit; either because the ISP pins are being used for other functions by the circuit, or because the microcontroller lacks an external clock source (which may be required for in-system programming). However, the connector design is, at least, flexible enough to allow the possibility if you want to take advantage of it.

At this point, you might be wondering why I chose to use SPI rather than, say, I2C (which requires fewer I/O lines and would allow a true "bus" configuration with a single set of signals routed to all peripherals) or CAN, which is better suited for unfriendly environments such as automotive applications. The first reason is code simplicity. CAN and I2C are both, by comparison with SPI, relatively complex protocols. For example, I2C uses bidirectional I/O lines and it's a little complicated to isolate an I2C device from the rest of the bus, because your isolation component needs to understand the state of the bus. I2C is also best suited for applications where a master device is programming registers or memory locations in a slave device. SPI is a slightly better protocol—with virtually no overhead—for peripherals that deliver a constant stream of data.

For the purposes of this chapter, we'll primarily be talking about controlling E2BUS peripherals directly from the parallel (Centronics) printer port of a PC-compatible running Linux. This is the easiest scenario to describe, and it illustrates all of the required techniques nicely. Following is a schematic for a fairly simple parallel port interface that allows you to connect up to eight SPI-style peripherals to a PC. By means of LEDs, the interface shows you which device is currently selected, and activity on the data input and output pins.

This circuit might appear unnecessarily complicated, but it's really quite simple. The eight data lines from the parallel port are used as select lines for the eight peripherals. These signals are buffered through 74HC244s, the outputs of which are tristated by the parallel port's _STROBE signal. The reason for the tristate control is to reduce the chance of spurious bus transactions while the SBC is performing power-on initialization. Note that this system assumes that the device(s) in use in your peripherals have their own pullup resistors on the select lines. An additional HC244 buffers the same signals to a row of indicator LEDs that show you which device is currently selected. A third HC244 buffers the control signals used for MISO, MOSI, and SCK, and additionally drives the _RESET line.

A side benefit of this circuit: If you use 5 V-tolerant, 3.3 V-capable devices where I have specified 74HC244s, you can use the design in Figure 4-3, virtually unmodified, to communicate between a standard 5 V-level PC parallel port and external devices that use 3.3 V I/Os.

You don't need to build this entire circuit to communicate with the projects in this chapter. If you only want to talk to one peripheral at a time, if you're exceedingly lazy, and if you're willing to take a bit of a risk on port compatibility, you can experiment with a quick-n-dirty cable wired as shown in Table 4-2. The left-hand column indicates the E2BUS pin number, and the right-hand number indicates which corresponding signal should be wired on a DB25M connector.

Be warned—there **is absolutely no protection** for your computer's parallel port if you use this circuit. If you accidentally short, say, a 24 V motor supply onto one of the parallel lines, you will need a new motherboard. I strongly warn you not to use this quick and dirty hack with a laptop computer, unless it's a disposable $50 laptop you bought off eBay!

Also be warned that the simple cable is substantially less tolerant of variations in the motherboard's parallel port implementation than the full E2BUS interface board. If you

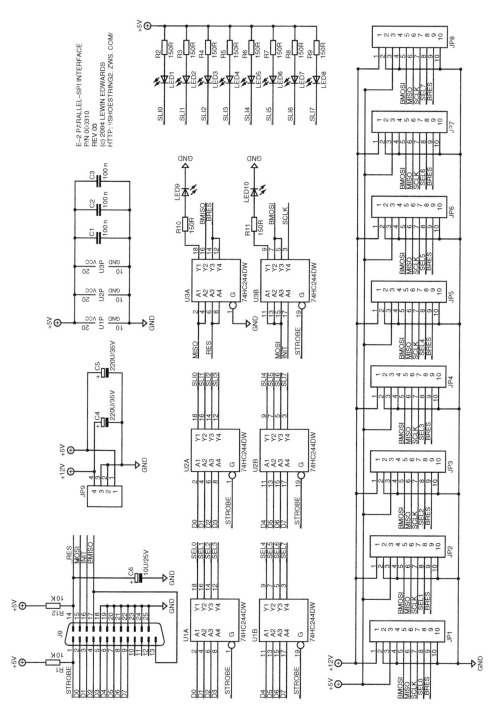

Figure 4-3: Parallel Port E2BUS Interface.

Table 4-2

	Name	Connect to
1	+12 V	External +12 VDC regulated supply
2	GND	+12 VDC ground return
3	+5 V	External +5 VDC regulated supply
4	GND	+5 VDC ground return
5	MOSI	Pin 15 of DB25M.
6	MISO	Pins 17 and 13 of DB25M.
7	SCK	Pin 16 of DB25M.
8	_SSEL	Pin 2 of DB25M.
9	_RESET	Pin 14 of DB25M.
10	GND	Ground, pins 18–25 of DB25M.

find yourself missing transmit or receive bits, or getting garbage data, try adding a rather strong pullup, say 1 K, to the SCLK and MOSI lines. If you still have problems, it may be possible to mitigate them by slowing down your data rates, but there will certainly be some trial and error waiting for you.

As I mentioned in the introduction to this chapter, the actual E-2 project isn't structured exactly as I have described in this section, and the principal reason is energy consumption. The PCM-5820 and its dependent peripherals are the greediest power hog in the entire submarine (these modules of the circuit pull considerably more current than both drive motors operating at full speed), and its brains aren't required most of the time on a typical E-2 voyage. For this reason, the master controller on the voyage is another AVR microcontroller—an ATmega128, to be exact. The peripheral select signals are generated by three GPIOs fed to a 74HC138 1-of-8 decoder. However, I originally started the project by connecting the peripherals directly to the SBC in the manner described in Figure 4-1, because it was the easiest way to debug the protocol and the peripherals themselves. For an early prototype, or for any laboratory fixture application that doesn't require battery power, you almost certainly want to do the same thing; it's much less challenging to debug the protocol and front-end interface issues in this configuration.

The ATtiny26L doesn't implement a full SPI interface in hardware, so the firmware in each peripheral needs to track the state of the select line and manually tristate its serial data output line when deselected. If any module happens to crash in an on-bus state, the entire bus could potentially be brought down. This design flaw could be mitigated to some degree by adding tristate buffers gated by the select line, or by migrating the peripherals to a different microcontroller that implements the full SPI interface in hardware. Also observe carefully that there is no reset generation circuitry on the individual peripheral modules; they rely on receiving an explicit software-generated reset from the attached SBC. A real-world design should implement an external reset generator with brownout detection, to ensure that all modules are reliably reset after a brownout or power-up event.

4.3 Host-to-Module Communications Protocol

The SPI specification only defines the bare outline of the communications protocol, including little more than the physical interface. This is a good thing and a bad thing. It's good, because you can make your protocols as simple as you like—and bad, because it means you have to specify and develop your own high-level protocols! The basic rules are as follows: Each slave device has an active-low slave select line (SS), a clock input (SCK), a data input (MOSI), and a data output (MISO). Note that the words "input" and "output" here are with reference to the slave device. It is fairly normal practice in schematics of SPI equipment to label the entire "output to slave(s)" net as MOSI and the "input from slave(s)" net as MISO. At this point we can sample the data stream out of the micro at MISO. Here's a sample waveform where the host is sending the code 0xFE to a peripheral (Figure 4.4). The top trace is MOSI and the bottom trace is SCK. Note how the pulses have rounded leading edges ("shark fins"). This trace was captured on a system connected using the quick and dirty cable as described previously.

The bit cell is approximately 9.6 μs, corresponding to a serial clock rate of 104.2 kHz. This is the fastest speed we can get out of the PCM-5820 using the code in e2bus.c with all timing delays commented out. Note that we're only using half the available bandwidth; it's entirely possible to implement a full-duplex protocol over the interface described in this section.

From a design perspective, you should observe also that for the projects described here, the Linux machine is always the bus master. This is a significant weak point in system reliability, because a crashed Linux box could potentially leave one or more peripheral modules in the "selected" state, listening to random noise coming down the bus. If you

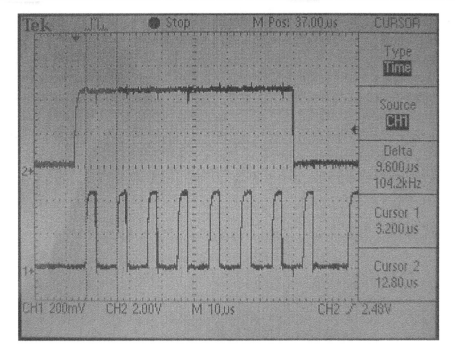

Figure 4-4: Example SPI Clock and Data Signals.

plan to implement a real system with this architecture, you should implement hardware and/or firmware interlocks to prevent such occurrences. For example, you could implement a timeout in the routine that monitors the SS line; if there is no SCK within a specified time period from SS going active, the peripheral should assume a crashed master, and go off-bus. Of course, this doesn't help you if the Linux box has pulled the master reset line low. You shouldn't use a configuration like this to control hardware that may need to be "safed" in event of a loss of control, unless you have some other external hardware that can overmonitor the control system and shut things down gracefully if the controller fails.

I have developed a simple piece of Linux code to do all the synchronous serial I/O you will need to talk to these projects. The meat of this code resides in five simple C functions. Note that these functions assume that your E2BUS interface is connected on the first parallel port. Also note that the timing they exhibit is quite sloppy, since we're not attempting to make Linux appear real time. You should not run this code inside a low-priority thread, because other things will preempt it and may cause spurious timeout problems.

Table 4-3 shows the basic function prototypes:

Table 4-3

Prototype	Description
int E2B_Acquire(void)	You must call this function before calling any other E2BUS functions. It attempts to get exclusive access to the first parallel port. It returns 0 for success or –1 for any error.
void E2B_Release(void)	You can call this function as part of your at-exit cleanup routines. It ensures that all devices are deselected, and releases the parallel port. If you exit without calling this function, the port will still be released implicitly as your task ceases to exist, but devices may still be selected.
void E2B_Reset(void)	Deselects all devices, asserts the reset line on the SPI bus for 250 ms, then pauses for an additional 250 ms before returning.
void E2B_Tx_Bytes(unsigned char *bytes*, int *count*, int *device*, int *deselect-after*)	Asserts the select line for the specified *device* (valid device numbers are 0–7), then clocks out the specified number of bytes one bit at a time. If *deselect-after* is nonzero, the device is deselected after the transmit operation is complete. Setting this argument to 0 allows you to read back a command response without having to set up a new SPI transaction.
void E2B_Rx_Bytes(unsigned char *bytes*, int *count*, int *device*, int *deselect-after*)	Works exactly the same as E2B_Tx_Bytes(), but receives data instead of transmitting it.

These functions, particularly E2B_Rx_Bytes and E2B_Tx_Bytes, are the low-level underpinnings of the E2BUS protocol.

On the device end, all the example circuits here share pretty much exactly the same code for serial transfer operations, though command processing details are naturally specific to each project. Incoming SPI data is received by the ATtiny26L's USART and processed by a very simple and hence robust state machine. When a device's SEL line is inactive, the state machine is in a quiescent mode (FSM_SLEEP); the MISO pin is set to input mode (to prevent it from driving the bus); clock and data from the USI are ignored, and USI interrupts are disabled. Asserting SEL pushes the state machine into a "listen for command byte" mode, resets the USI, and enables data receive interrupts. The first complete byte received generates an interrupt which causes a state transition. The destination state is determined by the value of the command byte received. The machine may transit through further states depending on whether the command requires additional data bytes or not. If the received command requires additional data, the system proceeds through intermediate states to receive these additional byte(s), and then executes the command before returning to quiescent mode.

If the destination state involves transmitting data back to the host, the data required for transmission is assembled for return to the host, and subsequent USART overflow (or rather, underflow) interrupts clock the data out a byte at a time. After the last reply byte is clocked out, the final underflow interrupt causes a transition back to the quiescent state.

Deasserting SEL at any time immediately disables the USART and tristates MISO. This completely aborts any data transfer or command in progress; any partially received command will be discarded, and partially transmitted data blocks will be forgotten.

4.3.1 Stepper Motor Controller

Stepper motors are useful for relatively low-speed, intermediate-torque drive and positioning applications, particularly where accurate sub-revolution rotor position control is necessary. Motors of this type are commonly used to drive the reels on electromechanical slot machines (one-armed bandits), to position floppy disk drive heads, operate trainable camera platforms, and to power the drive wheels of small mobile robots. In times of yore, they were also used to position hard disk heads, though such applications have long ago been taken over by voice-coil type mechanisms. Stepper motors are simple and cheap to use, and you don't need to have a fully closed-loop controller to use them accurately. Servomotors are much faster, but for guaranteeable positioning accuracy, you need to have a position encoder on the shaft to provide feedback on the actuator's position. By contrast, as long as you don't stray outside your system's nominal acceleration profile, a stepper-based system can reliably maintain its position indefinitely without recalibration.

There are several types of stepper motor, with varying electrical drive requirements. However, by far the most common type of motor to be found on the surplus market (or scavenged from unwanted computer equipment) is the four-pole unipolar type, so this is the type our circuit is designed to use. Without further ado, here's the schematic:

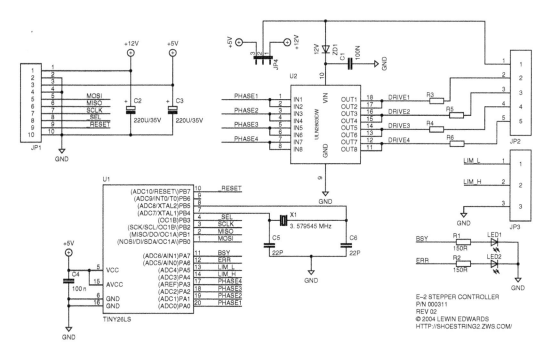

Figure 4-5: Stepper Motor Control Circuit.

Note: When faced with an unknown stepper motor of small to intermediate size, a very reliable gamble to play is as follows: if it has more than four wires, it's probably a four-pole unipolar motor, 0.9 degrees per step, and likely rated for either 5 V or 12 V operation. There are vast numbers of motors constructed with these characteristics.

Note: The alternate function for pin 1 is misprinted in Figure 4-5 as NOSI—it should be MOSI. This is an unimportant typographical error in the atmel.lbr library supplied by Cadsoft as part of the EAGLE package.

This project uses the ULN2803 octal high-voltage, high-current Darlington array to switch the stepper coils. This chip is readily available for around $0.75 in small quantities, and it is a handy solution for driving moderate loads. Until recently, one could often find this chip, or its close relatives, in commercial stepper motor applications such as inkjet printers and both sheet-fed and flatbed scanners. At present, however, it appears to be in decline as application-specific microcontrollers with high-current drivers on-chip take over its market space. On the subject of prices, you'll notice that I've specified an NTSC colorburst crystal as the clock source, despite the fact that the tiny26L is rated at up to 8 MHz for a 5 V supply voltage. I chose the 3.579545 MHz value, although it's not a nice integer to work with, because these crystals are available everywhere and are often cheaper than other speeds. Chances are you have several in your junkbox already, in fact. You'll also find that application notes for microcontrollers almost always give precalculated example timing constant values (e.g., for setting the baud rate of a UART) for this base clock speed.

Our example stepper controller module also has two active-low limit switch inputs. These are optionally used to signal end-of-travel in the increment and decrement step directions. Note that JP4, which selects between 5 V or 12 V drive for the stepper coils, is intended to be a wire link for factory configuration, rather than a user-changeable jumper. If you are using the device in 5 V drive mode, you should alter or remove ZD1; you can also omit C2, since it serves no function if you're driving the motor off the +5 V rail.

The controller operates in one of two modes: "drive" or "train." In drive mode, you simply specify a speed and direction, and the motor turns in that direction until commanded to stop. Optionally, you can request that it travel until either of the limit switches is triggered. Train mode is intended for positioning applications. In this mode, you command the stepper controller to seek to a specific offset from the current position, and it will automatically seek to that position while you carry out other tasks. The stepper will automatically cut off if it hits the high limit switch while seeking forwards, or the low limit switch while seeking backwards.

Note that the limit switches are permanently associated with specific seek directions. The "low" limit switch is only enforced for "backwards" seeking, and the "high" limit switch is only enforced for "forwards" seeking. The reasons for this are twofold: First, an external force—say, water rushing past a submarine's rudder—might turn the stepper past the make-point for the limit switch, before it reaches a mechanical stop. Second, switches are practically never perfect—in other words, the displacement required to make a contact isn't necessarily the displacement required to break it. You might need to push the arm of a microswitch two steps in to penetrate the oxide layer on its contacts; the first step in the other direction might leave the cleaned metal contact surfaces still touching. Or you might be using a reed switch—you need to bring the magnet to a certain proximity to close the switch, but a weaker field will suffice to hold the switch closed. In any of these sorts of cases, it could require one or more "extra" reverse steps to clear the limit condition.

The stepper controller accepts 8-bit command bytes, optionally followed by additional data. Similar serial reception code is used in other projects in this book, so it deserves a little additional study here. To begin with, please note that my choice of I/O pin assignments was by no means arbitrary. The AVR's pin-state-change interrupts are useful, but not very intelligent. On the tiny26L, there are only two such interrupts: PCINT0, which (if enabled) fires on state-changes for pins PB0-PB3, and PCINT1, which fires on state-changes for pins PA3, PA6, PA7, and PB4-PB7. When one of these interrupts fires, there is no direct way of determining which pin caused the interrupt; you have to maintain a shadow copy of the port registers and compare them to determine which pin(s) changed state.

Fortunately, when an alternate function is enabled for a pin, that pin will no longer generate state-change interrupts (note that there are a couple of exceptions to this rule). Even more fortunately, the three USI signals used for SPI-style communications are mapped to pins PB0-PB2. Thus, by configuring the USI in three-wire mode, PCINT0 will fire only if PB3 changes state. Since the USI in the tiny26L doesn't implement slave select logic in hardware, we need to do it in software—and as a result of all the discussion in the previous paragraph, it makes excellent sense to use PB3 as the SPI select line, since it has a state-change interrupt all to itself.

The entire meat of the stepper code is contained in three interrupt handlers: USI overflow, timer 0 overflow, and PCINT0. PCINT0 is probably the single most important function in the firmware—it is responsible for checking the state of PB3 and disabling the output driver on MISO (PB1) when the stepper controller is deselected (so we don't fight with anything else on the bus), or enabling it if _SSEL is asserted. When the device is deselected, this ISR also disables USI interrupts, because we don't care about other transactions that may be occurring on the bus, and having to service USI interrupts causes timing jitter in any step operation we happen to be running in the background. Here's the code in this handler:

```
; I/O pin change interrupt
; The only valid source of this interrupt is PB3, which is used as
; the 3-wire slave select line.
entry_iopins:
  push r0
      push r16
      push r17
      in r0, SREG

      ; Check state of select line, which is the only line that
      ; should have generated this interrupt.
      sbic PINB, PORTB_SEL
      rjmp usi_disable

      ; SEL line is LOW. Enable and reset USI and switch
      ;PB1 to output
      ldi r24, FSMS_RXCMD
      ldi r16, $00
      out USIDR, r16      ; Empty USI data register
      out USISR, r16      ; Clear USI status (including clock count!)
      sbi DDRB, PORTB_DO  ; set PB1 to output

      sbi USISR, USISIF   ; Clear start condition status
      sbi USISR, USIOIF   ; Clear overflow status
      sbi USICR, USIOIE   ; Enable USI overflow interrupts

      rjmp iopin_exit

      ; SEL line is HIGH. Disable USI and switch PB1 to input
      ; to take us off-bus
usi_disable:

      ; disable USI start and overflow interrupts
      cbi USICR, USISIE
      cbi USICR, USIOIE

      ; Disable output driver on PB1 (DO)
      cbi DDRB, PORTB_DO   ; set PB1 to input
iopin_exit:

      out SREG, r0
      pop r17
      pop r16
      pop r0
      reti
```

Actual stepping operations are performed in the timer 0 overflow interrupt. Timer 0, which has an 8-bit count register, is clocked through the prescaler at CK/256, which is approximately 14.053 kHz. When the overflow interrupt fires, the first thing the handler does is to reload the timer register with a step speed value. The default speed value is $00. Since timer 0 counts upwards, this means that by default the step speed is roughly 55 Hz, which is the slowest configurable speed. You can configure faster speeds by using the CMD_STEP_SETTICK command, followed by an 8-bit parameter that sets a new (larger) reload value. For instance, if you configure a reload value of $E0, the timer will overflow every 33rd ($21) tick instead of every 256th, thereby yielding a step speed of approximately 425 Hz. Theoretically, you could specify a reload value of $FF, resulting in an overflow on every tick and a 14.053 kHz step speed, but in practice there is an upper boundary on legal values for the timer reload figure. This boundary is set by the number of CPU instruction cycles required to service an incoming interrupt and make ready for the next, and it caps the step speed at about 7.1 kHz (reload value $FE) for the cheap NTSC colorburst clock crystal I specified. This shouldn't be a serious impediment: although many stepper motors are rated for as much as 10,000 steps/sec, real applications rarely exceed 2,000 steps/sec (300 rpm) due to the fact that the torque of a stepper motor rapidly decreases as step speed increases. The code for Timer 0 handler, along with the subroutines it calls, is as follows:

```
; Timer 0 overflow
entry_timer0:
      push r0
      push r16
      push r17
      in r0, SREG

      ; Reset TMR0 counter to start position for next tick
      lds r16, tick_speed
      out TCNT0, r16

      ; Update state of limit switch flags in machine
      ; status byte (for the benefit of the main thread only)

      sbr r25, (1 << LIM_H)
      sbic PINA, PORTA_LIM_H
      cbr r25, (1 << LIM_H)

      sbr r25, (1 << LIM_L)
      sbic PINA, PORTA_LIM_L
      cbr r25, (1 << LIM_L)
```

```
        ; Load current tick-command and see what we should be doing
        cpi r19, TICK_FWD
        breq tick_seek_fwd
        cpi r19, TICK_REV
        breq tick_seek_rev
        cpi r19, TICK_POWERDOWN
        breq tick_poweroff
        cpi r19, TICK_SEEK
        breq tick_seekto
        cpi r19, TICK_SEEK_FWEND
        breq tick_seek_fwdend
        cpi r19, TICK_SEEK_RVEND
        breq tick_seek_revend

        ; Note - TICK_SLEEP falls through to here
        rjmp tick_done

; Tick event - Seek forward, ignoring limit switch
tick_seek_fwd:
        rcall seek_fwd
        rjmp tick_done

; Tick event - Seek backward, ignoring limit switch
tick_seek_rev:
        rcall seek_rev
        rjmp tick_done

; Tick event - Seek forward, honoring limit switch
tick_seek_fwdend:
        sbis PINA, PORTA_LIM_H
        rjmp seekto_finished

        rcall seek_fwd
        rjmp tick_done

; Tick event - Seek backward, honoring limit switch
tick_seek_revend:
        sbis PINA, PORTA_LIM_L
        rjmp seekto_finished

        rcall seek_rev
        rjmp tick_done
```

```
; Tick event - Power down motor
tick_poweroff:
      andi r25, ~(1 << SEEKING)          ; Turn off busy flag
      in r16, PORTA
      andi r16, $F0
      out PORTA, r16
      ldi r19, TICK_SLEEP
      rjmp tick_done

; Tick event - Generic seek operation
tick_seekto:

      ; First check if the step count is 0 - if it is, then there's
      ; nothing left to do and we should go back to sleep.
      cpi r23, $00
      brne seekto_nz
      cpi r22, $00
      brne seekto_nz
      cpi r21, $00
      brne seekto_nz
      cpi r20, $00
      brne seekto_nz

      rjmp seekto_finished

seekto_nz:
      sbrc r25, DIRECTION
      rjmp seekto_fwd

      ; Seekto - REVERSE
      ; Check limit switch. If it's active, we stop.
      sbis PINA, PORTA_LIM_L             ; Check limit switch
      rjmp seekto_finished

      rcall seek_rev
      rjmp seekto_update_count

      ; Seekto - FORWARD

seekto_fwd:
      sbis PINA, PORTA_LIM_H             ; Check limit switch
      rjmp seekto_finished

      rcall seek_fwd
      rjmp seekto_update_count
```

```
seekto_update_count:
      dec r23
      cpi r23, $FF
      brne seekto_notz
      dec r22
      cpi r22, $FF
      brne seekto_notz
      dec r21
      cpi r21, $FF
      brne seekto_notz
      dec r20

   ; No terminal conditions have been encountered - continue stepping
; seekto_notz:
      rjmp tick_done

seekto_finished:
      ldi r19, TICK_SLEEP
      andi r25, ~(1 << SEEKING)            ; Turn off busy flag
      rjmp tick_done

tick_done:
      out SREG, r0
      pop r17
      pop r16
      pop r0
      reti

; SUBROUTINE - Seek forward one step
; Destroys R16, R17, SREG
; Updates X,Y
; Implicitly powers up motor and leaves it in powered state
seek_fwd:
      lds r16, stepper_phase
      inc r16
      andi r16, $03              ; Only the lower two bits interest us
      sts stepper_phase, r16     ; Store new current phase
      ldi r17, $01

sf_lp:
      cpi r16, $00
      breq sf_lp_done
      dec r16
      lsl r17
      rjmp sf_lp
```

```
sf_lp_done:
     in r16, PORTA
     andi r16, $F0
     or r16, r17
     out PORTA, r16

sf_update:
     inc r29                    ; increment step position
     brne sf_done
     inc r28
     brne sf_done
     inc r27
     brne sf_done
     inc r26

sf_done:
     ret

; SUBROUTINE - Seek backward one step
; Destroys R16, SREG
; Updates X,Y
; Implicitly powers up motor and leaves it in powered state
seek_rev:
     lds r16, stepper_phase
     dec r16
     andi r16, $03             ; Only the lower two bits interest us
     sts stepper_phase, r16    ; Store new current phase

     ldi r17, $01

sr_lp:
     cpi r16, $00
     breq sr_lp_done
     dec r16
     lsl r17
     rjmp sr_lp

sr_lp_done:
     in r16, PORTA
     andi r16, $F0
     or r16, r17
     out PORTA, r16
```

```
sr_update:
      ; Check for zero condition.
      sbic PINA, PORTA_LIM_L
      rjmp sb_notz
      clr r26
      clr r27
      clr r28
      clr r29
      ret

sb_notz:
      dec r29                          ; decrement step position
      cpi r29, $FF
      brne sr_done

      dec r28
      cpi r28, $FF
      brne sr_done
      dec r27
      cpi r27, $FF
      brne sr_done
      dec r26
      cpi r26, $FF
      brne sr_done

sr_done:
      ret
```

The most complex code segment, at least in terms of code volume, is the USI handler. This handler implements a simple state machine. The first byte received after the select line goes active is a command byte. This byte either causes the USI receive ISR to modify the system state directly, or to transit the ISR's state machine through further intermediate states either to receive more data, or to transmit a multibyte response back to the host. I won't reproduce the code here, because it's largely a very long switch … case statement implemented in assembly language.

The theoretical transfer speed between the stepper controller and its master is limited by a couple of factors. First, hardware absolutely limits the USI to transfer clock rates of $f_{CK}/4$ in three-wire slave mode. In our case, this is approximately 895 kHz, and we have to be careful that the SPI master doesn't exceed this speed. In the case of a PC parallel port master, it is unlikely (however, not entirely impossible) that you will be able to outrun the USI. The reason for this is tied up in the ancient PC architecture of twenty years ago, and the fact that the parallel interface is designed for backwards compatibility with 9-pindot-matrix printers run by slow 8-bit microcontrollers. Because of these compatibility issues, the parallel port registers are, conceptually or physically, on the other end of an ISA bridge; they are limited to ISA bus clock speeds (nominally 8 MHz) and may even have additional wait states inserted. Furthermore, the layers we traverse when making calls to change the parallel port registers add extra delays, the constant SMM interrupts on the Geode platform are stealing cycles from us regularly—and to cap it all off, the application layer in Linux is inherently so nonreal-time as to make the idea of accurately marking off 1 μS delay steps in userland quite silly indeed. You will observe, therefore, that the SPI code I provide in e2bus.c does not have explicit timing instructions throughout all of the state-changes. This code is tested on the 300 MHz Geode platform with Advantech's BIOS 1.23; if you need to rely on it to work faultlessly, it will need, at minimum, testing and requalification on other systems.

Table 4-4 shows a complete list of commands recognized by the stepper controller's firmware:

Table 4-4

Mnemonic	Value	Description
STEP_CMD_STOP	0	Aborts any current step operation. Note that the current step position, and the steps-remaining counter, are not altered—you can resume the step operation later, if necessary. The stepper motor coils remain powered.
STEP_CMD_SLEEP	1	Aborts the current step operation and de-energizes all stepper coils. The motor may move a step or two unpredictably when you re-energize it; a calibration may be necessary.
STEP_CMD_SETTICK	2	Set the step rate (larger values = faster rate). This command should be followed immediately by a one-byte step rate.
STEP_CMD_DRIVE_FWD	3	Starts the motor driving forwards at the configured step speed. The motor will continue to spin until another step command is received, ignoring the limit switches.
STEP_CMD_DRIVE_REV	4	Works the same as STEP_CMD_DRIVE_ FWD but steps backwards instead of forwards.

(continued)

Table 4-4 *(Continued)*

Mnemonic	Value	Description
STEP_CMD_STEP_FWD	5	Steps forwards a specified number of steps, or until the "high" limit switch is closed. This command should be followed immediately by a four-byte step count (MSB first).
STEP_CMD_STEP_REV	6	Works the same as STEP_CMD_STEP_ FWD but steps backwards instead of forwards, and monitors the "low" limit switch. The current step position is set to 00000000 when this switch is closed.
STEP_CMD_FWD_END	7	Steps forwards continuously until the "high" limit switch is asserted.
STEP_CMD_REV_END	8	Steps backwards continuously until the "low" limit switch is asserted. The current step position is set to 00000000 when this switch is closed.
STEP_CMD_READ_STATUS	254	Reads back the controller status byte and four-byte step position.
STEP_CMD_RESET	255	Performs a soft reset.

The status byte returned by STEP_CMD_READ_STATUS is formatted as shown in Table 4-5:

Table 4-5

Bit	Function
7	Unexpected interrupt or internal error detected.
6	Step error, e.g., attempt to seek beyond calibrated range (LED2 – ERR – tracks the state of this bit).
5	Reserved.
4	Reserved.
3	Reserved.
2	High limit switch asserted.
1	Low limit switch asserted.
0	Currently seeking to requested position (LED1 – BSY – tracks the state of this bit).

This is an extremely simple stepper control design; it is intended for low-speed positioning and simple low-torque drive applications only. The E-2 project uses these modules to position its rudder and dive planes, and to swivel a camera platform, neither of which are particularly demanding applications. For a little more understanding of the subtleties of stepper motor control, try running this little code snippet, which assumes that you have a stepper module connected as E2BUS device#0:

```c
#include <stdio.h>
#include "e2bus.h"
int main (int _argc, char *_argv[]) {
      unsigned char pkt[2];

      // Open port and start stepper motor
      if (E2B_Acquire()) {
            printf("Error opening E2BUS.\n");
            return -1;
      }
      E2B_Reset();

      pkt[0] = STEP_CMD_DRIVE_FWD;
      E2B_Tx_Bytes(pkt, 1, 0, 1);

      // Speed motor up gradually
      pkt[0] = STEP_CMD_SETTICK;
      for (pkt[1] = 0; pkt[1]<255; pkt[1]++){
            printf("Setting speed: %d\n", pkt[1]);
            E2B_Tx_Bytes(pkt, 2, 0, 1);
            sleep(1);
      }

      // Stop motor and de-energize it
      pkt[0] = STEP_CMD_SLEEP;
      E2B_Tx_Bytes(pkt, 1, 0, 1);

      return 0;
}
```

This will take a few minutes to complete its run. While it's proceeding, listen to and watch your motor. You will observe two things:

1. Certain step speeds are very noisy, but there will be a range of speeds—typically, the faster speeds—for which the motor is comparatively silent.

2. Depending on your stepper motor, at some point on the speed ramp, the motor will probably stop spinning and will simply begin to hum. Take a note of the approximate speed value when this happens; for the motors I am using (under no mechanical load), this is about 240.

The second point in particular is important, and needs elucidation. Try running this second code snippet:

```c
#include <stdio.h>
#include "e2bus.h"

int main (int _argc, char *_argv[])
{
        unsigned char pkt[2];

        // Open port and start stepper motor
        if (E2B_Acquire()) {
                printf("Error opening E2BUS.\n");
        }
        E2B_Reset();
        // Set the maximum speed your motor can withstand
        pkt[0] = STEP_CMD_SETTICK;
        pkt[1] = 230;
        printf("Setting speed: %d\n", pkt[1]);
        E2B_Tx_Bytes(pkt, 2, 0, 1);

        printf("Starting motor.\n");
        pkt[0] = STEP_CMD_DRIVE_FWD;
        E2B_Tx_Bytes(pkt, 1, 0, 1);

        sleep(2);

        // Stop motor and de-energize it
        pkt[0] = STEP_CMD_SLEEP;
        E2B_Tx_Bytes(pkt, 1, 0, 1);

        printf("Finished.\n");

        return 0;
}
```

Here we attempt to set a speed just shy of the fastest possible step rate we observed in the first test, while the motor is stopped—and then we try to start the stepper. You will observe, however, that it doesn't rotate at all—it just stalls and whines like an engineer awakened by his wife's alarm clock. (If you don't see this behavior, then increase the speed value a little, recompile, and try again).

To understand what's going on here, you need to think about the mechanics of the situation. Each position of the stepper's rotor is a stable mechanical state for a certain corresponding electrical (magnetic) state of the coils. This state can be visualized as a valley between two hills of a plot of net force vs. angle; the motor tries to seek the low point in the valley, where the clockwise and counterclockwise forces on the rotor are equal. If you put an external mechanical force on the rotor, in (say) a clockwise direction, you are pushing up one side of the hill. If you push all the way to the top, then when you release the rotor, it will fall down into the next valley, i.e., the next stable step position.

When we advance the step phase in software, we alter the electrical state of the coils, which creates a new position of mechanical equilibrium. Effectively, we move the two hills and valley along a quarter-phase, which means the motor is no longer sitting in the middle of the valley. Since the clockwise and counterclockwise forces on the rotor are no longer equal, the rotor turns until it is in an equilibrium position once again; the step operation is then complete. However, this process doesn't happen instantly, because the rotor—and whatever mechanical load it's driving—has inertia. The time required for the rotor to find its new equilibrium point depends, among other things, on the inertia (which is determined by the mechanical load on the motor) and the force exerted by the coils—which is directly proportional to coil current.

The above explains both why we observe unreasonable stepper noise at certain combinations of load and step rate, and the fact that we can't necessarily go from stationary to maximum achievable step speed instantly. At slow speeds and/or with light loads, the motor will snap to each new equilibrium point very quickly compared to the step rate, and will in fact stop at the bottom of the equilibrium valley waiting for the next change of magnetic state. This is *horrifically* wasteful of energy—for every single step, we're injecting enough energy to overcome static friction forces and impart some amount of angular momentum to the stepper shaft and its attached load. The motor responds quickly and reaches its new stable state, but our control software isn't ready to move on to the next step state yet. The motor's stored inertia causes it to start climbing the slope of the next "hill," which turns the stepper temporarily into a generator. The energy we just pumped in is shorted through the transient protection circuitry and turned into heat. The motor slumps back down to the stable position until the next step pulse comes along, whereupon the entire process is repeated. This, along with the possibility of simple mechanical resonances in the motor and other components at certain step speeds, creates a loud and objectionable noise.

The most efficient way to drive the motor is to transit to the next step state as soon as the rotor reaches its new equilibrium position; this way, you are always pushing the load in the desired direction. Knowing exactly when this occurs is difficult, and it leads nicely into the next issue: We're not exerting a constant force on the rotor; rather, we are kicking it periodically. In order for our kicks to do the most good, we have to time them to coincide with certain rotor positions. At slow speeds, the motor responds to the step pulses faster than we issue them (so the motor can be said to be "led" by the step pulses). However, as the speed increases, we start to rely increasingly on the fact that the motor's inertia will carry it to the equilibrium point by the time we kick it next. This is why you can't simply start the motor at its maximum possible step speed—starting from rest, the rotor won't have time to reach equilibrium point before we transit to the next state.

On the flip side of the coin, once the motor is whirling madly at high speeds, simply stopping the step pulses dead will very likely lead to mechanical overshoot. You probably won't observe that phenomenon on a bare motor—at least, not the small, light motors you're likely to be using—but it's a very real issue under load. The most common approach to this is to implement an "acceleration profile," which is a table of step rates describing how to get from one motor speed to another most efficiently and reliably. These sorts of tables are easily handled in small microcontrollers. Advanced stepper designs will also use higher drive current (in other words, steeper hills) to achieve faster speed changes.

Completely general solutions to these problems (and others not mentioned) are possible—for example, I should point out that as soon as the motor overshoots the current equilibrium position, it will start generating a back EMF that can be measured via a suitably delicate circuit. However, such complex solutions are not required for simple, low-speed stepper applications, and they lie beyond the scope of this text.

4.4 Speed-Controlled DC Motor with Tach Feedback and Thermal Cutoff

E-2's main propulsion system consists of two DC motors directly driving contra-rotating propellers. An underwater vehicle with a single propeller is subject to undesirable torque forces, especially if the vehicle has no significant keel. It's possible to counteract this by using a stator to straighten the water flow behind the propeller. It's also obviously possible to drive two propellers from a single motor using gears. However, using two independently controllable motors allows us to tighten the vehicle's turning circle by running the motors in opposite directions, if desired. It also lets the device limp home if one motor fails, or one propeller happens to foul something.

The textbook circuit configuration for controlling a reversible DC motor is the H-bridge, illustrated (representatively with bipolar transistors) in Figure 4-6:

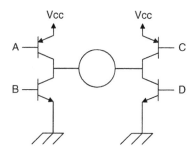

Figure 4-6: Standard H-Bridge Circuit.

This is very much a "lowest common denominator" circuit, and although you might build one on a breadboard for a very quick and dirty test of something, you would never want to field a device built around such a simple configuration. However, it's a good starting point to illustrate the basic principles. To run the motor in one direction, turn on the PNP transistor on one side of the bridge (say, at A), and the NPN transistor on the other side (say, at D). To run the motor in the opposite direction, turn on the opposite pair of transistors; B and C in our example. You can control the motor's speed by modulating the on-time of either or both of the active transistors.

There are numerous practical problems with such a simplistic design. Perhaps most importantly, there is no protection for the switching transistors from the inductive "kick-back" from the motor windings. You could mitigate this by putting a protection diode across the collector and emitter of each transistor. Also consider what would happen if you reverse the motor direction by switching from the configuration (A ON – D ON) to the configuration (B ON – C ON). The switching times of the individual BJTs or FETs you would be using are not exactly identical, so you run the risk that, for an instant, both sides of the bridge could be "on," thus shorting the power rails—and probably either burning out part of the driver circuit, blowing a fuse or just causing a momentary power glitch that could reset some or all of your system to an unknown state. You could work around this problem by ensuring that the firmware never goes directly from the "powered up—forward rotation" to "powered up—backward rotation" states; instead, it should switch both sides of the bridge off for a brief recovery period before changing directions.

Furthermore, there is also no intrinsic hardware protection to prevent a firmware bug from shorting the power rails directly through one side of the bridge (for example, by switching on A and B simultaneously due to a software error writing garbage values to an I/O latch)—you could solve this by providing some external logic providing "direction" and "enable" inputs that only allow the drivers to be turned on in permissible combinations. Finally, as the circuit stands, you have no way to diagnose the health of the switching circuit or gauge the current being drawn by the motor, so you can't detect a stalled rotor or shorted winding.

Rather than reinventing all these wheels and engineering a custom solution, we cut around these messy problems by using the National Semiconductor LMD18200T integrated H-bridge. This chip is not exactly cheap, at around $11.50 (in single-piece quantities). However, the price is well worth the engineering time saved. If nothing else, you would probably spend at least twice this amount on destroying MOSFETs while debugging your own circuit design. The LMD18200 also offers several useful bonus features, including an internal junction temperature watchdog that will signal to your microcontroller with a simple digital signal if the chip is overheating (and shut the drivers off if the H-bridge overheats), integral shoot-through protection, nonregenerative braking (this shorts the motor windings), and a current monitoring output that, with an appropriate shunt resistor and ADC, can be used to measure how much current is being drawn by the motor. In fact, we won't be using the latter feature, so you might prefer to use the slightly cheaper LMD18201, which is identical to the LMD18200 except that it doesn't have the handy drive-current-monitoring feature. The reason I specify the LMD18200 is simply because it seems to be stocked by more vendors than its cheaper sibling. The price difference is only a few pennies from the distributors I use, but maybe you'll come across a load of amazingly cheap LMD18201s in the surplus marketplace.

There is one more feature our circuit offers, which isn't always essential but is often useful—tachometer feedback. Without some feedback on the actual physical number of revolutions being executed per second, it is practically impossible to control the speed of a motor under varying load. The tach input of the board expects to see an active low pulse once per revolution. The sensor method I use on the E-2 is a Hall effect sensor mounted next to the motor shaft, and a tiny neodymium magnet glued to the shaft. You might prefer to use some other method, such as an optical sensor and a reflective (or dark) mark painted on the shaft. Some motors even have tach hardware of some sort built in; this is particularly common in small cooling fans, which frequently have an integral Hall effect sensor. With such motors, all you need to do is connect the wires properly, and you're done. However, you should note that the tach on these motors may not be reliable at anything less than 100% PWM duty cycle. These cooling fans are often designed to run continuously at full speed, with the tach providing feedback to the system that the fan hasn't stalled. The tach sensor is probably powered directly off the power input wire, and may not have enough of a decoupling capacitor connected to remain alive during the "off" portion of your PWM signal.

Note: Almost all fans output two tach pulses per revolution. Depending on what kind of tach sensor you employ, and how you mount it, your system may output only one pulse per revolution.

Enough talk, let's look at the schematic for our motor controller (Figure 4-7):

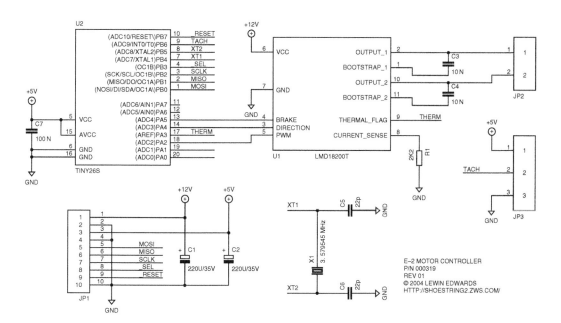

Figure 4-7: DC Motor Control Circuit.

The first thing you'll notice is that it's quite difficult for us to use the tiny26L's inbuilt PWM hardware, because its functions overlay the pins used by the USI. If you study the tiny26L's datasheet closely, you'll observe that we could use PB3 (which doubles as the OC1B output) as the PWM drive signal. However, this would mean either moving the SPI select functionality to a noninterrupt pin—unacceptable, because we have to respond to changes in the select state quickly so we can get off the bus—or enabling the second pin-change interrupt, PCINT1. The latter interrupt is a bit too "global" for comfort—it fires even when alternate hardware functions have been selected for the I/O pins. Extra care would be necessary to suppress these unwanted interrupts.

As a result, we have to use a software PWM drive scheme. Essentially, I had to choose between either software USI or software PWM, and in order to keep as much code as possible common between the various modules of this project, I chose to do the PWM in software. An unavoidable side effect of this is that serial I/O and tach pulses will cause minor glitches in the PWM output. A second side effect is that the PWM frequency is severely limited by the AVR's core clock speed. (The hardware PWM feature can be driven by an asynchronous clock generated by an on-chip PLL). Neither of these are serious problems for our application.

Observe also that we use the dedicated INT0 line for the tachometer input. The reason for this is that we're only interested in one edge of the tach signal. Although we've got enough free GPIOs to dedicate one of the pin-change interrupts to tach input, this interrupt would fire on both edges of the tach signal, which introduces unnecessary glitches into our other tasks. INT0 can be programmed to fire only on the edge of our choice.

Now let's study the firmware for this device. It's moderately complicated, because there is a lot going on, so let's study the major tasks separately. We'll first look at the task that handles the PWM:

```
; Timer 0 overflow - PWM motor output
; WARNING - This code must be bare-minimum optimized. It runs so
; frequently that it can easily take over the entire CPU.
entry_timer0:
      in r3, SREG
      sbi PORTA, PORTA_TEST

      ; Reset TMR0 counter to start position for next tick
      ldi r24, TMR0_RELOAD
      out TCNT0, r24
```

```
    ; Increment PWM count variable
    inc r22
    cpi r22, 100
    brne pwm_no_overflow

    ; If we've overflowed, reset to 0. Do a special-case check to see
    ; if the requested duty cycle is 0 - if so, turn off the PWM out-
    ; put

    ldi r22,0
    cp r1, r22
    brne pwm_duty_nonzero

    cbi PORTA, PORTA_PWM
    rjmp tmr0_done

pwm_duty_nonzero:
    sbi PORTA, PORTA_PWM
    rjmp tmr0_done

    ; If we didn't overflow the 100% counter, then compare to
    ; requested duty cycle.
    ; If count = duty cycle, turn off the PWM output. Note that the
    ; 100% special case is already handled by the overflow trapping
    ; above.

pwm_no_overflow:
    cp r22, r1
    brne tmr0_done
    cbi PORTA, PORTA_PWM

tmr0_done:
    cbi PORTA, PORTA_TEST
    out SREG, r3
    reti
```

There are a few parameters that need to be balanced here:

- The desired number of PWM "grayscales" between 0% (always off) and 100% (always on) duty cycle. We're going to use 100 levels, so the PWM value in use is actually the same number as the duty cycle percentage. There are a couple of reasons for choosing this value besides the mere elegance of specifying a percentage directly: firstly, since the AVR is an 8-bit micro, it's convenient for us to use a value that fits in an 8-bit variable with some wiggle room so we can avoid fencepost errors without having to insert any unwieldy special-case code for boundary conditions. Secondly, 100 steps is really more than enough for most drive motor applications; in fact, E-2 only adjusts the duty cycle in 10% increments.

> **Note:** Imagine you are building a fence 100 meters long. The posts along this fence are ten meters apart. How many posts do you need to buy? The instinctive response of ten is the canonical fencepost error.

- The target PWM frequency. 200 Hz is a workable (though noisy) frequency for motor driving. To give an admittedly oversimplified rule of thumb: ultrasonic frequencies are generally better for PWM applications, because they push the pulse noise out beyond our hearing range. However, the tiny26L is too slow to implement this in software. If we were using the hardware PWM, it wouldn't be a problem, but unfortunately we don't have that luxury, for reasons described previously. 200 Hz is a reasonable compromise.

- System clock speed and interrupt loading issues.

PWM in our design is implemented using Timer 0. This 8-bit timer increments with a selectable ratio (/1, /8, /64, /256 or /1024) of the system clock frequency, and interrupts when it rolls over from 0xFF to 0x00. The parameters we can vary here are the clock divisor and the value to be reloaded into the timer during the interrupt. In order to calculate these values, we need to work backwards from the 200 Hz figure. In each cycle of that 200 Hz signal, we need to have 100 sample points at which the PWM drive signal can be either off or on. This means we need a timer interrupt rate of 20 kHz. Given our system clock of 3.579545 MHz, the Timer 0 clock frequencies available to us are approximately 3.58 MHz, 447 kHz, 55.9 kHz, 14.0 kHz, or 3.50 kHz. The slowest timer speed that is faster than our 25 kHz target is 55.9 kHz. Dividing these frequencies we see that we need to divide the timer 0 clock source (by reloading the timer with some nonzero value) by 2.24 to get the right interrupt rate. The closest we can get is either a reload value of 0xFE (2 ticks per interrupt, 28.0 kHz) or 0xFD (3 ticks per interrupt, 18.6 kHz). We'll use 0xFE, because even though it represents a greater error with respect to our nominal target frequency, it's an error "in the right direction"—that is, towards better performance.

As an aside at this point, you'll notice by inspecting the subroutine that my code sets PA7 high on entry to the Timer 0 interrupt, and brings it low again just before returning from that ISR. The purpose of this strobe is so that we can observe on an oscilloscope the amount of CPU time being chewed up by the PWM function. Since this is the most frequently executed code path in the chip, it's instructive to have some means of measuring how big a timeslice it occupies. From quick inspection, rather less than 10% of the available CPU time is being occupied in the PWM ISR, which is quite acceptable. While we're talking about performance, though, also note how I've fine-tuned the register usage in this project to avoid having to save anything on the stack in the high-load ISRs. It gets harder and harder to do these kinds of down-to-the-last-nibble optimizations as you add more tasks to a system (simply because each task has a certain amount of state that needs to be stored). This is another useful argument in favor of breaking up a system's real-time responsibilities across multiple microcontrollers.

The second task in this module is tachometer measurement. This is handled by Timer 1. Timer 1 is a free-running up counter. Timer 1 has a richer selection of prescaler values than Timer 0, from /1 (3.58 MHz) all the way down to /16384 (218 Hz). What scaler value should we choose? It depends on the desired accuracy and anticipated range of the input signal. Let's say that we are going to use a motor with a maximum unloaded speed of 3,600 rpm and arbitrarily pick a divisor of CK/8192, or 437 Hz. To get a reasonably accurate measurement, we are actually going to perform a kind of running average: The tach interrupt increments a 16-bit counter, but does NOT permit it to roll over past 0xFFFF. Timer 2's overflow interrupt is allowed to fire eight times (2048 ticks, or in other words ~4.69 seconds). At the eighth interrupt, the tach counter is captured and reset; it can range from 0 (no motion) to 0xFFFE (838,000 rpm; exceedingly unlikely). For an unloaded motor of the type you're likely to be using, expect to see values around 274 (3,500 rpm). 0xFFFF values should be displayed by your user interface as "out of range," and 0 should probably be displayed as "tach failure," since these are extreme boundary conditions. Note that you can tinker with the dynamic range and sample rate of this measurement very simply by altering the tachometer divisor value in the Timer 1 interrupt. The actual code for the two tachometer-related ISRs is as follows:

```
; Timer 1 overflow - Tach sampler
entry_timer1:
     in r5, SREG

     inc r21
     cpi r21, TACH_DIVISOR
     brne tmr1_done

     ldi r21, $00
```

```
        ; Update RAM copy of counter
        sts tach_low, r30
        sts tach_high, r31

        ; Clear tach counter
        ldi r30, $00
        ldi r31, $00

tmr1_done:
        out SREG, r5
        reti

; Tachometer input handler (INT0)
; This interrupt fires once every revolution, and would typically
; be triggered by a stationary Hall effect sensor sensing a magnet on
; the shaft.
entry_int0:
    in r4, SREG

        ; Increment tach counter
        inc r30
        brne tach_done

        inc r31
        brne tach_done

        ; If tach has overflowed, peg it at $FFFF.
        ldi r30, $ff
        ldi r31, $ff

tach_done:
        out SREG, r4
        reti
```

Since this module isn't directly concerned with the actual motor speed, we just provide the above 16-bit counter result when queried for status—the host is expected to do the math to convert the raw count into a rotation speed in a unit acceptable to the end-user. The value can be calculated very simply by:

$$\text{speed in rpm} = (\text{tach value} * 60)/4.69.$$

The final task is the SPI interface management code. This code is very similar to the analogous portions of the stepper motor controller; a simple state machine which is stimulated by either the USI interrupt (for incoming data) or the pin-change interrupt that handles SPI slave selection and bus-on/-off events. The firmware supports the command codes shown in Table 4-6:

Table 4-6

Mnemonic	Value	Description
MTR_CMD_STOP	0	Stop motor.
MTR_CMD_FWD	1	Set PWM duty cycle and start motor spinning in the forward direction. The next byte following this command byte should be the duty cycle (0–100%).
MTR_CMD_REV	2	Set PWM duty cycle and start motor spinning in the reverse direction. The next byte following this command byte should be the duty cycle (0–100%).
MTR_CMD_READ_STATUS	254	Reads back current speed and tachometer status. The host can read back up to three bytes following this command; the first byte is the status/speed byte (lower 7 bits = current PWM duty cycle, upper bit is set if the thermal warning flag is active), and the next two bytes are the most recent tach count, low byte first.
MTR_CMD_RESET	255	Performs a soft reset.

4.5   Two-Axis Attitude Sensor Using MEMS Accelerometer

For a variety of reasons; navigation, hazard avoidance, and so on, it's desirable for a vehicle to be able to know its orientation with respect to the Earth. A ship, submarine or airplane has six degrees of freedom (land-bound vehicles generally have fewer). Three of these are rotational: rotation around an imaginary line from bow to stern (roll), rotation around an imaginary line perpendicular to the bow-stern line and parallel to the Earth's surface (pitch), and rotation in a plane parallel to the Earth's surface (yaw; turning the bow of the vehicle to point towards a new destination). The other three are translational, along the same axes just mentioned; respectively, surge (movement forwards or backwards), sway (movement from side to side) and heave (movement up or down).

Yaw is relatively difficult to measure directly, so let's discuss it first. One approach is to use a flux-gate sensor; an electronic compass, essentially. The difficulty with this is that every spot on the Earth's surface has more or less interference from local metallic deposits and other geographical features, so a compass needle doesn't always point at a known reference point (magnetic north). Magnetic north also moves about, and it doesn't coincide with the true geographic north pole. For short trips, a fixed variance setting can be looked up on a map of your area, and you can just ignore any errors caused by roaming about close to vast lodestone deposits! E-2's core electronic module doesn't directly measure yaw; it assumes that most of the vessel's motion vector is parallel to the bow-to-stern axis and hence uses GPS velocity data (while surfaced) to infer the direction the bow is pointing. If you want to try your hand at magnetic navigation methods, there are numerous kits containing flux-gate compass boards, intended for the hobbyist robotics market. Most of these incorporate some clever firmware to deal with variance issues.

Static roll and pitch, on the other hand, can easily be ascertained by measuring the gravity vector acting on the craft and comparing it to an imaginary reference vector at right angles to both the bow-stern and port-starboard axes of the vehicle. To perform this measurement task, we use an accelerometer.

At its simplest, an attitude or acceleration sensor is simply a pendulum. In fact, a reasonably useful two-dimensional attitude sensor can be constructed by simply taking a two-axis potentiometer assembly out of an off-the-shelf analog joystick, attaching a heavy weight to the joystick lever, and mounting the whole thing upside-down so that the weighted joystick can swing around freely. (Pay attention to align the axes of the sensor with the axes around which the sensor is expected to rotate). In cases where fine accuracy is not essential and it is desirable to connect this sensor directly to a PC, cannibalizing a joystick in this way is definitely the path of least resistance, not to mention an extremely fast way to construct a prototype. Apropos of the E-2 project, it is interesting to note that attitude and depth control in torpedoes of World War II vintage were actually controlled using a mechanically interlinked system of a pendulum and a manometer.

Despite its simplicity, there are a number of disadvantages to the simple pendulum method—it is bulky, and friction in the potentiometers and joystick bearings tends to makes the device insensitive to small accelerations. A better solution for some applications is to use free-turning weights or gyroscopes on exquisitely low-friction bearings, with some sort of optical or magnetic scale read-out, but these sorts of machines are expensive and relatively high-maintenance. The modern solution to this design problem is an integrated MEMS (MicroElectroMechanical System) part such as the Analog Devices' ADXL202 two-axis accelerometer. This particular device is only rated up to two gravities (approximately 19.7 ms^{-2}) of acceleration. This makes it suitable for assessing the overall attitude of a body, but not terribly useful for more demanding tasks. For truly challenging tasks like measuring deceleration during a car crash, or rocket takeoff forces, you need (at the very least) a part rated for much higher accelerations.

Before we go any further, please note that this chapter talks specifically about the older ADXL202JQC (commercial temperature grade) and ADXL202AQC (industrial temperature grade) parts, which were available in a 14-lead ceramic SOIC package. This variant has been discontinued by Analog Devices in favor of a ceramic 8-pin leadless chip carrier, the ADXL202JE (commercial) and ADXL202AE (industrial). However, the older part is still available in distribution channels and is quite widely used in hobbyist type applications because of the relative ease with which it can be hand-prototyped.

The best way to prototype with either part is, of course, with a small PCB or the evaluation board for the part. If this is not possible (note that the EVB is approximately three times the cost of the bare part in single-piece quantities), then an acceptable alternative for the older SOIC part is to glue the sensor to the non-coppered side of a piece of protoboard, and solder thin wires to the pins (wire-wrap type wire performs this duty very well). I prototyped the circuit in this chapter using this method; I used a two-part epoxy resin to glue the chip down. If you are using the more modern LCC part, however, life is more difficult. Figure 4-8 shows a picture of the older device (glued to a piece of prototype board), alongside a couple of samples of the LCC part, one of which has been turned upside-down to show the contact pattern on the underside.

Figure 4-8: Different ADXL202 Variants.

Those contacts are very fine gold deposits on the ceramic chip body; if you solder wires to them, you can quite easily pull the contacts right off the chip. Nevertheless, if that's your only prototyping option, it can be done—just be very careful not to apply any unnecessary stress, because those parts are expensive (about $15–20 each, in small quantities).

Mechanical package considerations of this sort aside, the ADXL202 is very easy to interface. The device outputs two square wave signals (one for each axis) with an identical period, T2, and a variable duty cycle with an on-time of T1. Note that although the two signals are guaranteed to have the same period, they are not guaranteed to start at the same time—they can have any amount of phase difference. The period can be configured from 1 to 10 ms by means of an external resistor (RSET in Analog Devices literature), selected according to the formula:

$$T_2 = R_{SET}/125000000,$$

where T_2 is in seconds, and R_{SET} is in ohms. I have chosen a 1 M resistor, which gives us a nominal period of 8 ms. Actual measurement of a real device in-circuit shows a period

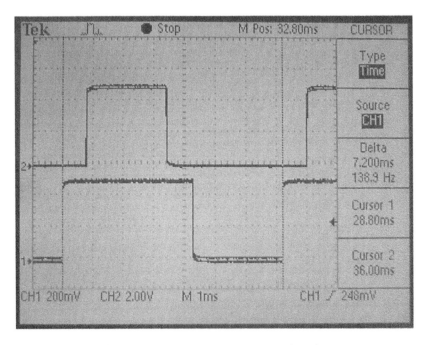

Figure 4-9: ADXL202 Output Signals.

of 7.2 ms, which is gratifyingly close to the mark. Rather than sketch the waveforms artificially, Figure 4-9 shows a picture of an actual scope trace showing both X (bottom) and Y (top) outputs for an ADXL202JQC. The device generating these signals was flat on my desk, which is approximately horizontal with reference to the Earth's surface.

The on-time period T_1 (the "hump" in the waveforms in Figure 4-9) is nominally supposed to be 0.5 T_2 when the acceleration on the axis in question is 0g. In practice, though, there is a wide deviation—as you can see from the measurements in Figure 4-9, where the accelerometer was known to be approximately horizontal.

Following is the schematic for our circuit (Figure 4.10):

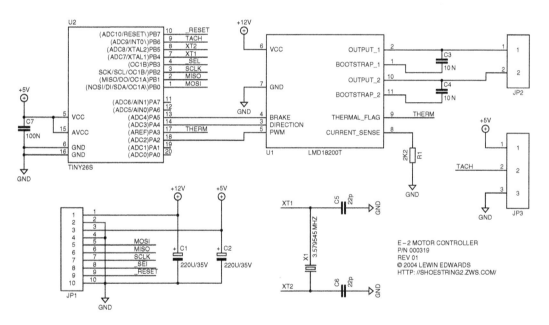

Figure 4-10: Accelerometer Schematic

The firmware supports only two command codes shown in Table 4-7:

Table 4-7

Mnemonic	Value	Description
ACL_CMD_READ_STATUS	254	Reads back six bytes of accelerometer status.
ACL_CMD_RESET	255	Performs a soft reset.

ACL_CMD_READ_STATUS returns three 16-bit words of status information; first, the measured value of T_2, then the T_1 value measured for the X axis, and finally the T_1 value measured for the Y axis. The high-order byte is transmitted first. Observe that we don't need to calibrate and transmit two copies of T_2, since it is known (by design) to be identical for both axes.

The meat of this project is contained in the Timer 1 and pin-change interrupt handlers. The interrupt handler for overflows in Timer 1, which runs at the full CPU clock speed, merely increments a high-order counter byte, thereby extending Timer 1's range to 16 bits. This corresponds to a maximum measurable T_2 of approximately 18 ms, with a theoretical resolution of 279 ns.

```
; Timer 1 interrupt
entry_tmr1:
      in r4, SREG

      inc r1

      out SREG, r4
      reti

; I/O pin change interrupt
; The only valid sources of this interrupt are PB3, which is used
; as the 3-wire slave select line, PA6 (X-input) and PA7 (Y-input)
entry_iopins:
   in r3, SREG

      ; Grab current timer value in case we need it later
   in r22, TCNT1

      ; Get last value of shadow register
   lds r26, porta_shadow

      ; Scan X-input. First, handle the case if it is high.
      sbis PINA, PORTA_X
      rjmp x_is_low

      sbrc r26, PORTA_X
      rjmp test_y         ; No change. Test Y-input.

      ori r26, PORTA_X

      ; +ve edge detected on X. We need to calculate T2
      ; by subtracting last-edge from current timer, and adding
      ; X-T1 to that result.
      mov r21, r22        ; timer lo byte
      lds r25, xle_lo
      sub r21, r25        ; r21 = intermediate val lo byte
      mov r24, r1         ; timer hi byte
      lds r25, xle_hi
```

```
      sbc r24, r25         ; r24 = intermediate val hi byte
      lds r25, x_t1_lo
      add r21, r25
      lds r25, x_t1_hi
      adc r24, r25
      sts t2_hi, r24
      sts t2_lo, r21

      rjmp update_x_edge

      ; X-input is low. Test for -ve edge.
x_is_low:
      sbrs r26, PORTA_X
      rjmp test_y

      ; -ve edge on X detected. We need to calculate X-T1
      ; by subtracting last-edge from current timer.
      mov r21, r22         ; timer lo byte
      lds r25, xle_lo
      sub r21, r25         ; r21 = T1 lo byte
      mov r24, r1                     ; timer hi byte
      lds r25, xle_hi
      sbc r24, r25         ; r24 = T1 hi byte
      sts x_t1_hi, r24
      sts x_t1_lo, r21

      ; Update last-edge timestamp for X axis
update_x_edge:
      sts xle_hi, r1
      sts xle_lo, r22

      ; Scan Y-input - First, handle the case if it is high.
test_y:
      sbis PINA, PORTA_Y
      rjmp y_is_low

      sbrc r26, PORTA_Y
      rjmp test_spi        ; No change. Go to SPI test.


      ; +ve edge detected on Y. We need to start calculating Y-T1.
      rjmp update_y_edge
y_is_low:
      sbrs r26, PORTA_Y
      rjmp test_spi        ; No change. Go to SPI test.

      ; -ve edge on Y detected. We need to calculate Y-T1
      ; by subtracting last-edge from current timer.
      mov r21, r22         ; timer lo byte
      lds r25, yle_lo
      sub r21, r25         ; r21 = T1 lo byte
      mov r24, r1                     ; timer hi byte
      lds r25, yle_hi
      sbc r24, r25         ; r24 = T1 hi byte
```

```
        sts y_t1_hi, r24
        sts y_t1_lo, r21

        ; Update last-edge timestamp for Y axis

update_y_edge:
        sts yle_hi, r1
        sts yle_lo, r22

test_spi:
        in r26, PINA
        sts porta_shadow, r26

        ; Check state of SPI select line.
        sbic PINB, PORTB_SEL
        rjmp usi_disable

        ; If PB1 is already an output, don't reset the USI. This
        ; special code is necessary so accelerometer interrupts
        ; don't mess with partially complete USI transactions.
        sbic DDRB, PORTB_DO
        rjmp iopin_done

        ; SEL line is LOW. Enable and reset USI and switch PB1 to output
        ldi r23, FSMS_RXCMD
        ldi r26, $00

        out USIDR, r26        ; Empty USI data register
        out USISR, r26        ; Clear USI status (including clock count!)
        sbi DDRB, PORTB_DO     ; set PB1 to output

        sbi USISR, USISIF     ; Clear start condition status
        sbi USISR, USIOIF     ; Clear overflow status
        sbi USICR, USIOIE     ; Enable USI overflow interrupts

        rjmp iopin_done

        ; SEL line is HIGH. Disable USI and switch PB1 to input to take us off-bus

usi_disable:
    ; disable USI start and overflow interrupts
        cbi USICR, USISIE
        cbi USICR, USIOIE

        ; Disable output driver on PB1 (DO)
        cbi DDRB, PORTB_DO        ; set PB1 to input

iopin_done:
        out SREG, r3
        reti
```

Some averaging or filtering is advisable on the PC end of this equation. Here's a simple program that takes out continuous readings and prints them to a single line on the console:

```
/*
   main.c
   Demonstration applet for E2BUS stepper interface code

   From "Open-Source Robotics and Process Control  Cookbook"
   Lewin A.R.W. Edwards (sysadm@zws.com)
*/

#include <stdio.h>

#include "e2bus.h"

int main (int _argc, char *_argv[])
{
   unsigned char pkt[6];
   int i=0,j;

   // Open port
   if (E2B_Acquire()) {
      printf("Error opening E2BUS.\n");
      return -1;
}
   E2B_Reset();
   printf("Reset complete, pausing...\n");
   sleep(1);
   printf("XXXXXXXXXXXXXXXXXXXXXXXXXXXXXXXXXXXXX");
   while (1)
   {
      usleep(750000);
      pkt[0]=ACL_CMD_READ_STATUS;
      E2B_Tx_Bytes(pkt, 1, 0, 0);
      E2B_Rx_Bytes(pkt, 6, 0, 1);
      for (j=0;j<37;j++) printf("\b");
      printf("Sample %-08.8X status %-02.2X%-02.2X,"
        "%-02.2X%-02.2X,%-02.2X%-02.2X",
        i, pkt[0],pkt[1],pkt[2],pkt[3],pkt[4],pkt[5]);
      fflush(stdout);
      i++;
   }
   return 0;
}
```

If you run this program, you'll see that even if the accelerometer is stationary, there's a certain amount of jitter in the output values. This is partly due to the irritatingly analog nature of the Universe, partly due to vibration of the accelerometer, and partly due to the fact that serial interrupts can slightly skew the time measurement task. For example, here is a sequence of three consecutive readings from my prototype:

```
6821, 2C0A, 3AD6
67D7, 2BFC, 3AF0
67E9, 2C46, 3A2C
```

Because this phenomenon is unavoidable, some averaging or filtering is desirable before working with the sensor output. Simple averaging is acceptable, but a Kalman filter is better; for more information on this topic, I recommend R.M. du Plessis, *Poor Man's Explanation of Kalman Filters or How I Stopped Worrying and Learned to Love Matrix Inversion* (1967), ISBN 0-9661016-0-X.

4.6 RS-422–Compatible Indicator Panel

This circuit is a bit of a departure from the rest of the content in this chapter, inasmuch as the appliance I describe here is not part of the E-2 project. The reason I have included this section is because the circuit and firmware illustrate several relevant and interesting points, including multidrop differential serial communications over relatively long distances and using the AVR's internal RC oscillator instead of an external crystal. This application also provides a nice example of how the sorts of systems in this chapter can be used in real-world situations.

I developed this device for a shipping center application used in two of a company's warehouses. The products shipped from these centers consist of standard and customized kits of individually packaged parts. A number of conveyor belts run through the warehouse area, past the various bins of parts. At the "start" end of each conveyor belt is a large matrix of 416 pigeonholes arranged as 26 rows by 16 columns. Before each shift, administrative staff stock these pigeonholes with pick-lists describing different standard subassemblies. Each pigeonhole has an indicator lamp (actually, an LED) over it. A central computer, connected to the company's order processing system, controls all these indicator lamps over a piece of Category 5e cable that runs approximately 800 feet from the computer room to the warehouse floor; the indicator panels show workers along the conveyor belt which pick-lists to gather for an individual order as it progresses down the line. As initially installed, all the panels were to repeat a single set of commands; however, it was desired to leave the functionality open-ended so that in future, more panels could be added to the same bus, but show a different set of signals (to process multiple orders simultaneously on the same line). For this reason, each panel has an 8-bit address; commands coming down the wire have an address field indicating the intended recipient. It's legal for multiple indicators to have the same address if you want them to repeat duplicate data. See Figure 4-11 for the schematic for our circuit.

The actual LEDs are omitted from this schematic for clarity's sake; they are wired in a simple matrix with the cathodes connected to the ULN2803s driving the column lines, and the anodes connected to the row lines.

I chose an ATmega16 part for this design purely for the large I/O and memory budget; although it would be possible to implement the project in a much smaller part, it was simply convenient and quick to pick the mega16. You'll observe that this project uses the AVR's on-chip clock generator rather than relying on an external crystal. Note that the internal RC oscillator in the AVR parts is factory-calibrated with a device-specific "fudge factor." This fudge factor is different for each supported oscillator speed. The specific calibration constants for each frequency are stored in a nonuser-accessible (probably OTP) area of the micro, and can be read out with the chip signature. They cannot be read directly by code running on the target device; you can only read them out with a device programmer like the STK500. You'll notice that the code I provide is set up for 1 MHz operation, but that I have also included commented-out initialization code for 8 MHz operation. If you want to run the project at 8 MHz, you need to do slightly more than just uncomment the faster initialization code, though—you need to make sure the chip is correctly initialized with the right oscillator fudge factor for 8 MHz. It's something of a design shortcoming in the AVR series, but when the device is configured for any RC oscillator mode, it automatically loads the 1 MHz calibration factor (the first calibration byte) into the processor's oscillator calibration register, regardless of what RC oscillator speed you have selected. If uncorrected, this can lead to considerable clock deviation from the expected speed. The oscillator might not even be reliable if it's grossly miscalibrated.

Atmel's official suggestion is to use an EEPROM or program flash location to store the factory calibration value, and copy it into the oscillator calibration register at power-up. If you run AVR Studio, select Tools—STK500/AVRISP/JTAG ICE—STK500/ AVRISP/JTAG ICE, and click the Advanced tab, you'll be able to read the signature bytes out of the chip using the Read Signature button. Select the desired speed from the drop-down list in the Oscillation Calibration byte section, and click "Read Cal. Byte." The appropriate calibration value will appear in the Value: box. (Remember—this value is specific to the individual chip you're looking at; you can't reuse this calibration value

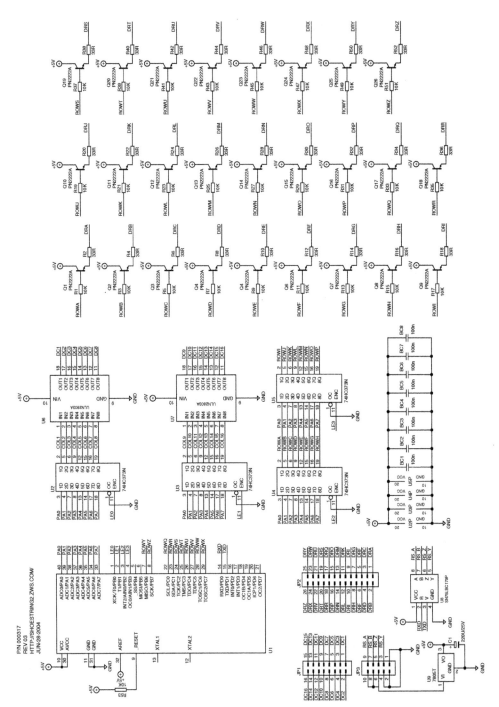

Figure 4-11: Schematic for RS-422-Compatible Indicator Panel.

in a different chip). You can now select a flash or EEPROM address using the fields under Write Address, and click "Write to Memory" to copy the calibration byte into flash or EEPROM, as illustrated by the following two screenshots (Figures 4-12 and 4-13):

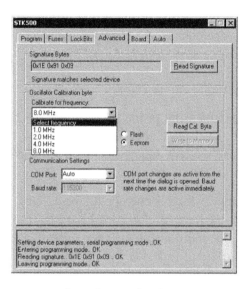

Figure 4-12: Select the Desired AVR Clock Speed.

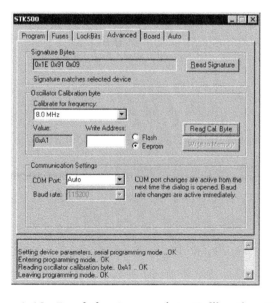

Figure 4-13: Read the Appropriate Calibration Byte.

While we're on the topic of burning chips, note that the ATmega16 ships with JTAG enabled by default. You need to set the fuses to disable JTAG in order to free up the associated I/O pins for use as general-purpose I/O in our application.

The indicator panel's serial interface operates at 2400 bps with 8 data bits, no parity and one stop bit. This panel is operated using a series of command strings, formatted as follows (note that these strings are case-sensitive):

- An attention character, '!'.

- An address character indicating which board(s) should hear the message. The code in this chapter has been hardcoded to use address 'A' (65). Refer to the sourcecode to change the unit address. If multiple indicator panels on the same bus have the same address, they will all respond to messages at the same time.

- A command character, which is one of: 'R' (reset; turn off all LEDs), '1' (turn on specified LED), '0' (turn off specified LED), 'B' (turn on blinking for a specified column), 'b' (turn off blinking for a specified column), or 'T' (start test mode; test mode runs until canceled with the 'R' command).

- The 1 and 0 commands require two additional bytes—a column identifier (A~P, corresponding to columns 0 through 15) and a row identifier (A~Z).

- The B and b commands require one additional byte identifying the column to blink (A~P, corresponding to columns 0 through 15).

- Example command strings:

    !AR – Turns off all LEDs on unit with address A.

    !A1BG – Turns on the LED (on unit with address A) at column 1, row G.

    !AT – Starts test mode on unit with address A.

    !A0FK – Turns off the LED (on unit with address A) at column 5, row K.

    !ABD – Starts blink mode for column 3 (on unit with address A).

    !AbD – Stops blink mode for column 3 (on unit with address A).

Since this is the most complex project, by code volume, included with this chapter, I am detailing the entire sourcecode in the text.

```
; Miscellaneous constants
.equ MY_ID      =65   ; ID number of this unit (default 'A')
.equ BLINK_RATE =30   ; Frames per blink-toggle

; Special serial Rx characters
.equ CHR_ATTENTION =$21   ; !
.equ CHR_RESET     =$52   ; R
.equ CHR_TESTMODE  =$54   ; T
.equ CHR_LEDOFF    =$30   ; 0
.equ CHR_LEDON     =$31   ; 1
.equ CHR_BLKON     =$42   ; B
.equ CHR_BLKOFF    =$62   ; b

; States for serial Rx state machine
.equ SRX_WAIT        =0 ; Wait for attention character
.equ SRX_ID          =1 ; Wait for unit ID
.equ SRX_GETCMD      =2 ; Wait for command byte
.equ SRX_ON_GETCOL   =3 ; Wait for column byte for LED-on command
.equ SRX_ON_GETROW   =4 ; Wait for row byte for LED-on command
.equ SRX_BLKON_COL   =5 ; Wait for column for blink-on command
.equ SRX_BLKOFF_COL  =6 ; Wait for column for blink-off command
.equ SRX_OFF_GETCOL  =7 ; Wait for column byte for LED-off command
.equ SRX_OFF_GETROW  =8 ; Wait for row byte for LED-off command


; Bits in flags
.equ FLAG_BLINK      =7 ; Blink flag, toggled every BLINK_RATE
frames
.equ FLAG_TESTMODE   =6 ; Nonzero = unit in test mode

; Variables in SRAM
.DSEG
.ORG 0x60
currentline:  .BYTE 1   ; Column# currently being driven
frameptr_lo:  .BYTE 1   ; Pointer to frame data for current column
frameptr_hi:  .BYTE 1
framecounter: .BYTE 1   ; Incremented each frame refresh
flags:        .BYTE 1
serialmode:   .BYTE 1   ; serial FSM code
tmpcol:       .BYTE 1   ; Temporary holding buffer for column#


.ORG 0x80
; BUGBUG Do not move this structure. The arithmetic that
; works with it will break if this 64-byte structure
; crosses a 256-byte boundary. For safety, leave it here.
; Each table entry is formatted as follows:
; BYTE   -   A-H
; BYTE   -   I-P
; BYTE   -   Q-X
```

```
; BYTE   -   Z (bit 7), Y (bit 6), blink (bit 0) - other bits
reserved, leave 0
framedata: .BYTE (4*16) ; Each 4 bytes is a column of LED data

; Interrupt vectors
; This must be the first thing in the executable flash image.

.CSEG
.ORG 0x0000
        jmp     entry_reset    ;Reset
        jmp     bad_irq        ;External interrupt request 0
        jmp     bad_irq        ;External interrupt request 1
        jmp     bad_irq        ;Timer/Counter2 Compare Match
        jmp     bad_irq        ;Timer/Counter2 Overflow
        jmp     bad_irq        ;Timer/Counter1 Capture Event
        jmp     bad_irq        ;Timer/Counter1 Compare Match A
        jmp     bad_irq        ;Timer/Counter1 Compare Match B
        jmp     bad_irq        ;Timer/Counter1 Overflow
        jmp     tc0_overflow   ;Timer/Counter0 Overflow
        jmp     bad_irq        ;Serial transfer complete
        jmp     usart_rx       ;USART Rx Complete
        jmp     bad_irq        ;USART data register empty
        jmp     bad_irq        ;USART Tx Complete
        jmp     bad_irq        ;ADC conversion complete
        jmp     bad_irq        ;EEPROM ready
        jmp     bad_irq        ;Analog comparator
        jmp     bad_irq        ;Two-wire serial interface
        jmp     bad_irq        ;External interrupt request 2
        jmp     bad_irq        ;Timer/Counter0 Compare Match
        jmp     bad_irq        ;Store Program Memory Ready

; Invalid exception handler
bad_irq: ldi r16, $BB
         out PORTA, r16
         rjmp bad_irq

; Power-on reset entry point
entry_reset:
        ; Initialize stack pointer to top of RAM
        ldi r16, high(RAMEND)
        out SPL, r16
        ldi r16, low(RAMEND)
        out SPH, r16

        ; Configure ports A,B,C as outputs
        ldi r16, $FF

        out DDRA, r16
        out DDRB, r16
        out DDRC, r16
```

```
        call clear_outputs

        ; clear status flags and serial FSM
        ldi r16, $00
        sts flags, r16
        sts serialmode, r16

        ; Configure PD1 (TxD) as output
        ldi r16,$02
        out DDRD, r16
;8MHz setup code
;       ldi r16, $CF
;       out UBRRL, r16
;       ldi r16, $00
;       out UBRRH, r16

        ; 1MHz clock setup
        ; Set up USART for 2400bps asynchronous mode
        ; Formula for calculating UBRR in this case is fosc/(16 * baud) - 1
        ldi r16, $19
        out UBRRL, r16
        ldi r16, $00
        out UBRRH, r16

        ; clear USART status
        out UCSRA, r16

        ; configure control/status register B
        ldi r16, (1 << RXCIE) + (1 << RXEN) + (1 << TXEN)
        out UCSRB, r16

        ; configure control/status register C
        ; 8 bits, no parity,
        ldi r16, (1 << URSEL) + (1 << UCSZ1)   + (1 << UCSZ0)

        ; set frame pointer to 0 and clear frame counter
        ldi r16, $00
        sts currentline, r16
        sts framecounter, r16

        ; Point X at start of framebuffer data
        ldi r26, low(framedata)
        ldi r27, high(framedata)
        sts frameptr_lo, r26
        sts frameptr_hi, r27

        ; clear timer counter
        ldi r16, $00
        out TCNT0, r16

        ; clear timer 0 interrupt flag
        ldi r16, $00
        out TIFR, r16
```

```
; 8MHz setup code
;     ldi r16, $03                        ; /64 prescaler
;     out TCCR0, r16
      ; 1MHz setup code
      ; set up timer 0 for fosc/8 (=125kHz)
      ; This corresponds to a line rate of ~977Hz, frame rate ~61Hz.

      ldi r16, $02                        ; /8 prescaler
      out TCCR0, r16

      ; enable timer 0 interrupt
      ldi r16, (1 << TOIE0)
      out TIMSK, r16

      ; enable interrupts
      sei

      call clearscreen
```

The USI-complete interrupt handles all the video write tasks:

```
; ISR
; Serial Rx complete
usart_rx:
      push r27
      push r26
      push r18
      push r17
      push r16
      push r0
      in r0, SREG

      ; Get received byte from USART
      in r16, UDR

      ; Act on the byte depending on the FSM state
      lds r17, serialmode

      cpi r17, SRX_WAIT
      brne rx_notwait

      ;Check for attention character
      cpi r16, CHR_ATTENTION
      breq rx_atn
      rjmp rx_done
rx_atn:
      ; Attention char received! Now wait for ID
      ldi r17, SRX_ID
      sts serialmode, r17
      rjmp rx_done
```

```asm
       ;Check for target ID
rx_notwait: cpi r17, SRX_ID
       brne rx_notid

       ; If this isn't our target, wait for the next command frame
       cpi r16, MY_ID
       breq targeted
       rjmp rx_finish
targeted:
       ; If we HAVE been targeted, we next expect a command byte
       ldi r17, SRX_GETCMD
       sts serialmode, r17
       rjmp rx_done

       ;Get command byte
rx_notid:    cpi r17, SRX_GETCMD
       brne rx_notcmd

       ; COMMAND - Reset
       cpi r16, CHR_RESET
       brne cmd_notres
       lds r16, flags
       andi r16, ~(1 << FLAG_TESTMODE)
       sts flags, r16
       call clearscreen
       rjmp rx_finish

       ; COMMAND - Test mode
cmd_notres: cpi r16, CHR_TESTMODE
       brne cmd_nottest
       lds r16, flags
       ori r16, (1 << FLAG_TESTMODE)
       sts flags, r16
       rjmp rx_finish

       ; COMMAND - Blink on
cmd_nottest:    cpi r16, CHR_BLKON
       brne cmd_notblkon
       ldi r17, SRX_BLKON_COL
       sts serialmode, r17
       rjmp rx_done

       ; COMMAND - Blink off
cmd_notblkon:   cpi r16, CHR_BLKOFF
       brne cmd_notblkoff
       ldi r17, SRX_BLKOFF_COL
       sts serialmode, r17
       rjmp rx_done
```

```
            ; COMMAND - LED on
cmd_notblkoff: cpi r16, CHR_LEDON
        brne cmd_notledon
        ldi r17, SRX_ON_GETCOL
        sts serialmode, r17
        rjmp rx_done

            ; COMMAND - LED off
cmd_notledon:   cpi r16, CHR_LEDOFF
        brne cmd_notledon
        ldi r17, SRX_OFF_GETCOL
        sts serialmode, r17
        rjmp rx_done

cmd_notledoff:

        rjmp rx_finish

        ;Get column byte for BLINK ON
rx_notcmd:   cpi r17, SRX_BLKON_COL
        brne rx_notblkon_col
        subi r16, $41               ; normalize column
        andi r16, $0F
        lsl r16                     ; multiply by 4
        lsl r16
        ldi r26, low(framedata)
        ldi r27, high(framedata)
        add r26, r16
        inc r26
        inc r26
        inc r26
        ld r16, X
        ori r16, $01
        st X, r16
        rjmp rx_finish

        ;Get column byte for BLINK OFF
rx_notblkon_col:cpi r17, SRX_BLKOFF_COL
        brne rx_notblkoff_col
        subi r16, $41               ; normalize column
        andi r16, $0F
        lsl r16                     ; multiply by 4
        lsl r16
        ldi r26, low(framedata)
        ldi r27, high(framedata)
        add r26, r16
        inc r26
        inc r26
```

```
        inc r26
        ld r16, X
        andi r16, $fe
        st X, r16
        rjmp rx_finish


        ;Get column byte for LED ON
rx_notblkoff_col:cpi r17, SRX_ON_GETCOL
        brne rx_noton_col
        subi r16, $41              ; normalize column
        andi r16, $0F
        sts tmpcol, r16
        ldi r17, SRX_ON_GETROW
        sts serialmode, r17
        rjmp rx_done


        ;Get row byte for LED ON and switch LED on
rx_noton_col:   cpi r17, SRX_ON_GETROW
        brne rx_noton_row
        subi r16, $41              ; normalize row


        ; First calculate column RAM offset
        lds r17, tmpcol
        lsl r17                    ; multiply by 4
        lsl r17
        ldi r26, low(framedata)
        ldi r27, high(framedata)
        add r26, r17


        ; Does desired LED lie in the current byte?
        cpi r16, $08
        brsh on_b2                 ; no->


        ldi r17, $01
b1_l:   cpi r16, $00
        breq on_done
        lsl r17
        dec r16
        rjmp b1_l


on_b2:inc r26                 ; Seek to next byte in framebuffer
        subi r16, $08         ; Chop off 8 from row#
        cpi r16, $08
        brsh on_b3
```

```
       ldi r17, $01
b2_1:  cpi r16, $00
       breq on_done
       lsl r17
       dec r16
       rjmp b2_1

on_b3:inc r26              ; Seek to next byte in framebuffer
       subi r16, $08       ; Chop off 8 from row#
       cpi r16, $08
       brsh on_b4

       ldi r17, $01
b3_1:  cpi r16, $00
       breq on_done
       lsl r17
       dec r16
       rjmp b3_1

       ; Rows Y and Z take special handling.
on_b4:inc r26
       subi r16, $08
       cpi r16, $00
       breq on_ry
       cpi r16, $01
       breq on_rz

       rjmp rx_finish      ; Column out of range.

on_ry:      ldi r17, $40
       rjmp on_done

on_rz:      ldi r17, $80
       rjmp on_done

       ; Finally we've finished this fandango and we write the video
byte

on_done: ld r16, X
       or r16, r17
       st X, r16
       rjmp rx_finish

       ;Get column byte for LED OFF
rx_noton_row:   cpi r17, SRX_OFF_GETCOL
       brne rx_notoff_col
       subi r16, $41            ; normalize column
```

```
      andi r16, $0F
      sts tmpcol, r16

      ldi r17, SRX_OFF_GETROW
      sts serialmode, r17
      rjmp rx_done

      ;Get row byte for LED OFF and switch LED off
rx_notoff_col: cpi r17, SRX_OFF_GETROW
      brne rx_notoff_row
      subi r16, $41          ; normalize row

      ; First calculate column RAM offset
      lds r17, tmpcol
      lsl r17                ; multiply by 4
      lsl r17
      ldi r26, low(framedata)
      ldi r27, high(framedata)
      add r26, r17

      ; Does desired LED lie in the current byte?
      cpi r16, $08
      brsh off_b2            ; no->

      ldi r17, $01
ob1_l:      cpi r16, $00
      breq off_done
      lsl r17
      dec r16
      rjmp ob1_l

off_b2:     inc r26    ; Seek to next byte in framebuffer
      subi r16, $08    ; Chop off 8 from row#
      cpi r16, $08
      brsh off_b3

      ldi r17, $01
ob2_l:    cpi  r16, $00
      breq off_done
      lsl r17
      dec r16
      rjmp ob2_l

off_b3:      inc r26    ; Seek to next byte in framebuffer
      subi r16, $08    ; Chop off 8 from row#
      cpi r16, $08
      brsh off_b4
```

```
        ldi r17, $01
ob3_l:     cpi  r16, $00
        breq off_done
        lsl r17
        dec r16
        rjmp ob3_l

        ; Rows Y and Z take special handling.
off_b4:     inc r26
        subi r16, $08
        cpi r16, $00
        breq off_ry
        cpi r16, $01
        breq off_rz

        rjmp rx_finish      ; Column out of range.

off_ry:     ldi r17, $40
        rjmp off_done

off_rz:     ldi r17, $80
        rjmp off_done

        ; Finally we've finished this fandango and we write the
video byte
off_done:     ld r16, x
        ldi r18, $FF
        eor r18, r18
        and r16, r17
        st x, r16
        rjmp rx_finish

rx_notoff_row:

        rjmp rx_done

        ; Finish transaction, return to SRX_WAIT
rx_finish:     ldi r17, SRX_WAIT
        sts serialmode, r17

        ; Clean up and exit ISR
rx_done:
        out SREG, r0
        pop r0
        pop r16
        pop r17
        pop r18
        pop r26
        pop r27
        reti
```

Video refresh functionality, including blinking of rows for which the blink bit is set, is handled in the timer 0 ISR:

```
; ISR
; Timer/counter 0 overflow
tc0_overflow:
      push r27
      push r26
      push r21
      push r20
      push r19
      push r18
      push r17
      push r16
      push r0
      in r0, SREG

      ; Turn off all column drivers
      ldi r16, $00
      out PORTA, r17
      nop
      ldi r16, $03    ; load zeros into U2, U3
      out PORTB, r16
      nop
      ldi r16, $00
      out PORTB, r16 ; disable load-enable for U2, U3

      ; Point X at framebuffer data
      lds r26, frameptr_lo
      lds r27, frameptr_hi

      ; Load data onto row drivers
      ld r18, X+
      out PORTA, r18
      nop
      ldi r19, $04
      out PORTB, r19
      nop
      ldi r19, $00
      out PORTB, r19
      nop
```

```
        ld r18, X+
        out PORTA, r18
        nop
        ldi r19, $04
        out PORTB, r19
        nop
        ldi r19, $00
        out PORTB, r19
        nop

        ld r18, X+
        out PORTC, r18

        ld r18, X+
        mov r21, r18
        andi r18, $C0
        out PORTB, r18

        mov r20, r18

        ; Check blink mode for this row
        andi r21, $01
        breq noblink
        lds r21, flags
        andi r21, (1 << FLAG_BLINK)
        brne noblink
        rjmp column_done

        ; Enable column driver
noblink: lds r18, currentline

        sbrc r18, 3    ; If the column modulo 16 is >= 8...
        rjmp sc_hi     ; ->

        ori r20, $01   ; Low bits are strobed with LE0
        rjmp sc_calc

sc_hi:ori r20, $02    ; High bits are strobed with LE1
sc_calc: andi r18, $07   ; mask off high bit of column#

        ; Shift left (R1) times to get the desired column latch byte
        ldi r17, $01
```

```
sc_c_lp: cpi r18, 0
        breq sc_c_done
        lsl r17
        dec r18
        rjmp sc_c_lp

sc_c_done:
        out PORTA, r17 ; put latch byte on PORTA
        nop
        nop
        out PORTB, r20
        nop
        nop
        andi r20, $C0
        out PORTB, r20

        ; Increment column
column_done:    lds r18, currentline
        inc r18
        andi r18, $0F
        sts currentline, r18

        ; Have we wrapped around to column 0?
        cpi r18, $00
        brne rfsh_cont

        ; Yes! Reset pointer and update blink count
        ldi r26, low(framedata)
        ldi r27, high(framedata)
        lds r18, framecounter
        inc r18
        ; See if we've had a blink overflow
        cpi r18, BLINK_RATE
        brne blink_ok

        lds r18, flags
        ldi r19, (1 << FLAG_BLINK)
        eor r18, r19
        sts flags, r18
        ldi r18, $00
blink_ok:
        sts framecounter, r18
```

```
rfsh_cont:

    ; store new frame pointer
    sts frameptr_lo, r26
    sts frameptr_hi, r27

    ; Force $80 into timer register to double frame-rate
    ldi r18, $80
    out TCNT0, r18

    out SREG, r0
    pop r0
    pop r16
    pop r17
    pop r18
    pop r19
    pop r20
    pop r21
    pop r26
    pop r27
    reti
```

This is the only project here that includes significant functionality in the main loop. This loop polls the test-mode bit, and when it is set by the serial ISR, the main loop calls the test-mode function. Test mode is provided mainly so the user can verify that all LEDs are wired and functioning correctly. In test mode, each entire column is illuminated sequentially (0–15), then each entire row is illuminated sequentially (A–Z). Any miswired or shorted row/column lines will become immediately apparent.

```
; Main program loop. Most of the functionality is actually in ISRs.
mainloop:
    lds r16, flags
    sbrs r16, FLAG_TESTMODE
    rjmp ml_nottest
    call testmode
    call clearscreen

ml_nottest:
    rjmp mainloop

; Test mode
; Destroys R16, R19, R26, R27
testmode:
```

```
; PHASE 1
; Walk a line of FFs across the (16) columns
call clearscreen
ldi r18, $10

; Point X at start of framebuffer data
ldi r26, low(framedata)
ldi r27, high(framedata)
ldi r20, $FF
ldi r21, $C0

w_f_0:
      lds r16, flags
      sbrs r16, FLAG_TESTMODE
      ret

      lds r16, framecounter
      cpi r16, $00
      brne w_f_0
      ; store new data
      call clearscreen
      st x+, r20
      st x+, r20
      st x+, r20
      st x+, r21

w_f_1:
      lds r16, flags
      sbrs r16, FLAG_TESTMODE
      ret

      lds r16, framecounter
      cpi r16, BLINK_RATE / 2
      brne w_f_1

      dec r18
      brne w_f_0

      ; PHASE 2
      ; Walk a line of FFs down the (26) rows
      call clearscreen
      ldi r18, 26
```

```
        ; Initialize data
        ldi r28, $01
        ldi r29, $00
        ldi r30, $00
        ldi r31, $00

w_r_0:
        lds r16, flags
        sbrs r16, FLAG_TESTMODE
        ret

        lds r16, framecounter
        cpi r16, $00
        brne w_r_0

        ; store new data
        ldi r26, low(framedata)
        ldi r27, high(framedata)
        ldi r25, $10

s_c_lp:     st x+, r28
        st x+, r29
        st x+, r30
        st x+, r31
        dec r25
        brne s_c_lp

w_r_1:
        lds r16, flags
        sbrs r16, FLAG_TESTMODE
        ret

        lds r16, framecounter
        cpi r16, BLINK_RATE / 2
        brne w_r_1

        ; shift "on-bit" one left through all 26 bits
        cpi r28, $00
        breq t_b1
        lsl r28
        brne t_done
        ldi r29, $01
        rjmp t_done
```

```
t_b1:    cpi r29, $00
    breq t_b2
    lsl r29
    brne t_done
    ldi r30, $01
    rjmp t_done

t_b2:    cpi r30, $00
    breq t_b3
    lsl r30
    brne t_done
    ldi r31, $40
    rjmp t_done

t_b3:    lsl r31

t_done:
    dec r18
    brne w_r_0

    rjmp testmode
```

At the very end of the code, we have a couple of miscellaneous subroutines to clear video RAM, and to reset all the output latches:

```
; SUBROUTINE
; Clear ALL framebuffer RAM
clearscreen:
    push r27
    push r26
    push r18
    push r16

    ldi r18, $00
    ldi r16, (4*16)
    ldi r26, low(framedata)
    ldi r27, high(framedata)

clslp:     st x+, r18
    dec r16
    brne clslp

    pop r16
    pop r18
```

```
    pop r26
    pop r27
    ret

; SUBROUTINE
; Sets all output latches to LOW state (ie turn off row/col
drivers) clear_outputs:
    push r16

    ; Set output drivers for ports A,B,C low
    ldi r16, $00
    out PORTA, r16
    out PORTB, r16
    out PORTC, r16

    ; Latch zeros into U2,U3,U4,U5
    ldi r16, $0f
    out PORTB, r16
    nop
    ldi r16, $00
    out PORTB, r16

    pop r16
    ret
```

CHAPTER 5

Installing and Using a Version Control System

Chris Keydel and Olaf Meding

In 1999 the FAA lost all of their code needed to control commercial flights between Chicago O'Hare and the regional airports. A disgruntled programmer left the agency and deleted all of the code from his computer. There were no backups. He did encrypt a copy on his home computer, which the FBI finally was able to access after six months of work to discover the encryption key.

Scary, isn't it?

In 2004 the Electronic Frontier Foundation announced their FTP site was hacked, and, by the way, they had no backups of a number of open source projects.

Is this acceptable? Of course not.

If you want one lousy resistor you have to go to the inventory room, fill out a form in triplicate, and deal with the scowls from the angry inventory clerk. Yet in all too many situations a developer can issue

rm − r *. *

...and completely delete the entire company's source code tree.

Accounting hasn't figured out that most of a company's value resides in the software. As a result it isn't locked down as are physical assets like that resistor. That has to change.

The first step in preserving the source is the use of a version control system (VCS). A VCS gets the source off of a developer's hard disk, and stores it in a common database that IT folks back up every day. A VCS chronicles the entire history of the ever-evolving code, saving the current configuration as well as every iteration from the past.

> *This chapter gives an overview of resources that come with a VCS, and which of those resources a developer really needs. They discuss issues like working with different kinds of file systems, plus poke holes in Microsoft's Common Source Code Control API.*
>
> *Here are a couple of other resources to add to those listed in this chapter:*
>
> *For a comparison of 16 VCSs see http://better-scm.berlios.de/comparison/comparison.html*
>
> *Subversion is the most popular VCS today: http://subversion.tigris.org*
>
> *Trac is an open source tool that hyperlinks info between subversion, bug-tracking software, and a wiki: http://trac.edgewall.org*
>
> *SCMbug integrates Bugzilla and Subversion. See http://www.mkgnu.net/?q=scmbug*
>
>   *—Jack Ganssle*

5.1 Introduction

Software configuration management (SCM or CM) is the discipline of managing and controlling the evolution of software (over time). It is many things to many (different) people and encompasses process, policy, procedure, and tools. CM is a cornerstone in software development, and thus has a very broad spectrum. The focus here is the embedded systems market and the software developer rather than a management perspective. Advice and concepts are more practical than theoretical.

Three central and key concepts of CM are versioning, tracking, and differencing. Versioning is the idea that every file or project has a revision number at any given state in the development process. Tracking is essential because it provides a central database to track and manage bug reports, enhancement requests, ideas, and user or customer feedback. Tracking is needed to maintain current and past releases and to improve future versions of a software product. Finally, differencing allows the developer to easily view the difference between any two revisions of a file or project. Many of the more advanced CM concepts integrate, manage, and track these core development activities.

Improving the quality of your code and boosting developer productivity are not mutually exclusive goals. In fact, the opposite is true. Both quality and productivity dramatically increase with the knowledge and the applications of the tools and concepts outlined below. The right tool will be fun to use while at the same time increase the engineer's confidence that a software project will achieve its goals. This chapter will show how basic and proven tools and procedures can be used to produce high quality software. From my own experience and by talking to colleagues I noticed that many CM tools, twenty years after their introduction, are still not being used to their full capabilities. This seems to be especially true in the Embedded Systems and PC (non-Unix) world. A few of the reasons are: small (often single-person) teams or projects, a wide variety of platforms and operating systems, budget constraints, including the cost of commercial tools, nonsupportive IS departments, prior negative experiences with hard to use CM tools, and lack of support from management. A trend in the development of software for embedded systems is the increased use of personal computers (PCs). More and more software development tools for PCs are becoming available and are being used. This is good news in that excellent and intuitive CM tools are also now available. This chapter seeks to outline such tool categories, including how they are used, along with some simple commonsense procedures. It is probably safe to assume that most source code including firmware for embedded systems and microcontrollers is edited and managed on PCs these days. Therefore, all concepts, ideas, and tools discussed below apply as well to these environments.

This chapter is divided into sections outlining version control overview, version control client issues, version control concepts explained, tips, bug tracking, non-configuration management tools, and finally a reference section. Other software development tools will be briefly mentioned and explained precisely because they too increase quality and productivity, but are usually not considered CM tools. The lack of their use or even knowledge of their existence, among many fellow professional software developers, motivated the inclusion. These tools include mirror software, backup software, World Wide Web browsers, and Internet News Groups.

5.2   The Power and Elegance of Simplicity

Before discussing version control (VC) systems in detail, it is worth noting that standard VC tools can be (and are) applied to all kinds of (text and binary) files and documents, including World Wide Web pages. The major and outstanding advantage of standard source code over any other kind of document is the fact that it is strictly and simply (ASCII) text based. This unique "feature" has two major advantages over other kinds of documents (HTML included), differencing and merging (explained in detail later). Differencing allows nonproprietary tools to show the difference between any two files. Secondly and equally important is that standard nonproprietary or generic tools can be used for merging a branch file back into the main trunk of development (also defined below). An unfortunate but minor exception is the line ending conventions used by prevailing operating systems, CRLF in PCs, CR in Macintosh, and LF in Unix.

Finally, many version control systems take advantage of the simple structure of standard text files by only storing deltas between file revisions and thus significantly reduce disk space requirements.

5.3   Version Control

Central to configuration management is the version control (VC) tool. VC captures and preserves the history of changes as the software (source code) evolves. VC empowers the user to re-create any past revision of any file or project with ease, 100% reliability, and consistency.

A common misconception is that VC is not needed for projects with only a single developer. Single developers benefit just as much as teams do. All the reasons for using VC also apply to single developers. Many VC systems are easy to set up and use and the VC learning curve is significantly smaller then any single side effect (see below) of not using a VC system.

Finally, how safe is the assumption that a single developer will maintain a certain software project throughout its entire life cycle? And even if this holds true, how likely is it that this developer will be available and able to remember everything in an emergency case? Corporate productivity and software quality are issues as well.

5.4   Typical Symptoms of Not (Fully) Utilizing a Version Control System

Here are some telltale symptoms that indicate only partial or perhaps wrong use of VC or the complete lack of a VC tool. These symptoms include manual copying of files, use of revision numbers as part of file or directory names, not being able to determine with 100% confidence what code and make files were used to compile a particular software release, and bugs considered fixed that are still being reported by customers. Also, checking in source code less frequently than multiple times per week and developers creating their own sub-directories for "testing" indicate serious problems with the (use of the) VC tool. Two other symptoms include the inability to determine when (date and release) a particular feature was introduced, and not being able to tell who changed or wrote a particular (any) section of code.

5.5 Simple Version Control Systems

Basic and single-developer VC systems have been around for a long time. One of the simplest, oldest, and most successful VC system is the Revision Control System (RCS). RCS is included with most Unix distributions and is available for many, if not all, other operating systems. Many modern and full-featured VC systems are based on RCS ideas and its archive files. The key idea is simple. Every source file has a corresponding archive file. Both the source file and its archive file reside in the same directory. Each time a file is checked in (see below) it is stored in its corresponding archive file (also explained below). RCS keeps track of file revisions (AKA versions) and usually includes tools for file differencing and file merging. However, the early RCS systems were not well suited for teams, because of the location of archive files and the inability to share them between developers. Also, there was no concept of projects, i.e., sets of related files, representing a software project. Good documentation is readily available and most implementations are free and easily available and can be downloaded via Internet FTP sites.

5.6 Advanced Version Control Systems

Advanced VC systems offer a wide variety of options and features. They fulfill the needs of large teams with many (thousands of) projects and files. The team members may be geographically distributed. Advanced VC systems support a choice of connectivity options for remote users and also internally store time in GMT or UTC format to support software developers across time zones. Security and audit features are included, so access can be limited to sensitive files and to provide audit trails. Promotion models are used to manage software in stages, from prototyping to development to final testing, for example. Parallel development models allow all developers to work on independent branches and full featured and automated merge tools ease the check-in process. Some VC systems employ virtual file systems that transparently point or redirect the editor or compiler directly into an SQL strength database. Another useful feature is the ability to rename or move files within the version controlled directory tree without losing the integrity of older revisions of a file that was moved or renamed. Powerful build tools are also frequently included to help automate the build (compile) process of large software projects. Event triggers alert the software team or a manager of key events, such as a check-in or a successful nightly build. It is not uncommon for large projects to employ a full-time dedicated configuration administrator.

5.7   What Files to Put Under Version Control

This is a subject of many debates. Many published CM papers include a full discussion of this topic, so there is no reason to expand on this topic here. In brief summary, agreement exists that all source code files needed to create the final executable must be added to the VC system. This includes, of course, make files. A good VC system should make it very easy to re-create the final executable file, so there should be no need to put binary intermediate files (object files, pre-compiled header files, etc.) or even the final executable file itself under VC. Library files, device drivers, source code documentation, any kind of written procedure, test scripts, test results, coding standard or style guides, special tools, etc., can and probably should be added to the VC system.

5.8   Sharing of Files and the Version Control Client

Most VC systems are true client/server applications. Obviously, in a team environment, all developers somehow must have both read and write or modify access to the file archive located on a central file server. So, you might ask yourself exactly how files are transferred from that central file server to your local PC? The VC system vendor often provides a VC client that provides a user interface to exchange source code between the PC (workstation) and the repository (explained below) on the central file server. However, this VC client may or may not be accessible from your PC and its operating system. And the PC may or may not have direct access to the server's file system (NFS in Unix or drive mappings with Microsoft LANs and Novel's Netware). Physical location of both the client PC and the file server affect and also limit the choice of available options. In summary, there are two variables—whether the VC client can run on your local PC and whether your PC can access the file server's file system. The two variables combine into the four choices or variations outlined below.

5.8.1   No Local Client and No Common File System

This is the least desirable scenario. An example would be a PC accessing an older mainframe. Or a much more likely situation is a geographically separated team with a limited budget. Here the only choice is to use Telnet or some kind of remote Unix shell to run the client directly on the central file server or on a local workstation (with respect to the file server) and then manually FTP the files between the file server and your PC. For example, login to the file server, checkout a file, FTP it to your PC, change it, FTP it back, and then finally perform a check in.

5.8.2 No Local Client but Common File System

This is similar to the first scenario. However, the PC can access files directly. The applications on the PC (editor, compiler, etc) can either access source files across the LAN or the user has to manually copy them. The first method is slow and will increase network traffic. The latter requires extra and manual steps after checkout and before check-in.

5.8.3 Local Client but No Common File System

The problem here is that few VC systems provide clients the capability of accessing the server over a LAN or across the Internet, via a network protocol like TCP/IP. Fewer VC systems still use secure methods of communications (public keys instead of un-encrypted passwords during login and encrypting file content instead of transferring text).

5.8.4 Local Client and Common File System

This is the most common and flexible scenario. The client often takes advantage of the common file system to access the repository on the file server. The client may not even be aware that the repository is on a separate file server. The VC client is much easier to implement by a VC provider, which in part may explain its popularity.

5.9 Integrated Development Environment Issues

Integrated development environments (IDEs) for software development predate graphical (non-character-based) windows and became very popular during this decade, the decade of the graphical user interface (GUI). Most IDEs provide an excellent and integrated source code editor with many hip features, such as automatically completing source code constructs while editing and tool tip type syntax help displayed as you type. In addition, they usually feature well-integrated debuggers, full and finely tunable compiler and linker control, and all kinds of easily accessible online help. What is often missing in such IDEs, or was added only as an afterthought, is an interface to a VC system. The Source Code Control (SCC) specification (see below) is one such integration attempt. Another major problem with many modern IDEs is proprietary, in some cases even binary, make file formats. These proprietary make files are much more difficult to put under version control. Problems with these make files include merging, sharing when strict locking (see below) is enforced, and the fact that each IDE typically has a different and proprietary format. In fact, make file formats often change between each release of an IDE from the same vendor. IDEs are a nice and convenient tool and there is no doubt that they increase developer productivity. However, they often produce a handful of proprietary files, containing make information, IDE configuration information, debug and source code browse information, etc. And it is not always obvious exactly what information these files contain and whether they need to be put under VC.

5.10 Graphical User Interface (GUI) Issues

A number of VC vendors provide GUI front-end VC clients. These graphical (versus text-based) clients are much more difficult to implement well and are therefore often expensive. They can easily exceed the budget of an embedded systems project. They occasionally appear clumsy and sometimes only support a subset of features compared to an equivalent text based (command line) VC system. Another problem is that they may not be available across operating systems and hardware platforms.

5.11 Common Source Code Control Specification

The Microsoft Common Source Code Control (SCC) application-programming interface (API) defines an interface between a software development IDE and a VC system. Many Microsoft Windows development environments (IDEs) interface to a VC system through the SCC specification. The VC system acts as an SCC service provider to an SCC aware IDE. This enables the software developer to access basic VC system functions, such as checkout and check-in, directly through the host environment. Once an SCC service provider is installed, the user interface of the IDE changes. Menus are added to the IDE for the operations offered by the SCC service provider. And file properties are extended to reflect the status of the files under source code control.

The problem with this approach is that often only a subset of VC functions are provided through the SCC interface. Therefore the standalone VC client is still needed for access to the more advanced functions offered by most VC systems. So why not always use the standalone VC client with access to all features?

The VCC specification is only available to Microsoft Windows users since this a Microsoft specification. Furthermore, the documentation for the SCC API is not publicly available. To receive a copy, you must sign a nondisclosure agreement. It should be noted that this is not the only way to interface IDEs to VC systems. Some VC vendors call their service interface extensions, because the VC system extends its functionality to an IDE or a standard file browser offered by many operating systems.

5.12   World Wide Web Browser Interface or Java Version Control Client

This sounds like a great idea. In theory, well-written Java code should run anywhere either as a standalone application, or in a web browser as an applet, or as a so-called plug-in. The VC server could be implemented as a dedicated web server, which manages the VC archive, repository or database (explained below). Another advantage of Java is build-in security. However, for various reasons there are very few Java applications of any kind available that achieve the goal of write once and run anywhere. A select few VC vendors do provide Java based VC systems, but their user interfaces appear sluggish and somewhat clumsy. It should be noted that web browsers rarely support the latest version of Java, so the plug-in must be written in an older version of Java. In summary, Java offers many attractive features. I would expect to see more Java based VC systems in the future as the Java and related technologies, such as Jini, mature.

5.12.1   Version Control Concepts Explained

Revision, Version, and Release

All three terms refer to the same idea: namely that of a file or project evolving over time by modifying it, and therefore creating different revisions, versions, and eventually releases. Revision and version are used interchangeably; release refers to the code shipped (or released) to customers.

Checkout, Check-in, and Commit

Checkout and check-in are the most basic and at the same time the most frequently used procedures. A checkout copies the latest (if not otherwise specified) version of a file from the archive to a local hard disk. The check-in operation adds a new file revision to that file's archive. A check-in is sometimes called a commit. The new version becomes by default, if not otherwise specified, either the head revision if you are working on the trunk, or a tip revision, if you are working on a branch (all explained below). When a file is checked into a revision, other then the head or tip revision, a new branch is created.

Lock, Unlock, Freeze, and Thaw

The lock feature gives a developer exclusive write (modify) access to a specific revision of a file. Some VC systems require (or strongly recommend) and enforce a strict locking policy. The advantage of strict locking is prevention of file merges and potential conflicts. These VC systems often also provide an administration module that allows for enforcing and fine-tuning of policies. An example of such a policy is allowing only experienced developers to make changes to sensitive files. This approach works well in smaller teams with good communication. Here, the team (management) decides beforehand who will be working on which file and when. Each team member knows or can easily tell who is working on what file. One major drawback is that if someone else already locked the file he needs, a developer must ask and then wait for another developer to finish and check in her changes. Frozen files cannot be locked and therefore cannot be modified until thawed. This may be a useful management tool to let developers know that certain files should not be changed, say right before a major release.

Differencing

To be able to tell the difference between two revisions of a file is of utmost importance to any software developer, even if for whatever reason no VC system at all is used. In general, the difference tool allows a developer to determine the difference between two files and, more specifically, between two revisions of the same file. Any VC system has a built in difference tool. However, there are many difference tools available. Some tools are text based and others use a graphical (visual) user interface. It is a good idea to have a selection of different difference tools handy, depending on the task to be performed. The build-in tool is often sufficient and convenient to access through a menu choice for quick checks. More powerful difference tools can compare entire directory trees and may format and present the output in a more readable form.

Difference tools help avoid embarrassment by pointing out the debug or prototype code you forgot to remove. Often there is an overlap between developing a new feature and debugging a problem recently discovered or reported. For example, if someone asks you to find a high profile bug while you are working on implementing a new feature. Using a difference tool before each check-in helps prevent forgetting to take out code not meant for an eventual release to customers, such as printing out the states of (debug) variables at run time.

Another important use for a difference tool is detecting all code changes between two revisions or entire set of files that introduced a new bug. For example, customer support enters a defect report into the bug tracking database (see below) stating that a bug exists in release 1.1 but the same problem does not exist in release 1.0. The difference tool will show all sections of code changes that most likely introduced the bug. Being able to quickly locate a bug is important for Embedded Systems developers, where hardware is often in flux, because it may not be initially obvious if the bug is caused by a software change, hardware change, or a combination of the two.

Build and Make

A powerful build tool plays a vital part in many CM setups and more advanced VC systems include flexible and powerful build tools. A make tool is standard on Unix and also shipped with most compilers. However the standard make tools are often less capable compared to full-featured build tools included with VC systems. Both build and make tools require some careful planning, but once correctly implemented provide the ability to build with ease any release of an application at any time. The goal is to ensure consistent and repeatable program builds. Another advantage is automation of large-scale projects. Many software manufacturers perform builds of entire software trees overnight. The results of these nightly builds are then ready for examination on the following workday.

Trunk, Branch, and Merge

Merging refers to the process of consolidating changes made to the same file by either different developers or at different times, i.e., revisions of a file. A merge is typically required to copy a change made on a branch back to the main trunk of development. Usually each change is in a different area of the file in question and the merging tool will be able to automatically consolidate all changes. Conflicts will arise for overlapping changes, however, and they must be consolidated manually.

There are three main reasons for branching. They are changing the past, protecting the main trunk from intermediate check-ins, and true simultaneous parallel development. Most, if not all, VC systems support branching as a way to change the past, that is, to create a branch to add bug fixes (but not necessarily features) to an already released software product. For example, a bug fix results in release 1.1, while the main development effort continues on release 2.0. A merge may be required if the bug was first fixed for release 1.1. In this case, a developer would either manually copy or, better, use a merge operation to copy the code representing the bug fix into release 2.0.

Some teams decide that all new development should take place on a branch to protect the main trunk from intermediate check-ins. This approach requires merging code from the branch to the main trunk of development after completing each specific feature (set). Often teams decide to check-in all code to their respective branch at the end of every workday to protect themselves against data loss. A major advantage of this approach is that the trunk contains only finished changes that compile correctly and that are ready to be shared within the team. This approach also rewards frequent merging, because the longer a developer delays a merge, the higher is the chance of encountering merge conflicts.

Not all VC systems support true simultaneous parallel development, however, and there is some controversy and reluctance by some programmers to use or take advantage of branching because of potential subsequent merge conflicts. Notably, VC systems requiring exclusive write access (see strict locking above) before a change can be made do not support parallel development. On the other side, VC systems that do encourage simultaneous and independent development make extensive use of branching. This can result in minor conflicts while merging changes made on a branch back to the main trunk of development. However, minor conflicts are a small price to pay for large teams in exchange for the ability to work in parallel and independently from each other. In general, branching depends on frequent and thorough developer communications and it is a good idea to implement a consistent naming convention for labeling (see below) branch revisions.

Files and Projects

Some VC systems strictly handle only individual files. At best, all related files are located in a single directory (tree). This is often sufficient for smaller projects. Large projects, however, are often organized across a wide and deep directory tree and often consist of sub-projects. Therefore, a VC project is defined as the set of files belonging to the corresponding source code project. And a source code project is defined as the set of files that are compiled into the target executable, i.e., device driver, dynamic link library (DLL), static library, program, etc. More advanced VC systems support projects or modules, so that there is a one to one relationship between a source code project and the corresponding VC project. This greatly simplifies VC operations such as check-in, checkout, and tagging (see below). For example, instead of tagging a specified number of individual files and directories belonging to a source code project, the user can just tag the corresponding VC project in a single transaction.

Checkpointing, Snapshots, Milestones, Tagging, and Labeling

The goal here is to be able to retrieve the state (certain revision number) of all files of a project in the future. This is a very important activity because it allows a developer to retrieve all files that comprise, for example, release 1.0 of a project. Or a test engineer could mark all files that were present and their revision number when a certain defect was found. Not all VC systems fully support this critical feature. There are two common implementations. The first is to record all revision numbers of all files involved. A drawback of this method is that the checkpoint, snapshot, or milestone information has to be stored somewhere in a special VC file or database. The second implementation of the critical feature is to mark all revision numbers of all files involved with a certain label or tag. A VC system that favors this method has to be able to store many thousands of labels for each revision of each file. The label information is stored inside the archive file. Though all four terms refer to the same idea, they are often used interchangeably for both methods. Note that some VC systems allow a particular label to be used only once in a file's history and some VC systems limit any one revision to a single label.

Archive, Repository, and Database

The archive is a file containing the most recent version of a source code file and a record of all changes made to it since that file was put under version control. Any previous version of a file can be reconstructed from the record of changes stored in the archive file. Sometimes the file tree containing the archive files is called the repository. Some VC systems use a database, instead of individual archive files, for recording the history of source code changes. With the possible exception of single user systems, the archive is most frequently stored on a central file server. Usually, all that is needed to re-create any particular version of a file is that file's archive. Backup of archive data is therefore vitally important. A network administrator usually performs a daily backup as part of the file server maintenance. This frees the software engineer from this time consuming task. Some VC systems specifically designed to support physically distributed teams use archive replication as a way to share archives. In this scenario, replication software resynchronizes the archive every few hours between the central file server and a remote location file server. The advantage of this approach is faster access for the remote team of developers. Drawbacks include the complexities of installing and maintaining the replication software and potential resynchronization conflicts between file servers.

Working Copy and Sandbox

The working copy contains a copy of each file in a project at a specified revision number, most commonly the latest revision number of the main trunk or of a branch. Working copies provide a way for each developer to work independently from other developers in a private workspace. The working copy is usually located on the developer's local hard disk and is created there through a checkout operation. A working copy is sometimes also called a sandbox. It is not uncommon to have multiple working copies per developer per project, with each working copy at a different trunk or branch release. I often create a new working copy immediately after checkpointing a major project release to double check and verify all file revisions. All files in the new working copy should be identical to all files in the working copy that created the new release (see Tips below).

Event Triggers

Many VC systems provide a means to execute one or more statements whenever a predefined event occurs. These events include all VC operations and more; for example, e-mailing a project manager each time a file is checked in.

Keywords

A keyword is a special variable that can be included within source code. This variable is embedded usually between two $ characters (for example $Author$) to represent textual information in a working copy. Keywords can optionally be expanded, i.e., replaced with their literal value, when a revision is checked or viewed. This is a very useful feature, because it allows the VC system to place information directly into the source code. At the same time keywords allow the developer to select what VC information, i.e., which keyword(s) should be inserted and where it should be located in the source code, usually at the beginning or at the end of a file.

5.12.2 Tips

Source Code File Directory Tree Structure

One key aspect of any larger software development project is the implementation of a directory structure and layout. This directory tree reflects the relationship of the source code files that define a software project. Most importantly, there should be no hard coded paths for include file statements in C/C++, import file statements in Java, or unit file statements in Pascal source code. Instead, a relative path statement should be used so that (multiple) working copies of files and projects can be created anywhere on a software developer's local hard disk. Placing shared source code files in the directory tree is an art, because each method has advantages and disadvantages and much depends on the specifics of a project's architecture. Therefore, it is not uncommon to adjust the source code file directory tree as the project grows. However, not all VC systems support renaming and moving of source code files well. An article by Aspi Havewala, called "The Version Control Process" (see References below) illustrates one nice way to organize shared source code and C/C++ style header files.

More Applications of the Difference Tool

For extra safety and protection of source code and its long term storage, I sometimes backup my archives on the central file server and my working files on my PC to a writeable CD. CDs are nice because they have a much larger shelf life then any other backup medium. However, for a number of reasons CD burners sometimes fail to make exact copies. This is where a tool such as Microsoft's WinDiff (see Resources below) or similar tools provide assurance and confidence in backups needed to remove obsolete files from your hard disk. WinDiff will read (and in the process verify) every single byte on the new CD in order to confirm that all files were copied correctly. Being able to remove obsolete code efficiently frees time for the developer and therefore increases productivity and also helps the software engineer to focus on the main project.

Another useful application of WinDiff is verifying what files were copied where during installation of new software (tools) to a critical PC. This not only provides extra insight on how the new tool works, it might also be a great help to debug incompatibility problems between various applications, which has been and continues to be a major problem for Microsoft operating systems. For example, installing a new compiler or other software development tool may consist of many files being copied into an operating system specific directory. To find out exactly what files were installed where, run a recursive directory listing and redirect the output into a file. For example, type "dir /s/a c:\ > before.dir", install the new tool, run another directory listing, but this time redirecting the output into a file called after.dir. Then use your favorite differencing tool or WinDiff to display the delta between before.dir and after.dir. You can use the same procedure to debug and verify your own installation procedure before you ship your product to a customer.

Heuristics and Visual Difference Tools

It is worth noting that GUI-based difference tools rely on heuristics rather than on exact algorithms to detect and display code that moved from one location to another. Heuristics are also used to differentiate between changes and additions for displaying each in a different color. For example, consider these two revisions of the same file:

```
Revision 1.1        Revision 1.2
line 1 code         line 1 code
line 2 code         line 2 code changed
line 3 code         a new line of code inserted here
line 4 code         line 3 code changed
line 5 code         line 4 code
                    another new line of code
                    line 5 code
```

A visual difference tool will have no problem detecting the new code inserted after line 4. However, how can the tool tell the difference between new code inserted and code changed between lines 2 and 3? The new line could either be part of the line 2 change or the line 3 change, and it could also be new code inserted! There is more than one answer here and that is why a heuristic is used to display change codes and colors. So watch out when using a visual difference tool when searching for code additions displayed in, say, a blue color. There may be unexpected surprises; in the above example, added code may be shown as changed code. Try the above example or variations of it with your favorite visual difference tool.

No Differencing Tool

On more than one occasion, for example when visiting a customer, I needed to compare two versions of a section of code, and the only tool available was a printer. No problem. I just printed out both sections and then held both copies overlapped against a bright light source. By just moving one of the pages slightly back and forth over the other I was able to easily identify the one line of code that changed. This will also work with a computer monitor displaying two revisions of the same file each in its own editor window. All you have to do is display the code in question in two editor windows one exactly on top of the other and then rotate quickly back and forth between the two windows. Your eyes will easily be able to detect the differing code (if any) as you cycle back and forth between the two windows.

Complete Working Copies

It is a good idea to include any file—i.e., libraries, device drivers, configuration (.ini) files, etc.—that are needed to compile, test, and run a project into the VC system. This ensures that a working copy once created by the VC system is immediately ready for use.

Multiple Working Copies

It is not uncommon to keep multiple working copies of a project on the local PC. Each working copy contains a different release of the project. This simplifies the creation of a list of all changes between a new release and a previous release if you have a working copy for each. The difference tool will point out all sections of code that changed between the two releases. This makes it much easier to document all changes for the new release. Multiple working copies are also very handy for locating bugs in prior releases. Furthermore, I often create a new working copy right after creating a new project release. I then compare (using the difference tool) the new release working copy with the working copy that created the new release. These two working copies had better be 100% identical. One reason that they may not be exactly alike is adding a new source code file to your project but forgetting to put that new file under version control.

Style Guide and Coding Convention

Both terms refer to the same document. One of the more important CM documents spelling out coding conventions and styles as a common language, so that all developers working in the same company or on the same project understand each other's source code. This document helps to keep all code in a consistent format and, therefore, increases the overall quality of source code. Many papers and books have been written on this subject and the important thing to remember is that each software project should have a coding convention. This document should be put under VC as it evolves. It is worth noting that source code should be written with humans in mind, not machines. And a style guide helps to achieve this goal.

5.13 Bug Tracking

Bug tracking (BT) encompasses the entire project timeline—from the past to the present and into the future. Being able to track an issue into the past enables customer service to determine if, for example, a certain bug is present in the version currently being owned by a customer. Looking at the present tells a manager what issues are currently open and who is eliminating what problem or implementing (working on) that hot new killer feature. And the element of future is represented by feature requests (wish list) or product ideas, both of which are forming the base for future releases, which in turn provide the revenue needed to maintain a software project.

BT is also known as defect tracking. Both terms are somewhat misleading because they imply problems. And yes, problem tracking is vital. However, just as important and worth tracking are enhancement requests, customer feedback, ideas developers have during development or testing, test results, sales force observations, etc. In short, any kind of issue that documents the behavior of past versions and that might help to improve a future product release. Tracking these issues is mainly a database exercise with emphasis on powerful database query capabilities. A BT application provides a central database to track the issues mentioned above. Ideally, the database must be easily accessible to any person involved in a software project, i.e., developers, testers, managers, customer service, and sales personnel. And yes, the user and customer should also have (limited) access to this database.

Significant recent improvements have been made to database (server) technology, local and wide area networks, and (thin) clients capable of running on any desktop PC. So, the technology is available (and has been available for years), yet BT software lags behind VC software and only a few packages offer tight or complete integration between the two. An example of good integration would be a single piece of client software that can access both the VC archive and the BT database. Another example would be the ability to easily locate the code that closed a certain bug report. Part of the reason for this might be that BT is not as clearly defined as VC. Effective BT requires a real client/server type database (versus a special purpose yet much simpler archive or repository typically used by VC). Another problem might be the wide variety of team requirements of this database, such as what fields a bug report should have (date, time, who reported a bug, status, etc.), management type reporting capability, ease of constructing queries, database server hardware requirements, and database administrator labor.

Equally important is the ability to determine if a specific defect (characteristic) or feature is present in a specific version of a product. This way customer service will be able to respond quickly to specific questions.

I have not personally worked very much with BT tools and my projects and productivity have suffered greatly. The fact that BT tools require some effort to set up and configure for your environment is no excuse for not using this essential tool. I am currently working on this problem and plan to present my findings in a future paper or presentation.

5.14 Non-Configuration Management Tools

5.14.1 Mirror Software

Most editors support renaming the current file with a .bak (for backup) extension before making a change. This, at a minimum, allows you to revert back to the last saved copy of a file in case you lost control or your PC crashes while editing source code. Taking this idea one step further, there are source code editors (for example, the Borland BC5 IDE) that automatically save your current file in two places. This feature makes it possible to recover from both a hardware failure and software crash. I fail to understand why this simple and easy to implement feature (saving the current file to a second configurable place, preferably a different hard drive) is not available in all source code editors. However, there are standalone third-party products that automatically mirror any file that changed. One such product is as AutoSave by Innovative Software (see Resources below).

5.14.2 Automated Backups

A software developer should not have to worry about backups of his or her daily work. Having to back up your work to floppy disks or some other removable media is counterproductive and cause for many problems. Central backup of software archives, repositories, and databases is a must-have, and there are many applications available for this. If there is no regular backup of this critical data, make sure you copy your resume file to a floppy disk. You may need it soon.

5.14.3 World Wide Web Browser

World Wide Web browsers are a must-have tool to access all kinds of freely available (product support) information, including many forms of technical support. This tool is included here for obvious reasons and for completeness.

5.14.4 Internet News Groups

Internet news groups date back to the earliest days of the Internet. These news groups—a more descriptive name would be discussion groups—use the network news transfer protocol (NNTP) to exchange messages. There are now two types of news servers, vendor specific and traditional. The vendor specific news servers are stand-alone servers with one or more news groups dedicated to the products offered by that company. All you need to access these news servers is the address of the server, usually advertised on the product support web page. Once connected it is straightforward to find the news group you need, because there are typically only tens of them with obvious names. Traditional news servers, on the other hand, are all interconnected with each other to form a worldwide network. They often contain tens of thousands of news groups, and are maintained by your Internet Service Provider (ISP). A typical news group is comp.lang.c for the "C" programming language. There are also moderated (censored) news groups that filter out all spam (Internet slang for unwanted advertisements) and other nonrelated offensive message posts. News groups are an excellent (and free!) source of help to any kind of question or problem you might have. You often find an answer to your a.m. posting in the afternoon and messages you post before you leave work are frequently answered overnight. You will discover that answers to your questions come from all corners of the world (Russia, Germany, New Zealand, and India, to name a few). Often, while browsing newsgroups, you see posted questions that you know the answer to and in turn help someone else out there to improve her code. News readers are available stand-alone and are also part of most World Wide Web browsers. Chances are that one is already installed on your PC and waiting for you to use it. Embedded systems software developers are often isolated or work in small teams. News readers open the door to the world in terms of help and support, and therefore are a critical tool for all professional software development.

5.15 Closing Comments

This concludes the overview of basic, yet most important, configuration management tools. All tools mentioned above apply just as much to embedded systems projects as they do to all professional software development projects. If you are using configuration control and tracking tools, you appreciate their value; if you are not, your intuition is probably telling you that something is missing. Many good commercial and free tools are available. And they are fun to use, if implemented correctly, because you intuitively know that you are doing the right thing to improve your source code and productivity! Finally, note that configuration management is not a destiny, but a continuously evolving process.

I would like to thank Mary Hillstrom for all of her loving support and Bill Littman for his generous advice and for proofreading this paper. Comment? Feel free to send me an e-mail.

Suggested Reading, References, and Resources

Brown, William J., McCormick, Hays W. III, and Thomas, Scott W. *AnitPatterns and Patterns in Software Configuration Management.* Wiley, 1999. Available at http://www.wiley.com/compbooks/catalog/32929-0.htm.

Fogel, Karl. *Open Source Development with CVS.* Coriolis, November 1999. An outstanding book describing both Open Source Development and the free Concurrent Version System (CVS) used by many (free) large scale and distributed software development teams. CVS is the successor of the well known RCS VC system. Karl convincingly makes the case that the two are intimately related. The CVS specific chapters of his book are copyrighted under the GNU General Public License and are available at http://cvsbook.red-bean.com. Available at http://www.coriolis.com/bookstore (search for Fogel).

McConnell, Steve. *Code Complete: A Practical Handbook of Software Construction.* Microsoft Press, 1993. This book is a classic and the book's title says it all. Very comprehensive and still in print. Available at http://mspress.microsoft.com/books/28.htm.

Havewala, Aspi. "The Version Control Process." *Dr. Dobb's Journal*, May 1999, p. 100. This article includes a section labeled "Finding a Home for Shared Code" illustrating a nice way to organize shared source code and C/C++ style header files. Available at http://www.ddj.com/articles/1999/9905/9905toc.htm.

Radding, Alan. "Join The Team." *InformationWeek*, October 4, 1999, p. 1a. A recent high level write-up on CM issues and available tools for parallel development and geographically dispersed teams. This is the kind of (rare) article I would like to see many more of in the trade rags. Available at http://www.informationweek.com/755/55adtea.htm.

Microsoft Corporation. WinDiff Software. WinDiff is included with the "Windows NT 4.0 Resource Kit Support Tools" that you can download for free. For more information and the download location search the Microsoft Knowledge Base for article Q206848 at http:// support.microsoft.com/support.

Innovative Software. AutoSave. For more information on this inexpensive tool check its distribution partner at http://www.v-com.com/.

comp.software.config-mgmt. News group carried by all traditional news servers. This newsgroup is very active and features up-to-date frequently asked questions (FAQ). It is intended to be a forum for discussions related to CM, both the bureaucratic procedures and the tools used to implement CM strategies. The FAQ is also available at http://www.faqs.org/ if your ISP does not provide a news server or if you do not otherwise have access to a news reader.

microsoft.public.visual.sourcesafe. News group, carried only by the vendor-specific news server at "msnews.microsoft.com". This newsgroup is dedicated to Microsoft's Visual Source Safe product. It does not appear to have someone maintaining a frequently asked questions (FAQ) list nor does it seem to have active MVPs. Briefly, MVPs are recognized by Microsoft for their willingness to share their knowledge on a peer-to-peer basis. MVPs never seem to tire of answering posts. However, most message posts do receive multiple answers within a few hours.

borland.public.teamsource. News group, carried only by the vendor specific news server at "forums.inprise.com". This newsgroup is dedicated to Borland's (or Inprise's) TeamSource product. It does not appear to have someone maintaining a frequently asked questions (FAQ) list. This newsgroup is very active and monitored by Borland personal and TeamB members. Briefly, TeamB members are unpaid volunteer professionals selected by Borland, they are not Borland employees, and they tend to be outspoken with their opinions on virtually all aspects of Borland's presence in the market, including but not limited to product development, marketing, and support. TeamB members never seem to tire of answering posts.

CHAPTER 6

Embedded State Machine Implementation

Martin Gomez

State machines have been around pretty much since the first air-breather muddily crawled ashore hundreds of millions of years ago. They've always been a part of the embedded landscape, though were usually used more by EEs-turned-programmers than by computer science types, since state machine design is a basic part of synchronous logic synthesis. But in recent years they've gained more traction, due to the emphasis on them in model-based design, and because of commercial products like IAR's VisualState.

State machines aren't for every application, but a tremendous number of systems can benefit from their use. Whenever a system undergoes a well-defined sequence of transitions, consider using a state machine. Traffic light sequencing is a classic application.

In this short but descriptive chapter Martin Gomez goes through the basics of state machine design, starting with the standard diagrams and state transition logic and taking the reader all the way through generating actual code.

He uses a great example, that of an airplane's landing gear mechanism. This is a case where failure is not an option. Coding the sequences using normal procedural approaches would result in a spaghetti mess that's hard to test and harder to prove correct. With a state machine it's almost trivial to prove correctness, and is simple to change in case a new requirement is found (like "don't deploy gear if altitude greater than 10,000 feet").

Provability includes showing that the system doesn't do things that it shouldn't do. Martin shows that this, too, is easy with a state machine.

He uses function pointers to advance between states, which is the most common approach.

Sometimes state machines can grow outrageously complicated as the number of transitions expand. In this case an extension—the hierarchical state machine—can provide useful layers of abstraction. See Miro Samek's book Practical Statecharts in C/C++: Quantum Programming for Embedded Systems *for more on those.*

—Jack Ganssle

Many embedded software applications are natural candidates for mechanization as a state machine. A program that must sequence a series of actions, or handle inputs differently depending on what mode it's in, is often best implemented as a state machine.

This chapter describes a simple approach to implementing a state machine for an embedded system. Over the last 15 years, I have used this approach to design dozens of systems, including a softkey-based user interface, several communications protocols, a silicon-wafer transport mechanism, an unmanned air vehicle's lost-uplink handler, and an orbital mechanics simulator.

6.1 State Machines

For purposes of this chapter, a state machine is defined as an algorithm that can be in one of a small number of states. A state is a condition that causes a prescribed relationship of inputs to outputs, and of inputs to next states. A savvy reader will quickly note that the state machines described in this chapter are Mealy machines. A Mealy machine is a state machine where the outputs are a function of both present state and input, as opposed to a Moore machine, in which the outputs are a function only of state.[1] In both cases, the next state is a function of both present state and input. Pressman has several examples of state transition diagrams used to document the design of a software product.

Figure 6-1 shows a state machine. In this example, the first occurrence of a slash produces no output, but causes the machine to advance to the second state. If it encounters a non-slash while in the second state, then it will go back to the first state, because the two slashes must be adjacent. If it finds a second slash, however, then it produces the "we're done" output.

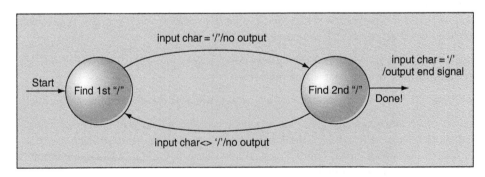

Figure 6-1: A Simple State Machine to Parse a Character String Looking For "//."

The state machine approach I recommend proceeds as follows:

- Learn what the user wants

- Sketch the state transition diagram

- Code the skeleton of the state machine, without filling in the details of the transition actions

- Make sure the transitions work properly

- Flesh out the transition details

- Test

6.2 An Example

A more illustrative example is a program that controls the retraction and extension of an airplane's landing gear. While in most airplanes this is done with an electrohydraulic control mechanism (simply because they don't have a computer on board), cases exist—such as unmanned air vehicles—where one would implement the control mechanism in software.

Let's describe the hardware in our example so that we can later define the software that controls it. The landing gear on this airplane consists of a nose gear, a left main gear, and a right main gear. These are hydraulically actuated. An electrically driven hydraulic pump supplies pressure to the hydraulic actuators. Our software can turn the pump on and off. A direction valve is set by the computer to either "up" or "down," to allow the hydraulic pressure to either raise or lower the landing gear. Each leg of the gear has two limit switches: one that closes if the gear is up, and another that closes when it's locked in the down position. To determine if the airplane is on the ground, a limit switch on the nose gear strut will close if the weight of the airplane is on the nose gear (commonly referred to as a "squat switch"). The pilot's controls consist of a landing gear up/down lever and three lights (one per leg) that can either be off, glow green (for down), or glow red (for in transit).

Let us now design the state machine. The first step, and the hardest, is to figure out what the user really wants the software to do. One of the advantages of a state machine is that it forces the programmer to think of all the cases and, therefore, to extract all the required information from the user. Why do I describe this as the hardest step? How many times have you been given a one-line problem description similar to this one: don't retract the gear if the airplane is on the ground.

Clearly, that's important, but the user thinks he's done. What about all the other cases? Is it okay to retract the gear the instant the airplane leaves the ground? What if it simply bounced a bit due to a bump in the runway? What if the pilot moved the gear lever into the "up" position while he was parked, and subsequently takes off? Should the landing gear then come up?

One of the advantages of thinking in state machine terms is that you can quickly draw a state transition diagram on a whiteboard, in front of the user, and walk him through it. A common notation designates state transitions as follows: < event that caused the transition >/< output as a result of the transition >.[2] If we simply designed what the user initially asked us for ("don't retract the gear if the airplane is on the ground"), what he'd get would look a bit like Figure 6-2. It would exhibit the "bad" behavior mentioned previously.

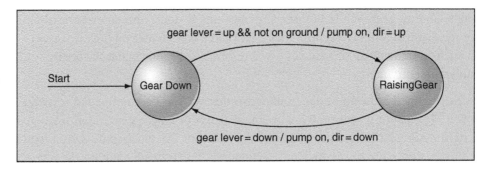

Figure 6-2: A State Machine Fragment that Does Only What the User Requested.

Keep the following in mind when designing the state transition diagram (or indeed any embedded algorithm):

- Computers are very fast compared to mechanical hardware—you may have to wait.

- The mechanical engineer who's describing what he wants probably doesn't know as much about computers or algorithms as you do. Good thing, too—otherwise you would be unnecessary!

- How will your program behave if a mechanical or electrical part breaks? Provide for timeouts, sanity checks, and so on.

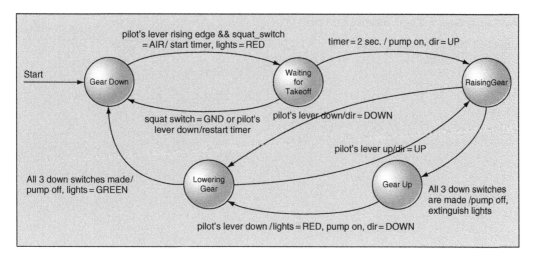

Figure 6-3: The Result of Finding Out What the User Really Wants.

We can now suggest the following state machine to the user, building upon his requirements by adding a few states and transitions at a time. The result is shown in Figure 6-3. Here, we want to preclude gear retraction until the airplane is definitely airborne, by waiting a couple of seconds after the squat switch opens. We also want to respond to a rising edge of the pilot's lever, rather than a level, so that we rule out the "someone moved the lever while the airplane was parked" problem. Also, we take into account that the pilot might change his mind. Remember, the landing gear takes a few seconds to retract or extend, and we have to handle the case where the pilot reversed the lever during the process. Note, too, that if the airplane touches down again while we're in the "Waiting for takeoff" state, the timer restarts—the airplane has to be airborne for two seconds before we'll retract the gear.

6.3 Implementation

This is a good point to introduce a clean way to code a finite state machine. The following is my implementation of the state machine in Figure 6-3.

Landing Gear Implementation

```
typedef enum {GEAR_DOWN = 0, WTG_FOR_TKOFF, RAISING_GEAR,
GEAR_UP, LOWERING_GEAR} State_Type;

/* This table contains a pointer to the function to call in each
state.*/
void (*state_table[]090 = {GearDown, WtgForTakeoff,
RaisingGear, GearUp, LoweringGear};

State_Type curr_state;

Main()
{
    InitializeLdgGearSM();

    /*  The heart of the state machine is this one loop.
        The function corresponding to the current state is called
        once per iteration. */
    while (1)
    {
        state_table[curr_state]();
        DecrementTimer();

        /* Do other functions, not related to this state machine.*/
    }
};

void InitializeLdgGearSM()
{
    curr_state = GEAR_DOWN;
    timer = 0.0;

    /* Stop all the hardware, turn off the lights, etc.*/
}

void GearDown()
{
    /* Raise the gear upon command, but not if the airplane is
    on the ground.*/

    if ((gear_lever == UP) && (prev_gear_lever == DOWN) &&
    (squat_switch == UP))
```

```
    {
        timer = 2.0;
        curr_state = WTG_FOR_TKOFF;
    };

    prev_gear_lever = gear_lever; /* Store for edge detection.*/
}

void RaisingGear()
{
    /* Once all 3 legs are up, go to the GEAR_UP state.*/
    if ((nosegear_is_up == MADE) && (leftgear_is_up == MADE) &&
    (rtgear_is_up == MADE))
    {
        Curr_state = GEAR_UP;
    };

    /* If the pilot changes his mind, start lowering
the gear.*/
    if (gear_lever == DOWN)
    {
        curr_state = LOWERING_GEAR;
    };
}

void GearUp()
{
    /* If the pilot moves the lever to DOWN, lower the gear.*/
    if (gear_lever == DOWN)
    {
        curr_state = LOWERING_GEAR;
    };
}

void WtgForTakeoff()
{
    /* Once we've been airborne for 2 sec., start raising
the gear.*/
    if (timer <=0.0)
    {
        curr_state = RAISING_GEAR;
    };
```

```
      /*If we touch down again, or if the pilot changes his
mind, start over.*/
      if ((squat_switch ==DOWN) || (gear_lever == DOWN))
      {
          timer = 2.0;
          curr_state = GEAR_DOWN;

          /* Don't want to require that he toggle the lever
             again this was just a bounce.*/
          prev_gear_lever = DOWN;
      };
}

void LoweringGear()
{
        if (gear_lever == UP)
        {
            curr_state = RAISING_GEAR;
        };
        if ((nosegear_is_down == MADE) && (Leftgear_is_down == MADE)
           && (rtgear_is_down == MADE))
        {
            curr_state = GEAR_DOWN;
        };
}
```

Let's discuss a few features of the example code. First, you'll notice that the functionality of each individual state is implemented by its own C function. You could just as easily implement it as a switch statement, with a separate case for each state. However, this can lead to a very long function (imagine 10 or 20 lines of code per state for each of 20 or 30 states). It can also lead you astray when you change the code late in the testing phase—perhaps you've never forgotten a break statement at the end of a case, but I sure have. Having one state's code "fall into" the next state's code is usually a no-no.

To avoid the switch statement, you can use an array of pointers to the individual state functions. The index into the array is curr_state, which is declared as an enumerated type to help our tools enforce correctness.

In coding a state machine, try to preserve its greatest strength, namely, the eloquently visible match between the user's requirements and the code. It may be necessary to hide hardware details in another layer of functions, for instance, to keep the state machine's code looking as much as possible like the state transition table and the state transition diagram. That symmetry helps prevent mistakes, and is the reason why state machines are such an important part of the embedded software engineer's arsenal. Sure, you could do the same thing by setting flags and having countless nested if statements, but it would be much harder to look at the code and compare it to what the user wants. The code fragment below fleshes out the RaisingGear( ) function.

The **RaisingGear()** Function

```
Void RaisingGear()
{
   /* Once all 3 legs are up, go to the GEAR_UP state.*/
   if (nosegear_is up == MADE) && (leftgear_is_up ==
      MADE) && (rtgear_is_up == MADE))
   {
      pump_motor = OFF;
      gear_lights = EXTINGUISH;
      curr_state = GEAR_UP;
   };

   /* If the pilot changes his mind, start lowering the gear.*/
   if (gear_lever == DOWN)
   {
      pump_direction = DOWN;
      curr_state = GEAR_LOWERING;
   };
}
```

Notice that the code for RaisingGear( ) attempts to mirror the two rows in the state transition table for the Raising Gear state.

As an exercise, you may want to expand the state machine we've described to add a timeout to the extension or retraction cycle, because our mechanical engineer doesn't want the hydraulic pump to run for more than 60 seconds. If the cycle times out, the pilot should be alerted by alternating green and red lights, and he should be able to cycle the lever to try again. Another feature to exercise your skills would be to ask our hypothetical mechanical engineer, "Does the pump suffer from having the direction reversed while it's running? We do it in the two cases where the pilot changes his mind." He'll say "yes," of course. How would you modify the state machine to stop the pump briefly when the direction is forced to reverse?

6.4   Testing

The beauty of coding even simple algorithms as state machines is that the test plan almost writes itself. All you have to do is to go through every state transition. I usually do it with a highlighter in hand, crossing off the arrows on the state transition diagram as they successfully pass their tests. This is a good reason to avoid "hidden states"—they're more likely to escape testing than explicit states. Until you can use the "real" hardware to induce state changes, either do it with a source-level debugger, or build an "input poker" utility that lets you write the values of the inputs into your application.

This requires a fair amount of patience and coffee, because even a mid-size state machine can have 100 different transitions. However, the number of transitions is an excellent measure of the system's complexity. The complexity is driven by the user's requirements: the state machine makes it blindingly obvious how much you have to test. With a less-organized approach, the amount of testing required might be equally large—you just won't know it.

It is very handy to include print statements that output the current state, the value of the inputs, and the value of the outputs each time through the loop. This lets you easily observe what ought to be the Golden Rule of Software Testing: don't just check that it does what you want—also check that it doesn't do what you don't want. In other words, are you getting only the outputs that you expect? It's easy to verify that you get the outputs that you expected, but what else is happening? Are there "glitch" state transitions, that is, states that are passed through inadvertently, for only one cycle of the loop? Are any outputs changing when you didn't expect them to? Ideally, the output of your prints would look a lot like the state transition table.

Finally, and this applies to all embedded software and not just to that based on state machines, be suspicious when you connect your software to the actual hardware for the first time. It's very easy to get the polarity wrong—"Oh, I thought a '1' meant raise the gear and a '0' meant lower the gear." On many occasions, my hardware counterpart inserted a temporary "chicken switch" to protect his precious components until he was sure my software wasn't going to move things the wrong way.

6.5   Crank It

Once the user's requirements are fleshed out, I can crank out a state machine of this complexity in a couple of days. They almost always do what I want them to do. The hard part, of course, is making sure that I understand what the user wants, and ensuring that the user knows what he wants—that takes considerably longer!

References

1. Hurst, S.L. *The Logical Processing of Digital Signals*. New York: Crane, Russak, 1978.

2. Pressman, Roger A. *Software Engineering: A Practitioner's Approach*, 3rd Edition. New York: McGraw-Hill, 1992.

CHAPTER 7

Firmware Musings

Jack Ganssle

When I wrote this chapter in 1999, peripheral datasheets were notorious for being incorrect and incomplete. Thus I advocated the iconoclastic notion of "hacking" drivers for I/O devices. That is, admit that the spec is unknown and use an iterative approach to discover a device's operation. Hack.

Now datasheets are even worse and peripherals more complex. Even memory has become another asset that requires gobs of code to handle, from programming flash to accessing serial EEPROM devices.

I remain amazed that the vendors, who must have written drivers to test and qualify their offerings, usually don't provide their customers with complete, canned driver solutions, the sort of thing Jean LaBrosse gives in Chapter 11 of this book. So hacking remains a staple of our profession.

Next I take on a number of memory issues, from predicting stack size and memory requirements to using dynamic memory allocation and then diagnostics. These issues are more relevant than when I first wrote about them. Malloc() remains the curse of our profession: it's a fabulous way to control RAM costs, but comes with perils most of us ignore. Memory fragmentation, unhandled exceptions, and the like cause too many embedded applications to crash like Windows machines.

Once one did diagnostics to find hard errors. Now things are changing. With geometries headed below 65 nm vendors are (only under pressure) admitting that their logic components, especially static RAMs, are susceptible to single-event upsets caused by cosmic rays. Systems operating in mile-high Denver sometimes exhibit much higher error rates than those at sea level.

Finally I talk about software prototypes. In one Dilbert cartoon the 23-year-old marketing droid gushes about how much the customer loved the new product. Dilbert informs him that it was a competitor's device with duct tape over the name. That doesn't dim the droid's enthusiasm, as he asks how quickly they can produce 1000 units.

> *Hardware people understand about prototypes. Design the PCB. Cover it with lots of green wires to fix design errors. Then they do the most important thing you can do with a prototype: they throw it away and reengineer the board to rid it of the mistakes. Show the boss a software prototype, though, and he's anxious to ship. He can't see the thousand or more virtual green wires that make the code barely functional, and completely unmaintainable. So check out my suggestions to thwart this dysfunctional behavior.*
>
> —*Jack Ganssle*

7.1 Hacking Peripheral Drivers

Experienced software engineers find no four-letter word more offensive than "hack." We believe that only amateurs, with more enthusiasm than skill, hack code.

Yet hacking is indeed a useful tool in limited circumstances.

This is not a rant against software methodologies—far from it. I think, though, a clever designer will identify risk areas and take steps to mitigate those risks *early* in a development program. Sometimes cranking code, maybe even lousy code, and diddling with it is the only way to figure out how to efficiently move forward.

No part of the firmware is more fraught with risks and unknowns than the peripheral drivers. *Don't* assume you are smart enough to create complex hardware drivers correctly the first time! Plan for problems instead of switching on the usual panic mode at debug time.

Before writing code, before playing with the hardware, build a shell of an executable using the tools allocated for the project. Use the same compiler, locator (if any), linker, and startup code. Create the simplest of programs, nothing more than the startup code and a null loop in main( ) (or its equivalent, when you're working in another language).

If the processor has internal chip-selects, figure out how to program these and include the setups in your startup code. Then, make the null loop work. This gives you confidence in the system's skeleton, and more importantly creates a backbone to plug test code into.

Next, create a single, operating, interrupt service routine. You're going to have to do this sooner or later anyway; swallow the bitter pill up front.

Identify every hardware device that needs a driver. This may even include memory, where (as with Flash) your code must do *something* to make it operate. Make a list, check it twice—LEDs, displays, timers, serial channels, DMA, communications controllers—include each component.

Surely you'll use a driver for each, though in some cases the driver may be segmented into several hunks of code, such as a couple of ISRs, a queue handler, and the like.

Next, set up a test environment for fiddling with the hardware. Use an emulator, a ROM monitor, or any tool that lets you start and stop the code. Manually exercise the ports (issue inputs and outputs to the device).

Gain mastery of each component by making it *do* something. Don't write code at this point—use your tool's input/output commands. If the port is a stack of LEDs, figure out how to toggle each one on and off. It's kind of fun, actually, to watch your machinations affect the hardware!

This is the time to develop a deep understanding of the device. All too often the documentation will be incomplete or just plain wrong. Bits inverted and transposed. Incorrect register addresses. You'll never find these problems via the normal design–code–inspect–debug cycle. Only playing with the devices—hacking!—with a decent debugging tool will unveil the peripheral's mysteries.

If you can't speak the hardware lingo, working with a part that has 100 "easy-to-set-up" registers will be impossible. If you are a hardware expert, dealing with these complex parts is merely a nightmare. Count on agony when the databook for a lousy timer weighs a couple of pounds.

Adopt a philosophy of creating a stimulus, then measuring the system's response with an appropriate tool.

Figures 7-1 and 7-2 illustrate this principle. The debugger's (in this case, driving an emulator) low-level commands configure the timer inside a 386EX. The response, measured on a scope, shows how the timer behaves with the indicated setup.

Using a serial port? Connect a terminal and learn how to transmit a single character. Again, manually set up the registers (carefully documenting what you did), using parameters extracted from the databook, using the tool's output command to send characters. Lots of things can go wrong with something as complicated as a UART, so I like to instrument its output with a scope. If the baud rate is incorrect, a terminal will merely display scrambled garbage; the scope will clearly show the problem.

Then write a shell of a driver in the selected language. Take the information gleaned from the databook and proven in your experiments to work, and codify it in code once and for all. Test the driver. Get it right!

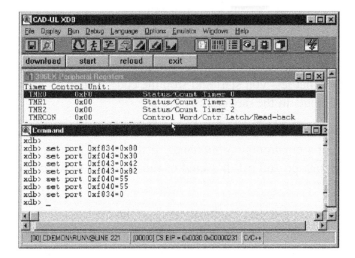

Figure 7-1: Hacking a Peripheral Driver.

Now you've successfully created a module that handles that hardware device.

Master one portion of a device at a time. On a UART, for example, figure out how to transmit characters reliably and document what you did, before you move on to receiving. Segment the problem to keep things simple.

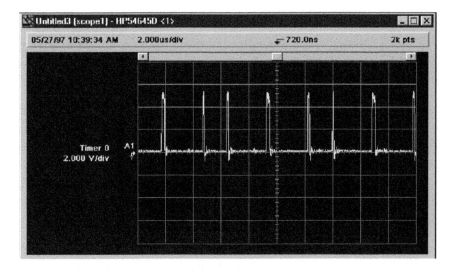

Figure 7-2: Hacking a Peripheral Driver.

If only we could live with simple programmed inputs and outputs! Most nontrivial peripherals will operate in an interrupt-driven mode. Add ISRs, one at a time, testing each one, for each part of the device. For example, with the UART, completely master interrupt-driven transmission before moving on to interrupting reception.

Again, with each small success immediately create, compile, and test code before you've forgotten the tricks required to make the little beast operate properly. Databooks are cornucopias of information and misinformation; it's astonishing how often you'll find a bit documented incorrectly. Don't rely on frail memory to preserve this information. Mark up the book, create and test the code, and move on.

Some devices are simply too complex to yield to manual testing. An Ethernet driver or an IEEE-488 port both require so much setup that there's no choice but to initially write a lot of code to preset each internal register. These are the most frustrating sorts of devices to handle, as all too often there's little diagnostic feedback—you set a zillion registers, burn some incense, and hope it flies.

If your driver will transfer data using DMA, it still makes sense to first figure out how to use it a byte at a time in a programmed I/O mode. Be lazy—it's just too hard to master the DMA, interrupt completion routines, and the part itself all at once. Get single-byte transfers working before opening the Pandora's box of DMA.

In the "make it work" phase we usually succumb to temptation and hack away at the code, changing bits just to see what happens. The documentation generally suffers. Leave a bit of time before wrapping up each completed routine to tune the comments. It's a lot easier to do this when you still remember what happened and why.

More than once I've found that the code developed this way is ugly. Downright lousy, in fact, as coding discipline flew out the window during the bit-tweaking frenzy. The entire point of this effort is to master the device (first) and create a driver (second). Be willing to toss the code and build a less offensive second iteration. Test that too, before moving on.

7.2 Selecting Stack Size

With experience, one learns the standard, scientific way to compute the proper size for a stack: Pick a size at random and hope.

Unhappily, if your guess is too small the system will erratically and maybe infrequently crash in horrible ways. And RAM is still an expensive resource, so erring on the side of safety drives recurring costs up.

With an RTOS the problem is multiplied, since every task has its own stack.

It's feasible, though tedious, to compute stack requirements when coding in assembly language by counting calls and pushes. C—and even worse, C++—obscures these details. Runtime calls further distance our understanding of stack use. Recursion, of course, can blow stack requirements sky-high.

Any of a number of problems can cause the stack to grow to the point where the entire system crashes. It's tough to go back and analyze the failure after the crash, as the program will often write all over itself or the variables, removing all clues.

The best defense is a strong offense. Odds are your stack estimate will be wrong, so instrument the code from the very beginning so you'll know, for sure, just how much stack is needed.

In the startup code or whenever you define a task, fill the task's stack with a unique signature such as 0x55AA (Figure 7-3). Then, probe the stacks occasionally using your debugger and see just how many of the assigned locations have been used (the 0x55AA will be gone).

Knowledge is power.

Also consider building a stack monitor into your code. A stack monitor is just a few lines of assembly language that compares the stack pointer to some limit you've set. Estimate the total stack use, and then double or triple the size. Use this as the limit.

Put the stack monitor into one or more frequently called ISRs. Jump to a null routine, where a breakpoint is set, when the stack grows too big.

Be sure that the compare is "fuzzy." The stack pointer will never *exactly* match the limit.

By catching the problem *before* a complete crash, you can analyze the stack's contents to see what led up to the problem. You may see an ISR being interrupted constantly (that is, a lot of the stack's addresses belong to the ISR). This is a sure indication of code that's too slow to keep up with the interrupt rate. You can't simply leave interrupts disabled longer, as the system will start missing them. Optimize the algorithm and the code in that ISR.

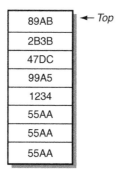

Figure 7-3: Proactively fill the stack with 0x55AA to find overrun problems. Note that the lower three words have been unused.

7.3 The Curse of Malloc( )

Since the stack is a source of trouble, it's reasonable to be paranoid and not allocate buffers and other sizable data structures as automatics. Watch out! Malloc( ), a quite logical alternative, brings its own set of problems. A program that dynamically allocates and frees lots of memory—especially variably-sized blocks—will fragment the heap. At some point it's quite possible to have lots of free heap space, but so fragmented that malloc( ) fails.

If your code does not check the allocation routine's return code to detect this error, it will fail horribly. Of course, detecting the error will also no doubt result in a horrible failure, but gives you the opportunity to show an error code so you'll have a chance of understanding and fixing the problem.

If you chose to use malloc( ), *always* check the return value and safely crash (with diagnostic information) if it fails.

Garbage collection—which compacts the heap from time to time—is almost unknown in the embedded world. It's one of Java's strengths and weaknesses, as the time spent compacting the heap generally shuts down all tasking. Though there's lots of work going on developing real-time garbage collection, as of this writing there is no effective approach.

Sometimes an RTOS will provide alternative forms of malloc( ), which let you specify which of several heaps to use. If you can constrain your memory allocations to standard-sized blocks, and use one heap per size, fragmentation won't occur.

One option is to write a replacement function of the form pmalloc (heap_number). You defined a number of heaps, each one of which has a dedicated allocation size. Heap 1 might return a 2000-byte buffer, heap 2100 bytes, and so on. You then constrain allocations to these standard-size blocks to eliminate the fragmentation problem.

When using C, if possible (depending on resource issues and processor limitations), always include Walter Bright's MEM package (www.snippets.org/mem.txt) with the code, at least for debugging. MEM provides the following:

- ISO/ANSI verification of allocation/reallocation functions

- Logging of all allocations and frees

- Verifications of frees

- Detection of pointer over- and under-runs

- Memory leak detection

- Pointer checking

- Out-of-memory handling

7.4 Banking

When asked how much money is enough, Nelson Rockefeller reportedly replied, "Just a little bit more." We poor folks may have trouble understanding his perspective, but all too often we exhibit the same response when picking the size of the address space for a new design. Given that the code inexorably grows to fill any allocated space, "just a little more" is a plea we hear from the software people all too often.

Is the solution to use 32-bit machines exclusively, cramming a full 4GB of RAM into our cost-sensitive application in the hopes that no one could possibly use that much memory?

Though clearly most systems couldn't tolerate the costs associated with such a poor decision, an awful lot of designers take a middle tack, selecting high-end processors to cover their posterior parts.

A 32-bit CPU has tons of address space. A 16-bitter sports (generally) 1 to 16 Mb. It's hard to imagine needing more than 16 Mb for a typical embedded app; even 1 Mb is enough for the vast majority of designs.

A typical 8-bit processor, though, is limited to 64k. Once this was an ocean of memory we could never imagine filling. Now C compilers let us reasonably produce applications far more complex than we dreamed of even a few years ago. Today the midrange embedded systems I see usually burn up something between 64k and 256k of program and data space—too much for an 8-bitter to handle without some help.

If horsepower were not an issue, I'd simply toss in an 80188 and profit from the cheap 8-bit bus that runs 16-bit instructions over 1 Mb of address space. Sometimes this is simply not an option; an awful lot of us design upgrades to older systems. We're stuck with tens of thousands of lines of "legacy" code that are too expensive to change. The code forces us to continue using the same CPU. Like taxes, programs always get bigger, demanding more address space than the processor can handle.

Perhaps the only solution is to add address bits. Build an external mapper using PLDs or discrete logic. The mapper's outputs go into high-order address lines on your RAM and ROM devices. Add code to remap these lines, swapping sections of program or data in and out as required.

7.5   Logical to Physical

Add a mapper, though, and you'll suddenly be confronted with two distinct address spaces that complicate software design.

The first is the *physical* space—the entire universe of memory on your system. Expand your processor's 64k limit to 256k by adding two address lines, and the physical space is 256k.

Logical addresses are the ones generated by your program, and thence asserted onto the processor's bus. Executing a MOV A,(0FFFF) instruction tells the processor to read from the very last address in its 64k logical address space. External banking hardware can translate this to some other address, but the code itself remains blissfully unaware of such actions. All it knows is that some data comes from memory in response to the 0FFFF placed on the bus. The program can never generate a logical address larger than 64k (for a typical 8-bit CPU with 16 address lines).

This is very much like the situation faced by 80x86 assembly-language programmers: 64k segments are essentially logical spaces. You can't get to the rest of physical memory without doing *something*; in this case reloading a segment register.

Conversely, if there's no mapper, then the physical and logical spaces are identical.

7.6 Hardware Issues

Consider doubling your address space by taking advantage of processor cycle types. If the CPU differentiates memory reads from fetches, you may be able to easily produce separate data and code spaces. The 68000's seldom-used function codes are for just this purpose, potentially giving it distinct 16-Mb code and data spaces.

Writes should clearly go to the data area (you're not writing self-modifying code, are you?). Reads are more problematic. It's easy to distinguish memory reads from fetches when the processor generates a fetch signal for every instruction byte. Some processors (e.g., the Z80) produce a fetch only on the read of the first byte of a multiple byte opcode; subsequent ones all look the same as any data read. Forget trying to split the memory space if cycle types are not truly unique.

When such a space-splitting scheme is impossible, then build an external mapper that translates address lines. However, avoid the temptation to simply latch upper address lines. Though it's easy to store A16, A17, et al. in an output port, every time the latch changes the *entire* program gets mapped out. Though there are awkward ways to write code to deal with this, add a bit more hardware to ease the software team's job.

Design a circuit that maps just portions of the logical space in and out. Look at software requirements first to see what hardware configuration makes sense.

Every program needs access to a data area that holds the stack and miscellaneous variables. The stack, for sure, must always be visible to the processor so calls and returns function. Some amount of "common" program storage should always be mapped in. The remapping code, at least, should be stored here so that it doesn't disappear during a bank switch. Design the hardware so these regions are always available.

Is the address space limitation due to an excess of code or of data? Perhaps the code is tiny, but a gigantic array requires tons of RAM. Clearly, you'll be mapping RAM in and out, leaving one area of ROM—enough to store the entire program—always in view. An obese program yields just the opposite design. In either of these cases a logical address space split into three sections makes the most sense: common code (always visible, containing runtime routines called by a compiler and the mapping code), mapped code or data, and common RAM (stack and other critical variables needed all the time).

For example, perhaps 0000 to 03FFF is common code. 4000 to 7FFF might be banked code; depending on the setting of a port it could map to almost any physical address. 8000 to FFFF is then common RAM.

Sure, you can use heroic programming to simplify the hardware. I think it's a mistake, as the incremental parts cost is minuscule compared to the increased bug rate implicit in any complicated bit of code. It *is* possible—and reasonable—to remove one bank by copying the common code to RAM and executing it there, using one bank for both common code and data.

It's easy to implement a three-bank design. Suppose addresses are arranged as in the previous example. A0 to A14 go to the RAM, which is selected when A15 = 1.

Turn ROM on when A15 is low. Run A0 to A14 into the ROM. Assuming we're mapping a 128k × 8 ROM into the 32k logical space, generate a fake A15 and A16 (simple bits latched into an output port) that go to the ROM's A15 and A16 inputs. However, feed these through AND gates. Enable the gates only when A15 = 0 (RAM off) and A14 = 1 (bank area enabled).

RAM is, of course, selected with logical addresses between 8000 and FFFF. Any address under 4000 disables the gates and enables the first 4000 locations in ROM. When A14 is a one, whatever values you've stuck into the fake A15 and A16 select a chunk of ROM 4000 bytes long.

The virtue of this design is its great simplicity and its conservation of ROM—there are no wasted chunks of memory, a common problem with other mapping schemes.

Occasionally a designer directly generates chip selects (instead of extra address lines) from the mapping output port. I think this is a mistake. It complicates the ROM select logic. Worse, sometimes it's awfully hard to make your debugging tools understand the translation from addresses to symbols. By translating *addresses* you can provide your debugger with a logical-to-physical translation cheat sheet.

7.7   The Software

In assembly language you control everything, so handling banked memory is not too difficult. The hardest part of designing remappable code is figuring out how to segment the banks. Casual calling of other routines is out, as you dare not call something not mapped in.

Some folks write a bank manager that tracks which routines are currently located in the logical space. All calls, then, go through the bank manager, which dynamically brings routines in and out as needed.

If you were foresighted enough to design your system around a real-time operating system (RTOS), then managing the mapper is much simpler. Assign one task per bank. Modify the context switcher to remap whenever a new task is spawned or reawakened.

Many tasks are quite small—much smaller than the size of the logical banked area. Use memory more efficiently by giving tasks two banking parameters: the bank number associated with the task, and a starting offset into the bank. If the context switcher both remaps and then starts the task at the given offset, you'll be able to pack multiple tasks per bank.

Some C compilers come with built-in banking support. Check with your vendor. Some will completely manage a multiple bank system, automatically remapping as needed to bring code in and out of the logical address space. Figure on making a few patches to the supplied remapping code to accommodate your unique hardware design.

In C or assembly, using an RTOS or not, be sure to put all of your interrupt service routines and associated vectors in a common area. Put the banking code there as well, along with all frequently used functions (when you're using a compiler, put the entire runtime package in unmapped memory).

As always, when designing the hardware carefully document the approach you've selected. Include this information in the banking routine so some poor soul several years in the future has a fighting chance to figure out what you've done.

And, if you are using a banking scheme, be sure that the tools provide intelligent support. Quite a few 8-bit emulators, for example, do have extra address bits expressly for working in banked hardware. This means you can download code and even set breakpoints in banked areas that may not be currently mapped into the logical address space.

But be sure the emulator works properly with the compiler or assembler to give real source-level support in banked regions. If the compiler and emulator don't work together to share the physical and logical addresses of every line of code and every global/static variable, the "source" debugger will show nothing more useful than disassembled instructions. That's a terrible price to pay; in most cases you'll be well advised to find a more debuggable CPU.

7.8 Predicting ROM Requirements

It's rather astonishing how often we run into the same problem, yet take no action to deal with the issue once and for all. One common problem that drives managers wild is the old "running out of ROM space" routine—generally the week before shipping.

For two reasons it's very difficult to predict ROM requirements in the project's infancy. First, too many of us write code before we've done a complete and thoughtful analysis of the project's size. If you're not estimating code size (in lines of code or numbers of function points or a similar metric), then you're simply not a professional software engineer.

Second, we're generally not sure how to correlate a line of C to a number of bytes of machine code. Historical data is most useful if you've worked with the specific CPU and compiler in the past.

Regardless, when you start coding, maintain a spreadsheet that predicts the project's size. As a professional you've done the best possible job estimating the functions' sizes (in LOC, lines of code). List this data.

Whenever you complete a function, append the incremental size of the executable to the spreadsheet. Figure 7-4 shows an example, including each function, with estimated and actual LOC counts, and compiled sizes.

Any idiot—or at least any idiot with an engineering degree—can then write an equation that creates an average size of an LOC in bytes, and another that predicts total system size based on estimated LOC.

Make sure your calculations do not include the bare system skeleton—the C startup code and a null main() function—since the first line of C brings in the runtime package.

7.9 RAM Diagnostics

Beyond software errors lurks the specter of a hardware failure that causes our correct code to die, possibly creating a life-threatening horror, or maybe just infuriating a customer. Many of us write diagnostic code to help contain the problem. Much of the resulting code just does not address failure modes.

Obviously, a RAM problem will destroy most embedded systems. Errors reading from the stack will surely crash the code. Problems, especially intermittent ones, in the data areas may manifest bugs in subtle ways. Often you'd rather have a system that just doesn't boot, rather than one that occasionally returns incorrect answers.

Module	Est LOC	Act LOC	Size
Skeleton	300	310	21,123
RTOS		3423	11,872
TIMER_ISR	50	34	534
ATOD_ISR	75	58	798
TOD	120	114	998
PRINT_E	80	98	734
COMM_SER	90		
RD_ATOD	40		
	Bytes/LOC	4.01	
	Est Size	36580	

Figure 7-4: A Spreadsheet that Predicts ROM Size.

Some embedded systems are pretty tolerant of memory problems. We hear of NASA spacecraft from time to time whose core or RAM develops a few bad bits, yet somehow the engineers patch their code to operate around the faulty areas, uploading the corrections over the distances of billions of miles.

Most of us work on systems with far less human intervention. There are no teams of highly trained personnel anxiously monitoring the health of each part of our products. It's our responsibility to build a system that works properly when the hardware is functional.

In some applications, though, a certain amount of self-diagnosis either makes sense or is required; critical life-support applications should use every diagnostic concept possible to avoid disaster due to a submicron RAM imperfection.

So, the first rule about diagnostics in general, and RAM tests in particular, is to clearly define your goals. Why run the test? What will the result be? Who will be the unlucky recipient of the bad news in the event an error is found, and what do you expect that person to do?

Will a RAM problem kill someone? If so, a very comprehensive test, run regularly, is mandatory.

Is such a failure merely a nuisance? For instance, if it keeps a cell phone from booting, if there's nothing the customer can do about the failure anyway, then perhaps there's no reason for doing a test. As a consumer I could care less why the damn phone stopped working . . . if it's dead, I'll take it in for repair or replacement.

Is production test—or even engineering test—the real motivation for writing diagnostic code? If so, then define exactly what problems you're looking for and write code that will find those sorts of troubles.

Next, inject a dose of reality into your evaluation. Remember that today's hardware is often very highly integrated. In the case of a microcontroller with on-board RAM, the chances of a memory failure that doesn't also kill the CPU is small. Again, if the system is a critical life-support application it may indeed make sense to run a test, as even a minuscule probability of a fault may spell disaster.

Does it make sense to ignore RAM failures? If your CPU has an illegal instruction trap, there's a pretty good chance that memory problems will cause a code crash you can capture and process. If the chip includes protection mechanisms (like the x86 protected mode), count on bad stack reads immediately causing protection faults your handlers can process. Perhaps RAM tests are simply not required, given these extra resources.

7.10 Inverting Bits

Most diagnostic code uses the simplest of tests—writing alternating 0x55 and 0xAA values to the entire memory array, and then reading the data to ensure that it remains accessible. It's a seductively easy approach that will find an occasional problem (like someone forgot to load all of the RAM chips), but that detects few real-world errors.

Remember that RAM is an array divided into columns and rows. Accesses require proper chip selects and addresses sent to the array—and not a lot more. The 0x55/0xAA symmetrical pattern repeats massively all over the array; accessing problems (often more common than defective bits in the chips themselves) will create references to incorrect locations, yet almost certainly will return what appears to be correct data.

Consider the physical implementation of memory in your embedded system. The processor drives address and data lines to RAM—in a 16-bit system there will surely be at least 32 of these. Any short or open on this huge bus will create bad RAM accesses. Problems with the PC board are far more common than internal chip defects, yet the 0x55/0xAA test is singularly poor at picking up these, the most likely failures.

Yet the simplicity of this test and its very rapid execution have made it an old standby that's used much too often. Isn't there an equally simple approach that will pick up more problems?

If your goal is to detect the most common faults (PCB wiring errors and chip failures more substantial than a few bad bits here or there), then indeed there is. Create a short string of almost random bytes that you repeatedly send to the array until all of memory is written. Then, read the array and compare against the original string.

I use the phrase "almost random" facetiously, but in fact it hardly matters what the string is, as long as it contains a variety of values. It's best to include the pathological cases, such as 00, 0xaa, 0x55, and 0xff. The string is something you pick when writing the code, so it is truly not random, but other than these four specific values, you fill the rest of it with nearly any set of values, since we're just checking basic write/read functions (remember: memory tends to fail in fairly dramatic ways). I like to use very orthogonal values—those with lots of bits changing between successive string members—to create big noise spikes on the data lines.

To make sure this test picks up addressing problems, ensure that the string's length is not a factor of the length of the memory array. In other words, you don't want the string to be aligned on the same low-order addresses, which might cause an address error to go undetected. Since the string is much shorter than the length of the RAM array, you ensure that it repeats at a rate that is not related to the row/column configuration of the chips.

For 64k of RAM, a string 257 bytes long is perfect: 257 is prime, and its square is greater than the size of the RAM array. Each instance of the string will start on a different low-order address. Also, 257 has another special magic: you can include every byte value (00 to 0xff) in the string without effort. Instead of manually creating a string in your code, build it in real time by incrementing a counter that overflows at 8 bits.

Critical to this, and every other RAM test algorithm, is that you write the pattern to all of RAM before doing the read test. Some people like to do nondestructive RAM tests by testing one location at a time, then restoring that location's value, before moving on to the next one. Do this and you'll be unable to detect even the most trivial addressing problem.

This algorithm writes and reads every RAM location once, so it's quite fast. Improve the speed even more by skipping bytes, perhaps writing and reading every 3rd or 5th entry. The test will be a bit less robust, yet will still find most PCB and many RAM failures.

Some folks like to run a test that exercises each and every bit in their RAM array. Though I remain skeptical of the need, since most semiconductor RAM problems are rather catastrophic, if you do feel compelled to run such a test, consider adding another iteration of the algorithm just described, with all of the data bits inverted.

7.11 Noise Issues

Large RAM arrays are a constant source of reliability problems. It's indeed quite difficult to design the perfect RAM system, especially with the minimal margins and high speeds of today's 16- and 32-bit systems. If your system uses more than a couple of RAM parts, count on spending some time qualifying its reliability via the normal hardware diagnostic procedures. Create software RAM tests that hammer the array mercilessly.

Probably one of the most common forms of reliability problems with RAM arrays is pattern sensitivity. Now, this is not the famous pattern problems of yore, where the chips (particularly DRAMs) were sensitive to the groupings of ones and zeroes. Today the chips are just about perfect in this regard. No, today pattern problems come from poor electrical characteristics of the PC board, decoupling problems, electrical noise, and inadequate drive electronics.

PC boards were once nothing more than wiring platforms, slabs of tracks that propagated signals with near-perfect fidelity. With very high-speed signals, and edge rates (the time it takes a signal to go from a zero to a one or back) under a nanosecond, the PCB itself assumes all of the characteristics of an electronic component—one whose virtues are almost all problematic. It's a big subject (read Howard Johnson and Martin Graham, *High Speed Digital Design—A Handbook of Black Magic*, PTR Prentice Hall, 1993 for the canonical words of wisdom on this subject), but suffice it to say that a poorly designed PCB will create RAM reliability problems.

Equally important are the decoupling capacitors chosen, as well as their placement. Inadequate decoupling will create reliability problems as well.

Modern DRAM arrays are massively capacitive. Each address line might drive dozens of chips, with 5 to 10 pF of loading per chip. At high speeds the drive electronics must somehow drag all of these pseudo-capacitors up and down with little signal degradation. Not an easy job! Again, poorly designed drivers will make your system unreliable.

Electrical noise is another reliability culprit, sometimes in unexpected ways. For instance, CPUs with multiplexed address/data buses use external address latches to demux the bus. A signal, usually named ALE (Address Latch Enable) or AS (Address Strobe), drives the clock to these latches. The tiniest, most miserable amount of noise on ALE/AS will surely, at the time of maximum inconvenience, latch the data part of the cycle instead of the address. Other signals are also vulnerable to small noise spikes.

Unhappily, all too often common RAM tests show no problem when hidden demons are indeed lurking. The algorithm I've described, as well as most of the others commonly used, trade off speed against comprehensiveness. They don't pound on the hardware in a way designed to find noise and timing problems.

Digital systems are most susceptible to noise when large numbers of bits change all at once. This fact was exploited for data communications long ago with the invention of the gray code, a variant of binary counting where no more than one bit changes between codes. Your worst nightmares of RAM reliability occur when all of the address and/or data bits change suddenly from zeroes to ones.

For the sake of engineering testing, write RAM test code that exploits this known vulnerability. Write 0xffff to 0x0000 and then to 0xffff, and do a read-back test. Then write zeroes. Repeat as fast as your loop will let you go.

Depending on your CPU, the worst locations might be at 0x00ff and 0x0100, especially on 8-bit processors that multiplex just the lower 8 address lines. Hit these combinations hard as well.

Other addresses often exhibit similar pathological behavior. Try 0x5555 and 0xaaaa, which also have complementary bit patterns.

The trick is to write these patterns back-to-back. Don't test all of RAM, with the understanding that both 0x0000 and 0xffff will show up in the test. You'll stress the system most effectively by driving the bus massively up and down all at once.

Don't even think about writing this sort of code in C. Any high-level language will inject too many instructions between those that move the bits up and down. Even in assembly the processor will have to do fetch cycles from wherever the code happens to be, which will slow down the pounding and make it a bit less effective.

There are some tricks, though. On a CPU with a prefetcher (all x86, 68k, etc.) try to fill the execution pipeline with code, so the processor does back-to-back writes or reads at the addresses you're trying to hit. And, use memory-to-memory transfers when possible. For example:

```
mov    si,0xaaaa
mov    di,0x5555
mov    [si],0xff
mov    [di],[si]        ; read ff00 from 0aaaa
                        ; and then write it
                        ; to 05555
```

DRAMs have memories rather like mine—after 2 to 4 milliseconds go by, they will probably forget unless external circuitry nudges them with a gentle reminder. This is known as "refreshing" the devices and is a critical part of every DRAM-based circuit extant.

More and more processors include built-in refresh generators, but plenty of others still rely on rather complex external circuitry. Any failure in the refresh system is a disaster.

Any RAM test should pick up a refresh fault—shouldn't it? After all, it will surely take a *lot* longer than 2–4 msec to write out all of the test values to even a 64k array.

Unfortunately, refresh is basically the process of cycling address lines to the DRAMs. A completely dead refresh system won't show up with the test indicated, since the processor will be merrily cycling address lines like crazy as it writes and reads the devices. There's no chance the test will find the problem. This is the worst possible situation: the process of running the test camouflages the failure!

The solution is simple: After writing to all of memory, just stop toggling those pesky address lines for a while. Run a tight do-nothing loop for a while (*very* tight … the more instructions you execute per iteration, the more address lines will toggle), and only then do the read test. Reads will fail if the refresh logic isn't doing its thing.

Though DRAMs are typically specified at a 2- to 4-msec maximum refresh interval, some hold their data for surprisingly long times. When memories were smaller and cells larger, each had so much capacitance that you could sometimes go for dozens of seconds without losing a bit. Today's smaller cells are less tolerant of refresh problems, so a 1- to 2-second delay is probably adequate.

7.12   A Few Notes on Software Prototyping

As a teenaged electronics technician I worked for a terribly undercapitalized small company that always spent tomorrow's money on today's problems. There was no spare cash to cover risks. As is so often the case, business issues overrode common sense and the laws of physics: all prototypes simply had to work, and were in fact shipped to customers.

Years ago I carried this same dysfunctional approach to my own business. We prototyped products, of course, but did so leaving no room for failure. Schedules had no slack; spare parts were scarce, and people heroically overcame resource problems. In retrospect this seems silly, since by definition we create prototypes simply because we expect mistakes, problems, and, well . . . failure.

Can you imagine being a civil engineer? Their creations—a bridge, a building, a major interchange—are all one-off designs that simply *must* work correctly the first time. We digital folks have the wonderful luxury of building and discarding trial systems.

Software, though, looks a lot like the civil engineer's bridge. Costs and time pressures mean that code prototypes are all too rare. We write the code and knock out most of the bugs. Version 1.0 is no more than a first draft, minus most of the problems.

Though many authors suggest developing version 1.0 of the software, then chucking it and doing it again, now correctly, based on what was learned from the first go-around, I doubt that many of us will often have that opportunity. Things are just too frantic, workforces too thin, and time-to-market pressures too intense. The old engineering adage "If the damn thing works at all, ship it," once only a joke, now seems to be the industry's mantra.

Besides—who wants to redo a project? Most of us love the challenge of making something work, but want to move on to bigger and better things, not repeat our earlier efforts.

Even hardware is moving away from conventional prototypes. Reprogrammable logic means that the hardware is nothing more than software. Slap some smart chips on the board and build the first production run. You can (hopefully) tune the equations to make the system work despite interconnect problems.

We're paid to develop firmware that is correct—or at least correct enough—to form a final product, first time, every time. We're the high-tech civil engineers, though at least we have the luxury of fixing mistakes in our creations before releasing the product to the cruel world of users.

Though we're supposed to build the system right the first time, we're caught in a struggle between the computer's need for perfect instructions, and marketing's less-than-clear product definitions. The B-schools are woefully deficient in teaching their students—the future product definers—about the harsh realities of working in today's technological environment. Vague handwaving and whiteboard sketches are not a product spec. They need to understand that programmers must be unfailingly precise and complete in designing the code. Without a clear spec, the programmers themselves, by default, must create the spec.

Most of us have heard the "but *that's* not what I wanted" response from management when we demo our latest creation. All too often the customer—management, your boss, or the end user—doesn't really know what they want until they see a working system. It's clearly a Catch-22 situation.

The solution is a prototype of the system's software, running a minimal subset of the application's functionality. This is not a skeleton of the final code, waiting to be fleshed out after management puts in their two cents. I'm talking about truly disposable code.

Most embedded systems do possess some sort of look and feel, despite the absence of a GUI. Even the light-up sneakers kids wear (which, I'm told, use a microcontroller from Microchip) have at least a "look." How long should the light be on? Is it a function of acceleration? If I were designing such a product, I'd run a cable from the sneaker to a development system so I could change the LED's parameters in seconds while the MBAs argue over the correct settings.

"Wait," you say. "We can't do that here! We *always* ship our code!" Though this is the norm, I'm running into more and more embedded developers who have been so badly burned by inadequate/incorrect specifications that even management grudgingly backs up their rapid prototyping efforts. However, any prototype will fail unless the goals are clearly spelled out.

The best prototype spec is one that models risk factors in the final product. Risk comes in far too many flavors: user interface (human interaction with the unit, response speed), development problems (tools, code speed, code size, people skill sets), "science" issues (algorithms, data reduction, sampling intervals), final system cost (some complex sum of engineering and manufacturing costs), time to market, and probably other items as well.

A prototype may not be the appropriate vehicle for dealing with all risk factors. For example, without building the real system it'll be tough to extrapolate code speed and size from any prototype.

The first ground rule is to define the result you're looking for. Is it to perfect a data reduction algorithm? To get consensus on a user interface? Focus with unerring intensity on just that result. Ignore all side issues. Build just enough code to get the desired result. Real systems need a spec that defines what the product does; a rapid prototype needs a spec that spells out what *won't* be in it.

More than anything you need a boss who shields you from creeping featurism. We know that a changing spec is the bane of real systems; surely it's even more of a problem in a quick-turn model system.

Then you'll need an understanding of what decisions will be made as a result of the prototype. If the user interface will be pretty much constant no matter what turns up in the modeling phase, hey—just jump into final product development. If you know the answer, don't ask the question!

Define the deadline. Get a prototype up and running at warp speed. Six months or a year of fiddling around on a model is simply too long. The raison d'être for the prototype is to identify problems and make changes. Get these decisions made early by producing something in days or weeks. Develop a schedule with many milestones where nondevelopers get a chance to look at the product and fiddle with it a bit.

For a prototype where speed and code size are not a problem, I like to use really high-level "languages" like Basic, Excel, Word macros. The goal is to get something going *now*. Use every tool, no matter how much it offends your sensibilities, to accomplish that mission.

Does your product have a GUI? Maybe a control panel? Look at products like those available from National Instruments and IoTech. These companies provide software that lets you produce "virtual instruments" by clicking and dragging knobs, displays, and switches around on a PC's screen. Couple that to standard data acquisition boards and a bit of code in Basic or C, and you can produce models of many sorts of embedded systems in hours.

The cost of creating a virtual model of your product, using purchased components, is immeasurably small compared to that of designing, building, and troubleshooting real hardware and software. Though there's no way to avoid building hardware at some point, count on adding months to a project when a new board design is required.

Another nice feature of doing a virtual model of the product is the certainty of creating worthless code. You'll focus on the real issues—the ones identified in your prototyping goals—and not the problems of creating documented, portable, well-structured software. The code will be no more than the means to the end. You'll toss the code as casually as the hardware folks toss prototype PC boards.

I mentioned using Excel. Spreadsheets are wonderful tools for evaluating the product's science. Unsure about the behavior of a data-smoothing algorithm? Fiddling with a fuzzy-logic design? Wondering how much precision to carry? Create a data set and put it in your trusty spreadsheet. Change the math in seconds; graph the results to see what happens. Too many developers write a ton of embedded code, only to spend months tuning algorithms in the unforgiving environment of an 8051 with limited memory.

Though a spreadsheet masks the calculations' speed, you can indeed get some sort of final complexity estimate by examining the equations. If the algorithm looks terribly slow, work within the forgiving environment of the spreadsheet to develop a faster approach. We all know, though too often ignore, the truth that the best performance enhancements come from tuning the algorithm, not the code.

Though the PC is a great platform for modeling, do consider using current company products as prototype platforms. Often new products are derivatives of older ones. You may have a lot of extant hardware and software—that works!—in a system on the shelf. Be creative and use every resource available to get the prototype up and running.

Toss out the standards manual. Use every trick in the book to get it done *fast*. Do code in small functions to get something testable quickly, and to minimize the possibility of making big mistakes.

There's a secret benefit to using cruddy "languages" for software prototypes: write your proto code in Visual Basic, say, and no matter how hard management screams, it simply cannot be whisked off into the product as final code. Clever language selection can break the dysfunctional last-minute conversion of test code to final firmware.

All of us have worked with that creative genius who can build anything, who pounds out a thousand lines of code a day, but who can never seem to complete a project. Worse—the fast coder who spends eons debugging the megabyte of firmware he wrote on a Jolt-driven all-nighter. Then there are the folks who produce working code devoid of documentation, who develop rashes or turn into Mr. Hyde when told to add comments.

We struggle with these folks, plead with them, send them to seminars, lead by example, all too often without success. Some of them are prima donnas who should probably get the ax. Others are really quite good, but simply lack the ability to deal with detail ... which is essential since, in a released product, every lousy bit must be right.

These are the ideal prototype developers. Bugs aren't a big issue in a model, and documentation is less than important. The prototype lets them exercise their creative zeal, while its limited scope means that problems are not important. Toss Twinkies and caffeine into their lair and stand back. You'll get your system fast, and they'll be happy employees. Use the more disciplined team members to get the bugless real product to market.

Part of management is effectively using people's strengths while mitigating their weaknesses. Part of it is also giving the workers a break once in a while. No one can crank out 70-hour weeks forever without cracking.

CHAPTER 8

Hardware Musings

Jack Ganssle

This chapter is a random collection of thoughts about working on hardware in embedded systems. It reflects my experience both as a designer and as a manager. Why would one write pages about a subject as prosaic as a resistor? Well, a resistor may be boring, but it's an essential component in any electronic system, embedded or not.

My observation is that recent graduates increasingly know less about basic components like resistors, capacitors, and inductors despite the fact that even something as simple as a wire or a track on a PCB exhibits all of the characteristics of these components. Try this experiment: show a new grad a ¼ watt 10 ohm resistor. Put it on a power supply and ask what will happen as one cranks the output to 1 volt, then 10, then 20. In a flurry of calculations you'll hear about Ohm's Law and I^2R. But watch their eyes when, at 20 volts the part starts to smoke, then bursts into flames.

In logic circuits resistors tie inputs to safe states. They're often used to terminate signals to minimize overshoot and other electromagnetic effects. An engineer must "get" the nuances of these seemingly simple parts, just as he understands all about the most complicated microprocessor.

As I write this I have an FPGA on my desk that has over 1500 pins (balls, really, as it's a BGA device) in a form factor no bigger than a square inch. That works out to almost 40 balls per inch in two axes. How would you probe this part? And worse, the balls are all sequestered underneath the package. They're soldered to the top side of the PCB. Need to probe a middle node? That's quite impossible. So it's even more important now than a few years ago to make your designs debuggable. Provide accessible nodes convenient for scoping. Because a perfectly wonderful bit of hardware that can't be debugged is simply junk.

Finally, I stand by my comments in praise of small processors. Though written eight years ago, these words are just as true today. In the intervening years analysts have told me that 8 and 16 bitters would soon be relegated to the scrap heap as Moore's Law makes 32 bit machines de rigor. Yet today the 16 bit market is undergoing a resurgence, largely inspired by TI's

MSP430 series, which offers a complete range of microcontrollers that operate with practically no power, and will give you plenty of change from a five dollar bill. The 32 bit world still suffers from code density problems (other than the Thumb subset of the ARM), and complete 32 bit microcontrollers remain somewhat rare.

The 8 bit world is even healthier as it continues to grow 15%–20% per year. In volume, some parts cost less than a dime. One company that makes flashlights discovered that an alternate action on-off switch costs more than a momentary switch coupled with a minimal 8051 used just to remember if the light was on or off!

—Jack Ganssle

8.1 Debuggable Designs

An unhappy reality of our business is that we'll surely spend lots of time—far too much time—debugging both hardware and firmware. For better or worse, debugging consumes project-months with reckless abandon. It's usually a prime cause of schedule collapse, disgruntled team members, and excess stomach acid.

Yet debugging will never go away. Practicing even the very best design techniques will never eliminate mistakes. No one is smart enough to anticipate every nuance and implication of each design decision on even a simple little 4k 8051 product; when complexity soars to hundreds of thousands of lines of code coupled to complex custom ASICs we can only be sure that bugs will multiply like rabbits.

We know, then, up front when making basic design decisions that in weeks or months our grand scheme will go from paper scribbles to hardware and software ready for testing. It behooves us to be quite careful with those initial choices we make, to be sure that the resulting design isn't an undebuggable mess.

8.2 Test Points Galore

Always remember that, whether you're working on hardware or firmware problems, the oscilloscope is one of the most useful of all debugging tools. A scope gives instant insight into difficult code issues such as operation of I/O ports, ISR sequencing, and performance problems.

Yet it's tough to probe modern surface-mount designs. Those tiny whisker-thin pins are hard enough to see, let alone probe. Drink a bit of coffee and you'll dither the scope connection across three or four pins.

The most difficult connection problem of all is getting a good ground. With speeds rocketing toward infinity the scope will show garbage without a short, well-connected ground, yet this is almost impossible when the IC's pin is finer than a spiderweb.

So, when laying out the PCB add lots of ground points scattered all over the board. You might configure these to accept a formal test point. Or, simply put holes on the board, holes connected to the ground plane and sized to accept a resistor lead. Before starting your tests, solder resistors into each hole and cut off the resistor itself, leaving just a half-inch stub of stiff wire protruding from the board. Hook the scope's oversized ground clip lead to the nearest convenient stub.

Figure on adding test points for the firmware as well. For example, the easiest way to measure the execution time of a short routine is to toggle a bit up for the duration of the function. If possible, add a couple of parallel I/O bits just in case you need to instrument the code.

Add test points for the critical signals you know will be a problem. For example:

- Boot loads are always a problem with downloadable devices (Flash, ROM-loaded FPGAs, etc.). Put test points on the critical load signals, as you'll surely wrestle with these a bit.

- The basic system timing signals all need test points: read, write, maybe wait, clock, and perhaps CPU status outputs. All system timing is referenced to these, so you'll surely leave probes connected to those signals for days on end.

- Using a watchdog timer? Always put a test point on the time-out signal. Better, use an LED on a latch. You've got to know when the watchdog goes off, as this indicates a serious problem. Similarly, add a jumper to disable the watchdog, as you'll surely want it off when working on the code.

- With complex power-management strategies, it's a good idea to put test points on the reset pin, battery signals, and the like.

When using PLDs and FPGAs, remember that these devices incorporate all of the evils of embedded systems with none of the remedies we normally use: the entire design, perhaps consisting of tens of thousands of gates, is buried behind a few tens of pins. There's no good way to get "inside the box" and see what happens.

Some of these devices do support a bit of limited debugging using a serial connection to a pseudo-debug port. In such a case, by all means add the standard connector to your

PCB! Your design will not work right off the bat; take advantage of any opportunity to get visibility into the part.

Also plan to dedicate a pin or two in each FPGA/PLD for debugging. Bring the pins to test points. You can always change the logic inside the part to route critical signal to these test points, giving you some limited ability to view the device's operation.

Similarly, if the CPU has a BDM or JTAG debugging interface, put a BDM/JTAG connector on the PCB, even if you're using the very best emulators. For almost zero cost you may save the project when/if the ICE gives trouble.

Very small systems often just don't have room for a handful of test points. The cost of extra holes on ultra-cheap products might be prohibitive. I always like to figure on building a real, honest, prototype first, one that might be a bit bigger and more expensive than the production version. The cost of doing an extra PCB revision (typically $1000 to $2000 for 5-day turnaround) is vanishingly small compared to your salary!

When management screams about the cost of test points and extra connectors, remember that you do not have to load these components during the production run. Install them on the prototypes, leaving them off the bill of materials. Years later, when the production folks wonder about all of the extra holes, you can knowingly smile and remember how they once saved your butt.

8.3 Resistors

When I was a young technician, my associates and I arrogantly believed we could build anything with enough 10k resistors and duct tape. Now it seems that even simple electronic toys use several million transistors encased in tiny SMT packages with hundreds of hairlike leads; no one talks about discrete components anymore. Yet no matter how digital our embedded designs get, we can never avoid certain fundamental electrical properties of our circuits.

For example, somehow the digital age has an ever-increasing need for resistors—so many, in fact, that most "discrete" resistors are now usually implemented in a monolithic structure, like an SIP, not so different from the ICs they are tied to.

Too often we spend our time carefully analyzing the best way to use a modern miracle of integration only to casually select discrete components because they are, well, *boring*. Who can get worked up over the lowly carbon resistor? You can't even buy them one at a time any more. At Radio Shack they come paired in bright decorator packages for an outrageous sum.

Back when I was in the emulator business we dealt with a lot of user target systems that, because of poor resistor choices, drove the tools out of their minds. Consider one typical example: a unit based on an 8-MHz 80188, memory and I/O all connected in a carefully thought-out manner. Power and ground distribution were well planned; noise levels were satisfyingly low. And yet . . . the only tool that seemed to work for debugging code was a logic analyzer. Every emulator the poor designer tested failed to run the code properly. Even a ROM emulator gave erratic results.

Though the emulator wouldn't run the user's code, it did show an immediate service of the non-maskable interrupt—which wasn't used in the system.

> **Note:** When things get weird, always turn to your emulator's trace feature, which will capture weirdness like no other tool.

A little further investigation revealed that the NMI input (which is active high on the 188) was tied low through a 47k resistor.

Now, the system ran fine with a ROM and processor on the board. I suppose the 47k pull-down was at least technically legitimate. A few microamps of leakage current out of the input pin through 47k yields a nice legal logic zero. Yet this 47k was too much resistance when any sort of tool was installed, because of the inevitable increase in leakage current.

Was the design correct because it violated none of Intel's design specs? I maintain that the specs are just the starting point of good design practice. Never, ever, violate one. Never, ever, assume that simply meeting spec is adequate.

A design is *correct* only if it reliably satisfies all intended applications—including the first of all applications, debugging hardware and software. If something that is technically correct prevents proper debugging, then there is surely a problem.

Pull-down resistors are often a source of trouble. It's practically impossible to pull down an LS input (leakage is so high the resistor value must be frighteningly low). Though CMOS inputs leak very little, you must be aware of every potential application of the circuit, including that of plugging tools in. The solution is to avoid pull-downs wherever possible.

In the case of a critical edge-triggered (read "really noise sensitive") input such as NMI, you simply should never pull it low. Tie it to ground. Otherwise, switching noise may get coupled into the input. Even worse, every time you lay out the PC board, the magnitude of the noise problem can change as the tracks move around the board.

Be conservative in your designs, especially when a conservative approach has no downside. If any input must be zero all of the time, simply tie it to ground and never again worry about it. I think folks are so used to adding pull-ups all over their boards that they design in pull-downs through the force of habit.

Once in a while the logic may indeed need a pull-down to deal with unusual I/O bits. Try to come up with a better design.

(The only exception is when you plan to use automatic test equipment to diagnose board faults. ATE gear injects signals into each node, so you'll often need to use a resistor pull-down in place of a ground. Use a small—really small, like 220 ohms—value.)

Though pull-downs are always problematic, well-designed boards use plenty of pull-up resistors—some to bias unused inputs, others to deal with signals and busses that tristate, and some to put switches and other inputs into known one states.

The biggest problem with pull-ups is using values that are too low. A 100k pull-up will in fact bias that CMOS gate properly, but creates a circuit with a terribly high impedance. Why not change to 10k? You buy an order of magnitude improvement in impedance and noise immunity, yet typically use no additional current since the gate requires only microamps of bias.

Vcc from a decent power supply is essentially a low-impedance connection to ground. Connect a 100k pull-up to a CMOS gate and the input is 100k away from ground, power, and everything else—you can overcome a 100k resistance by touching the net with a finger. A 10k resistor will overpower any sort of leakage created by fingers, humidity, and other effects.

Besides, that low-impedance connection will maintain a proper state no matter what tools you use. In the case of NMI from the example above, the tools weakly pulled NMI high so they could run stand-alone (without the target); the 47k resistor was too high a value to overcome this slight amount of bias.

If you are pulling up a signal from off-board, by all means use a very low value of resistance. The pull-up can act as a termination as well as a provider of a logic one, but the characteristic impedance of any cable is usually on the order of hundreds of ohms. A 100k pull-up is just too high to provide any sort of termination, leaving the input subject to cross coupling and noise from other sources. A 1k resistor will help eliminate transients and crosstalk.

Remember that you may not have a good idea what the capacitance of the wiring and other connections will be. A strong pull-up will reduce capacitive time constant effects.

8.4   Unused Inputs

Once upon a time, back before CMOS logic was so prevalent, you could often leave unused inputs dangling unconnected and reasonably expect to get a logic one. Still, engineers are a conservative lot, and most were careful to tie these spare pins to logic one or zero conditions.

But what exactly is a logic one? With 74LS logic it's unwise to use Vcc as an input to any gate. Most LS devices will happily tolerate up to 7 volts on Vcc before something fails, while the input pins have an absolute maximum rating of around 5.5 volts. Connecting an input to Vcc creates a circuit where small power glitches that the devices can tolerate may blow input transistors. It's far better (when using LS) to connect the input to Vcc through a resistor, thus limiting input current and yielding a more power-tolerant design.

Modern CMOS logic in most of its guises has the same absolute maximum rating for Vcc as for the inputs, so it's perfectly reasonable to connect input pins directly to Vcc—if you're sure that production will never substitute an LS equivalent for the device you've called out.

CMOS does require that every unused input be pulled to a valid logic zero or one to avoid generating an SCR latchup condition.

Fast CMOS logic (like 74FCT) switches so quickly, even at very low clock rates, that glitches with Fourier components into billions of cycles per second are not uncommon. Reduce noise susceptibility by tying your logic zeroes and ones directly to the power and ground planes.

And yet . . . one must balance the rules of good design with practical ways to make a debuggable system. A thousand years ago circuits used vacuum tubes mounted on a metal chassis. All connections were made by point-to-point wiring, so making engineering changes during prototype checkout must have been pretty easy. Later, transistors and ICs lived on PC boards, but incorporating modifications was still pretty simple. Now we're faced with whisker-thin leads on surface-mount components, with 8- and 10-layer boards where most tracks are buried under layers of epoxy and out of reach of our X-Acto knives. If we tie every unused input, even on our spare gates, to a solid power or ground connection, it'll be awfully hard to cut the connection free to tie it somewhere else. Lifting the pins on those spare gates might be a nightmare.

One solution is to build the prototype boards a little differently than the production versions. I look at a design and try to identify areas most likely to require cutting and pasting during checkout. A prime example is the programmable device—PALs or FPGAs or whatever. Bitter experience has taught me that probably I'll forget a crucial input to that PAL, or that I'll need to generate some nastily complex waveform using a spare output on the FPGA.

Some engineers figure that if they socket the programmable logic, they can lift pins and tack wires to the dangling input or output. I hate this solution. Sometimes it takes an embarrassing number of tries to get a complex PAL right—each time you must remove the device, bend the leads back to program it, and then reinstall the mods. (An alternative is to put a socket in the socket and lift the upper socket's leads.) When the device is PLCC or another, non-DIP package, it's even harder to get access to the pins.

So I leave all unused inputs on these devices unconnected when building the prototype, unfortunately creating a window of vulnerability to SCR latchup conditions. Then it's easy to connect mod wires to the unconnected pins. When the first prototype is done I'll change the schematic to properly tie off the unused inputs so prototype 2 (or the production unit) is designed correctly.

In years of doing this I have never suffered a problem from SCR latchup due to these dangling pins. The risk is always there, lurking and waiting for an unusual ESD or perhaps even a careless ungrounded finger biasing an input.

I do tie spare gate inputs to ground, even with the first run of boards. It just feels a little too dangerous to leave an unconnected 74HC74 lead dangling. However, if at all possible, I have the person doing the PCB layout connect these grounds on the bottom layer so that a few quick strokes of the X-Acto knife can free them to solve another "whoops."

In designs that use through-hole parts, by all means leave just a little extra room around each chip so you can socket the parts on the prototype. It's a lot easier to pull a connected pin from a socket than to cut it free from the board.

8.5 Clocks

For a number of years embedded systems lived in a wonderful era of compatibility. Just about all the signals on any logic board were relatively slow and generally TTL compatible. This lulled designers into a feeling of security, until far too many of us started throwing digital ICs together without considering their electrical characteristics. If a one is 2.4 volts and a zero 0.7, if we obey simple fanout rules, and as long as speeds are under 10 MHz or so, this casual design philosophy works pretty well. Unfortunately, today's systems are not so benign.

In fact, few microprocessors have *ever* exclusively used TTL levels. Surprise! Pull out a data sheet on virtually any microprocessor and look at the electrical specs page—you know, the section without coffee spills or solder stains. Skip over those 300 tattered pages about programming internal peripherals, bypass the pizza-smeared pinout section, and really look at those one or two pristine pages of DC specifications.

Most CPUs accept TTL-level data and control inputs. Few are happy with TTL on the clock and/or reset inputs. Each chip has different requirements, but in a quick look through the data books I came up with the following:

- 8086: Minimum Vih on clock: Vcc – 0.8

- 386: Minimum Vih on clock: Vcc – 0.8 at 20 MHz, 3.7 volts at 25 and 33 MHz

- Z80: Minimum Vih on clock: Vcc – 0.6

- 8051: Minimum Vih on clock and reset: 2.5 volts

In other words, connect your clock and maybe reset input to a normal TTL driver, and the CPU is out of spec. The really bad news is that these chips are manufactured to behave far better than the specs, so often they'll run fine despite illegal inputs. If only they failed immediately on any violation of specifications! Then, we'd find these elusive problems in the lab, long before shipping a thousand units into the field.

Fully 75% of the systems I see that use a clock oscillator (rather than a crystal) violate the clock minimum high-voltage requirement. It's scary to think we're building a civilization around embedded systems that, well, may be largely misdesigned.

If you drive your processor's clock with the output of a gate or flip-flop, be sure to use a device with true CMOS voltage levels. 74HCT or 74ACT/FCT are good choices. Don't even consider using 74LS without at least a heavy-duty pull-up resistor.

Those little 14-pin silver cans containing a complete oscillator are a good choice . . . if you read the data sheet first. Many provide TTL levels only. I'm not trying to be alarmist here, but look in the latest DigiKey catalog—they sell dozens of varieties of CMOS *and* TTL parts.

Clocks must be clean. Noise will cause all sorts of grief on this most important signal. It's natural to want to use a The venin termination to more or less match impedance on a clock routed over a long PCB trace or even off board. Beware! The venin terminations (typically a 220-ohm resistor to +5 and a 270 to ground) will convert your carefully crafted CMOS level to TTL.

Use series damping resistors to reduce the edge rate if noise is a problem. A pull-up might help with impedance matching if the power supply has a low impedance (as it should).

A better solution is to use clock-shaping logic near the processor itself. If the clock is generated a long way away, use a CMOS hysteresis circuit (such as a 74HCT14) to clean it up. The extra logic adds delay, though. If your system requires clock synchronization, then use a special low-skew clock driver made for that purpose.

In slower systems—under 20 MHz or so—I prefer to design circuits that don't depend on a synchronous clock. What happens if you change to a second sourced processor with slightly different timing? Keep lots of margin.

Never drive a critical signal such as clock off board without buffering. There are a very few absolutely critical signals in any system that must be noise-free. Examine your design and determine what these are, and take appropriate steps. Clock, of course, is the first that comes to mind. Another is ALE (Address Latch Enable), used on processors with a multiplexed address/data bus. A tiny bit of noise on ALE can cause your address register to latch in the middle of a data cycle, driving an incorrect address to the memories.

OK—so now your voltage levels are right. Go back to the data sheet and make sure the clock's timing is in spec.

The 8088 requires a 33% clock duty cycle. Sure, it's a little odd, but this is a fundamental rule of nature to 8088 designers. Other chips have tight duty cycle requirements as well.

Rise and fall times are just as important, though difficult to design for. Some chips have *minimum* rise/fall time requirements! It's awfully hard to predict the rise/fall time for a track routed all over the board. That's one attraction of microprocessors with a clock-out signal. Provide a decent clock-input to the chip, connect nothing to this line other than the processor, and then drive clock-out all over the board.

Motorola's 68HC16 pulls a really neat trick. You can use a 32,768-Hz standard watch crystal to clock the device. An internal PLL multiplies this to 16 MHz or whatever, and drives a clock output to feed to the rest of the board. This gets around many of the clock problems and gives a "free" accurate time-of-day clock source.

8.6 Reset

The processor's reset input is another source of trouble. Like clocks, some processors have unusual input voltage requirements for reset. Be wary.

Other chips require synchronous circuits. The old Z280 had a very odd timing spec, clearly spelled out in the documentation, that everyone ignored only to find massive troubles getting the CPU to start. I think every single Z280 design in the world suffered from this particular ill at one time or another.

Sometimes slew rate is an issue. The old RC startup circuit generates a long ramp that some processors cannot tolerate. You might want to feed it into a circuit with hysteresis, like a Schmidt Trigger, to clean up the ramp.

The more complex CPUs require a long time after power-up to stabilize their internal logic. Reset cannot be unasserted until this interval goes by. Further complicating this is the ramp-up time of the system power supply, as the CPU will not start its power-up sequence until the supply is at some predefined level. The 386, for example, requires 2^{19} clock cycles if the self-test is initiated before it is ready to run.

Think about it: in a 386 system four events are happening at once. The power supply is coming up. The CPU is starting its internal power-up sequence. The clock chip is still stabilizing. The reset circuit is getting ready to unassert reset. How do you guarantee that everything happens to spec?

The solution is a long time delay on reset, using a circuit that doesn't start timing out until the power supply is stable. Motorola, Dallas, and others sell wonderful little reset devices that clamp until the supply hits 4.5 volts or so. Use these in conjunction with a long time constant so the processor, power supply, and clocks are all stable before reset is released.

When Intel released the 188XL they subtly changed the timing requirements of reset from that of the 188. Many embedded systems didn't function with this "compatible" part simply because they weren't compliant with the new chip's reset spec. The easy solution is a three-pin reset clamp.

The moral? Always read the data sheets. Don't skip over the electrical specifications with a mighty yawn. Those details make the difference between a reliable production product and a life of chasing mysterious failures.

One of my favorite bumper stickers reads "Question Authority." It's a noble sentiment in almost all phases of life...but not in designing embedded systems. Obey the specifications listed in the chip vendors' datasheets!

If you've read many annual reports from publicly held companies, you know that the real meat of their condition is contained in the notes. This is just as true in a chip's data sheet. It seems no one specifies sink and source current for a microprocessor's output, but the specification of the device's Vol and Voh will always reference a note that gives the test condition. This is generally a safe maximum rating.

With watchdog timers and other circuits connected to reset inputs, be wary of small timing spikes. I spent several frustrating days working with an AMD part that sometimes powered up oddly, running most instructions fine but crashing on others. The culprit was a subnanosecond spike on the reset input, one too fast to see on a 100-MHz scope.

Homemade battery-backed-up SRAM circuits often contain reset-related design flaws. The battery should take over, maintaining a small bias to the RAM's Vcc pins, when main power fails. That's not enough to avoid corrupting the memory's contents, though.

As power starts to ramp down, the processor may run crazy for a while, possibly creating errant writes that destroy vast amounts of carefully preserved data in the RAM. The solution is to clamp the chip's reset input *as soon as power falls below the part's minimum Vcc* (typically 4.75 volts on a 5-volt part).

With reset properly asserted, Vcc now at zero, and the battery providing a bit of RAM support, be sure that the chip select and write lines to the RAM are in guaranteed "idle" states. You may have to use a small pull-up resistor tied to the battery, but be wary of discharging the battery through the resistor when the system is operating normally.

And be sure you can actually pull the line up despite the fact that the driver will experience Vcc's from +5 to zero as power fails. The cleanest solution is to avoid the problem entirely by using a RAM with an active high chip select, which you clamp to zero as soon as Vcc falls out of spec.

Despite our apparent digital world, the harsh reality is that every component we use pushes electrons around. Electrical specifications are every bit as important to us as to an analog designer. This field is still electronic engineering filled with all of the tradeoffs associated with building things electronic. Ignore those who would have you believe that designing an embedded system is nothing more than slapping logic blocks together.

8.7 Small CPUs

Shhhh! Listen to the hum. That's the sound of the incessant information processing that subtly surrounds us, that keeps us warm, washes our clothes, cycles water to the lawn, and generally makes life a little more tolerable. It's so quiet and keeps such a low profile that even embedded designers forget how much our lives are dominated by data processing. Sure, we rail at the banks' mainframes for messing up a credit report while the fridge kicks into auto-defrost and the microwave spits out another meal.

The average house has some 40 to 50 microprocessors embedded in appliances. There's neither central control nor networking: each quietly goes about its business, ably taking care of just one little function. This is distributed processing at its best.

Billions and billions of 4- to 16-bit micros find their way into our lives every year, yet mostly we hear of the few tens of millions that reside on our desktops.

Now, I'd never give up that zillion-MIP little beauty I'm hunched over at the moment. We all crave more horsepower to deal with Microsoft's latest cycle-consuming application. I'm just getting tired of 32-bit hype for embedded applications. Perhaps that 747 display controller or laser printer needs the power. Surely, though, the vast majority of applications do not.

A 4-bit controller that formed the basis for a calculator started this industry, and in many ways we still use tiny processors in these minimal applications. That is as it should be: use appropriate technology for the job at hand.

Derivatives of some of the earliest embedded CPUs still dominate the market. Motorola's 6805 is a scaled up 6800 which competed with the 8080 back in the embedded Dark Ages. The 8051 and its variants are based on the almost 20-year-old 8048.

8051s, in particular, have been the glue of this industry, corresponding to the analog world's old 741 op amp or the 555 timer. You find them *everywhere*. Their price, availability, and on-board EPROM made them the natural choice for applications requiring anywhere from just a hint of computing power to fairly substantial controllers with limited user interfaces.

Now various vendors have migrated this architecture to the 16-bit world. I can't help but wonder if this makes sense, as scaling a CPU, while maintaining backward compatibility, drags lots of unpleasant baggage along. Applications written in assembly may benefit from the increased horsepower; those coded in C may find that changing processor families buys the most bang for the buck.

Microchip, Atmel, and others understand that the volume part of the embedded industry comes from tiny little CPUs scattered with reckless abandon into every corner of the world. These are cool parts! The smaller members offer a minimum amount of compute capability that is ideal for simple, cost-sensitive systems. Higher-end versions are well suited for more complicated control applications.

Designers seem to view these CPUs as something other than computers. "Oh, yeah, we tossed in a couple of PIC16s to handle the microswitches," the engineer relates, as if the part were nothing more than a PAL. This is a bit different from the bloodied, battered look you'll get from the haggard designer trying to ship a 68030-based controller. The micro-controller is easy to use simply because it is stuffed into easy applications.

L.A. Gear sells sneakers that blink an LED when you walk. A PIC16C5x powers these for months or years without any need to replace the battery. Scientists tag animals in the wild with expendable subcutaneous tracking devices powered by these parts. In Chapter 7 I mentioned the benefit of adding small CPUs just to partition the code. There are other compelling reasons as well.

A friend developing instruments based on a 32-bit CPU discovered that his PLDs don't always properly recover from brown-out conditions. He stuffed a $2 controller on the board to properly sequence the PLD's reset signals, ensuring recovery from low-voltage spikes. The part cost virtually nothing, required no more than a handful of lines of code, and occupied the board space of a small DIP. Though it may seem weird to use a full computer for this trivial function, it's cheaper than a PAL.

Not that there's anything wrong with PALs. Nothing is faster or better at dealing with complex combinatorial logic. Modern super-fast versions are cheap (we pay $12 in singles for a 7-nanosecond 22V10) and easy to use, and their reprogrammability is a great savior of designs that aren't quite right. PALs, though, are terrible at handling anything other than simple sequential logic. The limited number of registers and clocking options means you can't use them for complicated decision making. PLDs are better, but when speed is not critical a computer chip might be the simplest way to go.

As the industry matures, lots of parts we depend on become obsolete. One acquaintance found the UART his company depended on was no longer available. He built a replacement in a PIC16C74, which was pin-compatible with the original UART, saving the company expensive redesigns.

In the good old days of microcomputing, hardware engineers also wrote and debugged all of the system's code. Most systems were small enough that a single, knowledgeable designer could take the project from conception to final product. In the realm of small, tractable problems like those just described, this is still the case. Nothing measures up to the pride of being solely responsible for a successful product; I can imagine how the designer's eyes must light up when he sees legions of kids skipping down the sidewalk flashing their L.A. Gears at the crowds.

Part of the recent success of these parts comes from the aggressive use of Flash and One-Time Programmable (OTP) program memory. OTP memory is simply good old-fashioned EPROM, though the parts come without an erasure window. That small quartz opening typical of EPROMs and many PLDs is very expensive to manufacture. You can program the memory on any conventional device programmer, but, since there's no window, you can never erase it. When it's time to change the code, you'll toss the part out.

Intel sold OTP versions of their EPROMs many years ago, but they never caught on. A system that uses discrete memory devices—RAM, ROM, and the like—has intrinsically higher costs than one based on a microcontroller. In a system with $100 of parts, the extra dollar or two needed to use erasable EPROMs (which are very forgiving of mistakes) is small.

The dynamics are a bit different with a minimal system. If the entire computer is contained in a $2 part, adding a buck for a window is a huge cost hit. OTP starts to make quite a bit of sense, assuming your code will be stable.

This is not to diminish Flash memory, which has all of the benefits of OTP, though sometimes with a bit more cost.

Using either technology, the code *can* be cast in concrete in small applications, since the entire program might require only tens to hundreds of statements. Though I have to plead guilty to one or two disasters where it seemed there were more bugs than lines of code, a program this small, once debugged and thoroughly tested, holds little chance of an obscure bug. The risk of going with OTP is pretty small.

You can't pick up a magazine without reading about "time to market." Managers want to shrink development times to zero. One obvious solution is to replace masked ROMs with their OTP equivalents, as producing a processor with the code permanently engraved in a metalization layer takes months... and suffers from the same risk factors as does OTP. The masked part might be a bit cheaper in high volumes, but this price advantage doesn't help much if you can't ship while waiting for parts to come in.

Part of the art of managing a business is to preserve your options as long as possible. Stuff happens. You can't predict everything. Given options, even at the last minute, you have the flexibility to adapt to problems and changing markets. For example, some companies ship multiple versions of a product, differing only in the code. A Flash or OTP part lets them make a last-minute decision, on the production floor, about how many of a particular widget to build. If you have a half million dollars tied up in inventory of masked parts, your options are awfully limited.

Part of the 8051's success came from the wide variety of parts available. You could get EPROM or masked versions of the same part. Low-volume applications always took advantage of the EPROM version. OTP reduces the costs of the parts significantly, even when you're only building a handful.

Microcontrollers do pose special challenges for designers. Since a typical part is bounded by nothing more than I/O pins, it's hard to see what's going on inside. Nohau, Metalink, and others have made a great living producing tools designed specifically to peer inside of these devices, giving the user a sort of window into his usually closed system.

Now, though, as the price of controllers slides toward zero and the devices are hence used in truly minimal applications, I hear more and more from people who get by without tools of any sort. While it's hard to condone shortchanging your efficiency to save a few dollars, it's equally hard to argue that a 50-line program needs much help. You can probably eyeball it to perfection on the first or second iteration. Again, appropriate technology is the watchword; 5000 lines of assembly language on a 6805 will force you to buy decent debuggers... and, I'd hope, a C compiler.

You can often bring up a microcontroller-based design without a logic analyzer, since there's no bus to watch. Some people even replace the scope with nothing more than a logic probe.

An army of tool vendors supply very low-cost solutions to deal with the particular problems posed by microcontrollers. You have options—lots of them—when using any reasonable controller—far more than if you decide to embed a SPARC into your system.

Some companies cater especially to the low end. Most do a great job, despite the low cost. I recently looked at Byte Craft's array of compilers for microcontrollers from Microchip, Motorola, and National. Despite the limited address spaces of some of these parts, it's clear a decent C compiler can produce very efficient code.

One friend cross-develops his microcontroller code on a PC. Using C frees him from most processor dependencies; compile-time switches select between the PC's timer/UART, etc., and that contained in the controller. He manages to debug more than 80% of the code with no target hardware.

Working in a shop using mostly midrange processors, I'm amazed at the amount of fancy equipment we rely on, and am sometimes a bit wistful for those days of operating out of a garage with not much more than a soldering iron, a logic probe, and a thinking cap. Clearly, the vibrant action in the controller market means that even small, under- or uncapitalized businesses still can come out with competitive products.

8.8 Watchdog Timers

I'm constantly astonished by the utter reliability of computers. While people complain and fume about various PC crashes and other frustrations, we forget that the machine executes millions of instructions per second, even when sitting in an idle loop. Smaller device geometries mean that sometimes only a handful of electrons represent a one or zero. A single-bit failure, for a fleetingly transient bit of time, is disaster.

Yet these failures and glitches are exceedingly rare. Our embedded systems, and even our desktop computers, switch trillions of bits without the slightest problem.

Problems can and do occur, though, due more often to hardware or software design flaws than to glitches. A watchdog timer (WDT) is a good defense for all but the smallest of embedded systems. It's a mechanism that restarts the program if the software runs amok.

The WDT usually resets the processor once every few hundred milliseconds unless reset. It's up to the firmware to reinitialize the watchdog timer, restarting the timing interval. The code tickles the timer frequently, restarting the countdown interval. A code crash means the timer counts down without interruption; at time-out, hardware resets the CPU, ideally bringing the system back on-line.

The first rule of watchdog design is to drive the CPU's reset input, not an interrupt (such as NMI). A WDT time-out means that something awful happened, something that may have left the CPU in an unpredictable scrambled state. Only RESET is guaranteed to bring the part back on-line.

The non-maskable interrupt is seductive to some designers, especially when the pin is unused and there's a chance to save a few gates. For better or worse, NMI—and all other interrupt inputs—is not fail-safe. Confused internal logic will shut down NMI response on some CPUs.

On other chips a simple software problem can render the non-maskable interrupt unusable. The 68 K, for example, will crash if the stack pointer assumes an odd value. If you rely on the WDT to save the day, driving an interrupt while SP is odd results in a double bus fault, which puts the CPU in a dead state until it's reset.

Next, think through the litigation potential of your system. Life-threatening failure modes mean you've got to beware of simple watchdog timers! If a single I/O instruction successfully keeps the WDT alive, then there's a real chance that the code might crash but continue to tickle the timer. Some companies (Toshiba, for example) require a more complex sequence of commands to the timer; it's equally easy to create a PLD yourself that requires a fiendishly complex WDT sequence.

It's also a very bad idea to put the WDT reset code inside of an interrupt service routine. It's always intriguing, while debugging, to find your code crashed but one or more ISRs still functioning. Perhaps the serial receive routine still accepts characters and echoes them to the sender. After all, the ISR by definition runs independently of the rest of the code, so will often continue to function when other routines die. If your WDT tickler stays alive as the world collapses around the rest of the code, then the watchdog serves no useful purpose.

This problem multiplies in a system with an RTOS, as a reliable watchdog monitors all of the tasks. If some of the tasks die but others stay alive—perhaps tickling the WDT—then the system's operation is at best degraded.

In this case write the WDT code as its own task, driven by a timer. All other tasks send messages to the watchdog process, indicating "I'm alive." Only when the WDT activity sees that all tasks that should have checked in are indeed operating does it service the watchdog. If you use RTOS-supplied messaging to communicate the tasks' health—rather than dreaded though easy global variables—there's little chance that errant code overwriting RAM can create a false indication that all's OK.

Suppose the WDT does indeed find a fault and resets the CPU. Then what? A simple reset and restart may not be safe or wise.

One system uses very high-energy gamma rays to measure the thickness of steel. A hardware problem led to a series of watchdog time-outs. I watched, aghast, as this system cycled through WDT resets about once a second, *each time opening the safety shield around the gamma ray source!* The technicians were understandably afraid to approach close enough to yank the power cord.

If you cannot guarantee that the system will be safe after the watchdog fires, then you simply must add hardware to put it in a reasonable, nondangerous, mode.

Even units that have no safety issues suffer from poorly thought-out WDT designs. A sensor company complained that their products were getting slower. Over time, and with several thousand units in the field, response time to user inputs degraded noticeably. A bit of research showed that their system's watchdog properly drove the CPU's reset signal, and the code then recognized a warm boot, going directly to the application with no indication to the users that the time-out had occurred. We tracked the problem down to a floating input on the CPU that caused the software to crash—up to several thousand times per second. The processor was spending most of its time resetting, leading to apparently slow user response.

If your system recovers automatically from a WDT time-out, add an LED or status display so users—or at least the programmers!—know that the system had an unexpected reset. Don't use a bit of clever watchdog code to compensate for software or hardware glitches.

Should embedded systems have a reset switch?

It seems almost traditional to put a reset switch on the back panel of an embedded system. When something horrible happens, hit the reset and retry! Doesn't this make the customer feel that we don't trust our own products? Electronic systems never had reset switches until the introduction of the microprocessor. Why add them now?

A reset switch is no substitute for flaky hardware. It's pretty easy (or, at least possible) to design robust, reliable microprocessor circuits. Any failure is most likely to be a hard fault that a simple reset will not cure.

This argument implies that a reset switch is mostly useful to cure software bugs. We have a choice of writing 100% reliable code or adding some sort of an escape hatch for the user. I hereby proclaim, "We shall all now write correct code."

The problem is now cured.

OK, so perhaps a bug just might creep in once in a while. My feeling is that a reset switch is still a mistake. It conveys the message that no one really trusts the product. It's much better to include a very robust watchdog timer that asserts a good, hard reset when things fall apart. The code might still be unreliable, but at least we're not announcing to the world that bugs are perhaps rampant. Remember when Microsoft eliminated the Unexpected Application Error message from Windows 3.1 . . . by renaming it?

No watchdog is perfect, but even a simple one will catch 99% of all possible code crashes. Combine this percentage with the (ideally) low probability of a software crash, and the watchdog failure rate falls to essentially zero.

8.9 Making PCBs

In the bad old days we created wire-wrapped prototypes because they were faster to make than a PCB, and a lot cheaper. This is no longer the case. Except for the very smallest boards, the cost of labor is so high that it's hard to get a wire-wrapped prototype made for less than $500 to several thousand dollars. Turnaround time is easily a week.

Cheap autorouting software means any engineer can design a PCB in a matter of a couple of days—and you'll have to do this eventually anyway, so it's not wasted time. Dozens of outfits will convert your design to a couple of PCBs in under a week for a very reasonable price. How much? Figure $1000–1500 for a 50-square-inch 4- to 6-layer board, with one-week turnaround.

It's magic. Modem your board design to the vendor, and days later FedEx delivers your custom design, ready for assembly and test.

PCBs are much quieter, electrically, than their wire-wrapped brethren. With fast rise times and high clock rates, noise is a significant problem even in small embedded designs. I've seen far too many cases of "Well, it doesn't work reliably, but that's probably due to the wire wrap. It'll probably get better when we go to PC." These are clearly cases where the prototype does not accomplish its prime objective: identify and fix all risk factors.

Always build your prototype on a PCB, never on wirewrap or other impedance-challenged technologies. And figure on using a multilayer design, with unadulterated power and ground planes. Modern logic is just too fast, too noisy, and too intolerant of ground bounce and other impedance issues to try to mix power and signals on any PCB layer.

The best source for information about speed and noise issues on PC boards is *High Speed Digital Design—A Handbook of Black Magic*, by Howard Johnson and Martin Graham (1993, PTR Prentice Hall). This is a must-read for all digital engineers. If you felt that your college electromagnetics was a flunk-out course, one you squeaked through, fear not. The authors do use plenty of math, but their prose descriptions are so lucid you'll gain a lot of insight by just reading the words and skipping over the equations.

Design your prototype PCB with room for mistakes. Designing a pure surface-mount board? These usually use tiny vias (the holes between layers) to increase the density. *Think* about what happens during the prototyping phase: you'll make design changes, inevitably implemented by a maze of wires. It's impossible to run insulated wire through the tiny holes! Be sure to position a number of unusually large vias (say, 0.031") around the board that can act as wiring channels between the component and circuit sides of the board.

Add pads for extra chips; there's a good chance you'll have to squeeze another PAL in somewhere. My latest design was so bad I had to glue on five extra chips. Guess who felt like an idiot for a few days....

Always build at least two copies of each prototype PCB. One may lag the other in engineering modifications, but you'll have options if (when) the first board smokes. Anyone who has been at this for a while has blown up a board or two.

I generally buy three blank prototype PCBs, assemble two, and use the third to see where tracks run. Though sometimes you'll have to go back to the artwork to find inner tracks, it sure is handy to have the spare blank board on the bench during debug.

It's scary how often the firmware group receives a piece of "functional" prototype hardware from the designers accompanied by nothing more than the schematics—schematics that are usually incomprehensible to the software folks, made even more abstruse by massive use of PLDs and similar functional blocks plopped down on the page, with perhaps hundreds of connections. They are documentation black holes—every signal goes in, and presumably something comes out, but without the designer's suite of design tools even the brightest firmware person will never make sense of the design.

Where does one draw the line between the responsibilities of the hardware designers and those of the firmware folks? Should the designers include device drivers? Seems reasonable to me, since surely they did indeed at least hack together a bit of code to test each device. Why not structure the development plan to make this test code part of the framework of the final software? The hardware tends to be so complex now that it's unfair to give "naked iron" to the software people. At the very least, deliver low-level drivers with well-defined interfaces.

If you live and breathe hardware only, do talk to your software counterparts. You may be surprised to learn that all too often your cool new product makes debugging the code practically impossible. Poor design decisions might seriously affect the firmware schedule. All embedded people must understand that their creation does not exist in isolation; the code and the chips all function together, to form the seamless gestalt that (you hope) delights the user.

8.10   Changing PCBs

After spending a couple of months writing code, it's a bit of a shock to come back to the hardware world. Fixing bugs is a real pain! Instead of a quick edit/compile, you've got to break out a soldering iron, wire, parts, and then manipulate a pin that might be barely visible.

PALs, FPGAs, and PLDs all ease this process to some extent. Many changes are not much more difficult than editing and recompiling a file. It is important to have the right tools available: your frustration level will skyrocket if the PAL burner is not right at the bench.

FPGAs that are programmed at boot time via a ROM download usually have a debugging mechanism—a serial connection from the device to your PC, so you can develop the logic in a manner analogous to using a ROM emulator. Be sure to put the special connector on your design, and buy the little adapter and cable. Burning ROMs on each iteration is a terrible waste of time.

PLDs often come like EPROMs, in ceramic packages with quartz erasure windows. These are great . . . if you were clever enough either to socket the parts, or to have left room around the part for a socket.

On through-hole designs I generally have the technicians load sockets for every part on the prototype. I want to replace suspected failed devices quickly, without spending a lot of time agonizing over "Is it really dead?"

Sockets also greatly ease making circuit modification. With an 8-layer board it's awfully hard to know where to cut a track that snakes between layers and under components. Instead, remove the pin from the socket and wire directly to it.

You can't lift pins on programmable parts, as the device programmer needs all of them inserted when reburning the equations. Instead, stack sockets. Insert a spare socket between the part and the socket soldered on the board. Bend the pins up on this one. All too often the metal on the upper socket will, despite the bent-out pin, still short to the socket on the bottom. Squish the metal in the bottom socket down into the plastic to eliminate this hard-to-find problem.

Surface-mount parts are much more problematic. Get a good set of dental tools and a very fine soldering iron, so you can pry up pins as needed. You'll need a bright light with magnifier, a steady hand, and abstinence from coffee. A decent surface-mount rework machine (such as from Pace Electronics) is essential; get one that vectors hot air around the IC's pins. Don't even try to use conventional solder on fine-pitch parts; use solder paste instead, and keep it fresh (usually it's best stored in a fridge).

Since SMT is so tough, I always make prototype boards with tracks on the outer layers. Sure, the final version might reverse this (power and ground outside to reduce emissions), but reverse the layering during debug. It's easy to cut tracks with an X-Acto knife.

Every engineer needs at least two X-Acto knives. One is for fingernail cleaning, cutting open envelopes, and tossing at the dartboard. The other is only for PCB work and always has a new, sharp blade. Keep 50 or100 spare blades in your drawer, since PCB work invariably breaks the very sharp and very essential pointy end off in no time.

8.11 Planning

Engineers have managers, who "run" projects, ensuring that resources are available when needed, negotiate deadlines and priorities with higher-ups, and guide/mentor the developers toward producing a decent product on time. Planning is one of any manager's main goals. Too often, though, managers do planning that more properly belongs to the engineers. You know more about what your project needs than your boss ever will; it's silly, and unfair, to expect him to deal with all of the details.

There are many great justifications for a project running late. In engineering it's usually impossible to predict all of the technical problems you'll encounter! However, lousy planning is simply an unacceptable, though all too common, reason.

I think engineers spend too much time *doing*, and not enough time *thinking about doing*. Try spending two hours every Monday morning planning the next week and the next month. What projects will you be working on? What's their status? *What is the most important thing you need to do to get the projects done?* Focus on the desired goal, and figure out what you need to do to get there. Do you need to order parts? Tools? Does some of your test equipment need repair or calibration?

Find the critical paths and do what's required to clear the road ahead. Few engineers do this effectively; learn how, and you'll be in much higher demand.

When you're developing a rush project (all projects are rush projects . . .), the first design step is a block diagram of the each board. From this you'll create the schematic, then do a PCB layout, create a bill of materials, and finally, order parts for the prototype.

Not. The worst thing you can do is have a very expensive quick-turn PCB arrive, with all of the components still on back order. The technicians will snicker about your "hurry up and wait" approach, and management will be less than thrilled to spend heavily for fast-turn boards that idle away the weeks on a shelf.

Buy the parts first, before your design is complete. Surely you'll know what all of the esoteric parts are—the CPU, odd analog components, sensors, and the like. These are likely to be the hardest and slowest to get, so put them on order immediately.

The nickel and dime components, such as gates and PALs, resistors and capacitors, are hard to pin down until the schematic is complete. These should mostly be in your engineering spares closet. Again, part of planning is making sure your lab has the basic stuff needed for doing the job, from soldering irons to engineering spares. Make sure you have a good selection of the sort of components your company regularly uses, and avoid the temptation to use new parts unless there's a good reason.

CHAPTER 9

Closed Loop Controls, Rabbits and Hounds

John M. Holland

No embedded World Class Designs book would be complete without a description of control theory, which is a subject often and surprisingly ignored in college curricula and barely touched on in most engineering texts.

Perhaps you've seen the control theory demonstrations. Take a standard attaché case. Put one wheel on a corner. A processor drives the wheel, and balances the case on edge, 45 degrees to perpendicular, using feedback from some sort of attitude sensor like Lewin Edwards describes in Chapter 4.

That seems like a simple problem to the initiated. Just write some code that drives the motor in the right direction depending on the sensed errors. Try it, and you'll quickly find yourself overwhelmed by the physical issues like inertia, damping, and overshoot that are present in any mechanical system. I suspect it's quite impossible to make the case balance on the wheel unless you change the algorithm to one derived from control theory, like PID.

In this chapter John Holland does a fabulous job describing the very nontrivial behavior of the PID algorithm. He manages this without any calculus (other than one hastily abandoned derivative), quite a feat considering the second term is "integral."

PIDs were around long before computers. They're easily implemented using opamps.

John extends the conventional PID approach by adding a predictive term that helps the system stabilize more quickly. He shows how one can enhance that by adding more refinements, such as a first or second derivative term if one has a priori knowledge that such a configuration mirrors the physical reality of the system.

This chapter is the best explanation I have seen of the PID algorithm, and of how each term plays into the overall system response. A couple of drawings using opamps, instead of complicated logic analogs, are especially illuminating.

A bit of PID code is included, written in a variant of Basic or Fortran. It's great pseudo-code that translates directly to C.

Finally, John concludes with a section about the mechanical issues that affect the PID loop in a robot, and gives hints about tuning the various parameters.

If you have access to Matlab or a similar program it's easy and enlightening to implement a software PID, and then to play with the parameters and see how the response changes.

Old-timers will get quite a kick out of John's "flashback."

—Jack Ganssle

Any good athlete will tell you that the key to an exceptional performance is to imagine the task ahead and then to practice until the body can bring this imagined sequence into reality. Similarly, generals create "battle plans," and business people create "business plans" against which they can measure their performance. In each case there is a master plan, and then separate plans for each component. An athlete has a script for each limb, the general for each command, and so forth. All of these plans must be kept in synchronization.

Controlling a robot or other complex intelligent machine requires that it have a plan or model of what it expects to accomplish. Like all such plans, it is quite likely to require modification within moments of the beginning of its execution, but it is still essential.

Like the commander or executive who must modify a plan to allow for changing situations, a robot must also be able to modify its plan smoothly, continuously, and on the fly. If circumstances cause one part of the plan to be modified, then the plans for all other parts must be capable of adjusting to this.

For the plan to have any meaning, physical hardware will need to track to the plan to the best of its ability. Thus we trespass into the fiefdom of Controls. This subject is deep and broad, and engineers spend their entire careers mastering it. In short, it is a great subject to trample on. The fact is that it is quite possible to create decent controls for most applications using little more than common sense algorithms.

The purpose of a control system is to compare the plan to reality, and to issue commands to the servos or other output devices to make reality follow the plan. The desired position of the robot or one of its servos is often referred to as the *rabbit* and the position of the actual robot or servo is referred to as the *hound*. This terminology comes from the dog racing business, where a mechanical rabbit is driven down the track ahead of the pack of racing canines. A closed loop system is one that measures the error between the rabbit and hound, and attempts to minimize the gap without becoming erratic.

Classical control theory has been in existence for a very long time. Engineering courses have traditionally taught mathematical techniques involving poles and zeros and very abstract mathematics that beautifully describe the conditions under which a control can be designed with optimal performance. Normally, this involves producing a control system that maintains the fastest possible response to changing input functions without becoming underdamped (getting the jitters). It is perhaps fortunate that the professor who taught me this discipline has long ago passed from this plane of existence, else he would most probably be induced to cause me to do so were he to read what I am about to say.

Control theory, as it was taught to me, obsessed over the response of a system to a step input. Creating a control that can respond to such an input quickly and without overshooting is indeed difficult. Looking back, I am reminded of the old joke where the patient lifts his arm over his head and complains to the doctor, "It hurts when I do this, Doc," to which the doctor replies, "Then don't do that."

With software controls, the first rule is never to command a control to move in a way it clearly cannot track.

If you studied calculus, you know that position is the integral of velocity, velocity is the integral of acceleration, and acceleration is the integral of jerk. Calculus involves determining the effect of these parameters upon each other at some future time.

In real-time controls this type of prediction is not generally needed. Instead, we simply divide time into small chunks, accumulating the jerk to provide the acceleration, and accumulating the acceleration to provide the velocity, etc. This set of calculations moves the rabbit, and if all of these parameters are kept within the capability of the hardware, a control can be produced that will track the rabbit.

Furthermore, the operator who controls the "rabbit" at a dog race makes sure that the hounds cannot pass the rabbit, and that it never gets too far ahead of them. Similarly, our software rabbit should adjust its acceleration if the servo lags too far behind.

A common problem with applying standard control theory is that the required parameters are often either unknown at design time, or are subject to change during operation. For example, the inertia of a robot as seen at the drive motor has many components. These might include the rotational inertia of the motor's rotor, the inertia of gears and shafts, rotational inertia of its tires, the robot's empty weight, and its payload. Worse yet, there are elements between these components such as bearings, shafts, and belts that may have spring constants and friction loads.

Notwithstanding the wondrous advances in CAD (computer aided design) systems, dynamically modeling such highly complex systems reliably is often impractical because of the many variables and unknowns. For this reason, when most engineers are confronted with such a task, they seek to find simpler techniques.

Flashback . . .

As a newly minted engineer, my first design assignment was a small box that mounted above the instrument panel of a carrier borne fighter plane. The box had one needle and three indicator lights that informed the pilot to pull the nose up or push it down as the multi-million dollar aircraft hurtled toward the steel mass of the ship. As I stood with all of the composure of a deer caught in the landing lights of an onrushing fighter, I was told that it should be a good project to "cut my teeth on"!

The design specification consisted of about ten pages of Laplace transforms that described the way the control should respond to its inputs, which included angle-of-attack, air speed, vertical speed, throttle, etc. Since microprocessors did not yet exist, such computers were implemented using analog amplifiers, capacitors, and resistors, all configured to perform functions such as integration, differentiation, and filtering. Field effect switches were used to modify the signal processing in accordance to binary inputs such as "gear-down."

My task was simply to produce a circuit that could generate the desired mathematical function. The problem was that the number of stages of amplifiers required to produce the huge function was so great that it would have been impossible to package the correct mathematical model into the tiny space available. Worse yet, the accumulation of tolerances and temperature coefficients through so many stages meant the design had no chance of being reproducible. With my tail between my legs, I sought the help of some of the older engineers.

To my distress, I came to realize that I had been given the task because, as a new graduate, I could still remember how to manipulate such equations. I quickly came to the realization that none of the senior engineers had used such formalities in many years. They designed more by something I can best describe as enlightened instinct. As such, they had somehow reprogrammed parts of their brains to envision the control process in a way that they could not describe on paper!

Having failed miserably to find a way out, I took a breadboarding system to the simulator room where I could excite my design with the inputs from a huge analog simulator of the aircraft. I threw out all the stages that appeared to be relatively less significant, and eventually discovered a configuration with a mere three operational amplifiers that could produce output traces nearly identical to those desired. It passed initial testing, but my relief was short lived. My next

---

assignment was to produce a mathematical report for the Navy showing how the configuration accomplished the desired function!

At this point I reverted to the techniques I had honed in engineering school during laboratory exercises. I fudged. I manipulated the equations of the specification and those of the desired results to attempt to show that the difference (caused by my removal of 80% of the circuitry) was mathematically insignificant. In this I completely failed, so when I had the two ends stretched as close to each other as possible, I added the phrase "and thus we see that" for the missing step.

A few months later my report was returned as "rejected." Knowing I had been exposed, I began planning for a new carrier in telemarketing. I opened the document with great apprehension. I found that the objections had to do with insufficient margins and improper paragraph structures. I was, however, commended for my clear and excellent mathematical explanation!

9.1 Basic PID Controls

At the simple extreme of the world of controls is the humble PID control. PID stands for Proportional, Integral, and Derivative. It is a purely reactive control as it only responds to the system error (the gap between the rabbit and the hound). PID controls were first introduced for industrial functions such as closed loop temperature control, and were designed using analog components. Amazingly, to this day, there are still a very significant number of analog PID controls sold every year. Because of its simplicity, many robot designers have adapted software PID controls for servo motor applications. Figure 9-1 shows a simplified block diagram of a PID controller. Since some of the

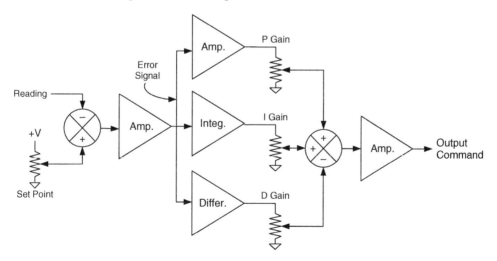

Figure 9-1: A Classic Analog PID Control.

terms of a PID do not work optimally in motor controls, we will consider the classical examples of PIDs in temperature control, and then work our way toward a configuration more capable of driving motors.

The PID control is actually three separate controls whose outputs are summed to determine a single output signal. The current reading of the parameter being controlled is subtracted from the command (in this case a set point potentiometer) to generate an error signal. This signal is then presented to three signal processors.

9.1.1 The Error Proportional Term

The first processor is a straight amplifier. This "proportional" stage could be used by itself but for one small problem. As the reading approaches the set point, the error signal approaches zero. Therefore, a straight proportional amplifier can never close to the full command.

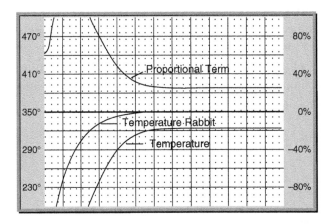

Figure 9-2: Critically Damped Proportional Control with Classic P Droop.

Remember that there is a lag between the time the control generates an output and the reading reflects it. Therefore, if the gain of a proportional stage is increased too much in an attempt to minimize the error, the system will begin to oscillate. The minimum obtainable steady-state error that can be achieved is called the P Droop (see Figure 9-2).

9.1.2 The Error Integral Term

What is needed at this point is a way of eliminating the P Droop. By placing an integrator in the loop, the system can slowly but continuously increase or decrease the output as long as there is any error at all. When the error is eliminated the integral term stops changing. At steady state, the error will go to zero and the integral term will thus replace the proportional term completely. Zero error may sound good, but remember it took quite some time to achieve it.

Since the integration process adds an additional time delay to the loop, if its gain is too high it is prone to induce a lower frequency oscillation than that of the *proportional term* (P-term). For this reason, the P-term is used to provide short-term response, while the I-term (integral term) provides long term accuracy.

The Achilles heel of the integral term is something called integral wind-up. During long excursions of the set point command, the integral term will tend to accumulate an output level that it will not need at quiescence. As the set point is approached by the reading, it will tend to overshoot because the algorithm cannot quickly dump the integral it has accumulated during the excursion. This problem is referred to as *integral wind-up*. Generally, the integral term is better suited to temperature controls than to motor controls, especially in mobile robots where the servo target is constantly being revised.

In motor control applications the speed command often varies so rapidly that the integral term is of little value in closing the error. For this reason, I do not recommend the integral term in any control where the power will normally go to zero when the rabbit reaches its destination (such as a position commanded servo on a flat surface).

In applications where an integral is appropriate, and where the rabbit tends to stay at set point for extended periods, an integral hold off is usually helpful. This simple bit of logic watches the rabbit and when its absolute derivative becomes small (the rabbit isn't changing much) it enables the integral term to accumulate. When the rabbit is in transition, the integral is held constant or bled off.

Generally, introducing sudden changes into the calculation of the PID (such as dumping the integral) is not an optimal solution because the effects of such excursions tend to destabilize the system.

Figure 9-3 shows a well-tuned PID control using only these P and I terms. The total power, proportional, and integral terms correspond to the scale on the right side of the graph, while the scale for the rabbit and temperature are on the left.

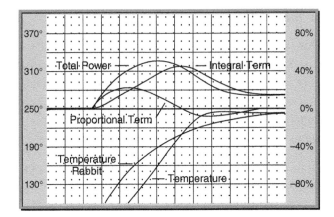

Figure 9-3: A Well-Tuned PID Using Only Proportional and Integral Terms.

Notice that the temperature rabbit is an exponential function that approaches the set point of 250 degrees. The trick here is that while the temperature clearly overshoots the rabbit, it almost perfectly approaches the desired quiescent state. Overshooting the rabbit causes the proportional term to go negative, decelerating the approach to set point. The integral term "winds up" during the ascent of the rabbit, but is cancelled out by the negative excursion of the proportional term as the temperature overshoots the rabbit. There is no *integral hold-off* in this example.

So why would we want anything more than this? The answer is that the performance is only optimal for the precise rabbit excursion to which the gains were tuned. Had we elected to send the rabbit to 350 degrees with these same gains, there would have been a significant overshoot because the integral would have wound up excessively. Since robot servos must operate over a wide range of excursions and rates, a more robust control is very desirable.

9.1.3 The Error Derivative Term

With the tendency of the P-term and I-term to cause oscillations, it is natural to look for a counterbalance. The purpose of the D-term is just that. The D-term is a signal that is proportional to the derivative or rate-of-change of the error. This term is subtracted from the other terms, in order to suppress changes in the error. Thus, the whole PID equation that must be coded is merely:

$$P_t = (K_p \cdot E) + (K_i \cdot \int E) - (K_d \cdot dE/dt)$$

Where P_t is the total power command, E is the current error, and K_p, K_i, and K_d are the proportional, integral, and derivative gains. The algorithm is implemented by repeatedly calling a subroutine on a timed basis. The frequency of execution depends on the time frame of the response. The higher the frequency, the faster the servo can respond, but the more computer resources that are absorbed. Generally the PID rate for a mobile robot operating at less than 3 mph will be from 10 Hz to 100 Hz. A single execution of the PID is often called a tick (as in the tick of a clock).

The proportional term is a simple, instantaneous calculation. The integral term, however, is produced by multiplying the current error by the integral gain and adding the result to an accumulator. If the error is negative, the effect is to subtract from the accumulator.

The derivative can be calculated by simply subtracting the error that was present during the previous execution of the algorithm from the current error. In some cases, this may be a bit noisy and it may be preferable to average the current calculation with the past few calculations on a weighted basis.

Unfortunately, the D-term can also induce instability if its effective gain is greater than unity, and it amplifies system noise. I have found this term to be of some limited value in temperature control applications, but of much less value at all in controlling servo motors. In fact, purely reactive controls such as PIDs tend in general to be of limited use with most servo motors.

9.2 Predictive Controls

The reactive controls we have been discussing are driven entirely by error. This is a bit like driving a car while looking out the back window. Your mistakes may be obvious, but their realization may come a bit too late. By running controls strictly from the error signal, we are conceding that an error must occur. Indeed, there will always be a bit of error, but wouldn't it be nice if we could guess the required power first, and only use the remaining error to make up for relatively smaller inaccuracies in our guess?

Predictive controls do just this. Predictive controls do not care about the error, but simply watch the rabbit and try to predict the amount of power required to make the servo track it. Since they watch the rabbit, and are not in the feedback loop, they are both faster and more stable than reactive controls.

9.2.1 The Rabbit Term

Looking at the temperature rabbit curve of Figure 9-3, we begin to realize that there are two relationships between the rabbit and the required power that can be readily predicted. The first relationship is that for any steady-state temperature, there will be a certain constant power required to overcome heat loss to the environment and maintain that temperature. Let's call this the rabbit term, and in the simplest case it is the product of the rabbit and the rabbit gain. For a temperature control this is the amount of power required to maintain any given temperature relative to ambient. For the drive motor of a robot, this will be the amount of power that is required to overcome drag and maintain a fixed speed.

The power required may not bear a perfectly linear relationship with the set point over its entire range. This rabbit gain will vary with factors such as ambient temperature, but if the operating temperature is several hundred degrees, and the ambient variation is only say 20 degrees, then this power relationship can be assumed to be a constant. If the ambient temperature varies more appreciably, we could take it into account in setting the gain. *Note that in some cases the relationship may be too nonlinear to get away with a simple gain multiplier. In these cases an interpolated look-up table or more complex mathematical relationship may be useful.*

9.2.2 The Rabbit Derivative and Second Derivative Terms

The second relationship we can quickly identify has to do with the slope or derivative of the rabbit. For a position servo, the rabbit derivative is the *rabbit velocity*, and the rabbit second derivative is the *rabbit acceleration.*

If the rabbit is attacking positively, we will need to pour on the coals to heat the furnace up or to accelerate the robot. This term is thus proportional to the derivative of the rabbit, and we will call its gain the *rabbit derivative gain* or *rabbit velocity gain.*

If the rabbit is accelerating, then we will need even more power. This term is most commonly found in position controls and its gain is thus called the *rabbit acceleration gain.* If we perfectly guess these gains, then the servo will track the rabbit perfectly, with the minimum possible delay.

The big advantage to the rabbit derivative terms as opposed to the error derivative term is that, because they work from the nice clean rabbit command, they do not amplify system response noise.

9.3 Combined Reactive and Predictive Controls

Unfortunately, a predictive control of this type has no way of correcting for variables such as ambient temperature or payload. It is also unlikely that our perfectly selected gains will be as perfect for every unit we build. It is possible to learn these gains during operation, but this will be discussed later.

It is therefore not practical to expect exact performance from a predictive control. Yet, the prediction may buy us a guess within say 10% of the correct value, and its output is instantaneous and stable. For this reason, it is often optimal to combine terms of a conventional PID with those of a predictive control. Using our analog control as a metaphor, consider the configuration shown in Figure 9-4.

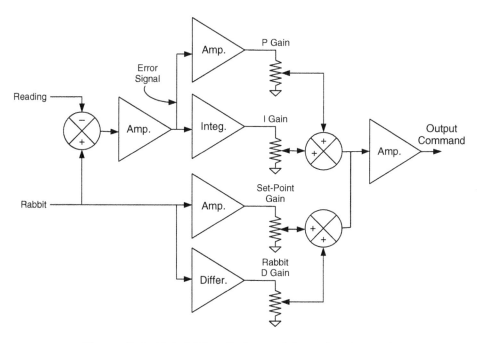

Figure 9-4: Hybrid Predictive and Reactive Control.

In this configuration, the instantaneous rabbit value is multiplied by a gain to produce the *rabbit term*. This term is sometimes called the *set point gain* because at set point (quiescence) this term should produce the bulk of the output needed.

The second new function is a differentiator that provides a term proportional to the rate of change of the rabbit. When added to the *rabbit term*, this term gives us a much better guess at the power required during transitions of the rabbit. Additionally, we have retained the *proportional* and *integral* terms driven by the error to make up for any inaccuracy in the prediction. The result is an extremely capable yet simple control architecture suitable for a wide range of applications.

9.4   Various PID Enhancements

Since the standard PID control algorithm is far from ideal for many types of controls, software engineers have produced an almost endless array of variations on the basic theme. In most cases, these modifications amount to replacing the gain constants we have been discussing with gain tables or simple functions.

9.4.1   Asymmetric Gains

One of the most common problems encountered is that many systems are asymmetric in their response to a power command. For example, a heater may be adequately powerful to quickly increase the temperature of a mass with say 30% power, but when the same amount of power is removed, the temperature may not drop nearly as fast as it had increased. This asymmetric response is obviously due to the fact that cooling is not merely the result of removing power, but rather the result of air convection.

The drive power requirement for a heavy robot is also very asymmetric. While a good deal of forward power is usually required to accelerate the robot, only a small amount of reverse power is required to decelerate it at the same rate. If a symmetric PID is used to control such a motor, it may be impossible to obtain adequate forward responsiveness without causing the robot to somersault when deceleration occurs. Things get even more exciting when a robot goes over the top of a ramp and starts down the back side.

For asymmetric loads, it is often useful to provide two gains for each term. One gain is used for positive term inputs, while the other is used for negative inputs.

9.4.2   Error Band Limits

Another useful modification to reactive PID terms is to limit the error range over which they respond proportionally. If the error is within this band, then the error is multiplied by the gain, otherwise the appropriate error limit is used. This is particularly useful with the integral term to prevent wind-up. The example below is coded in Visual Basic, and utilizes many of the techniques discussed thus far.

In all these cases, you will be governed by the principal of enlightened instinct. To become a Zen master of such techniques, you must understand the causes of problems, and the basics of physics and dynamics, but in the end the solution is often half theory and half instinct. If the choice is made to be rigidly mathematical in the solution, you may still end up guessing at a lot of parameters, and producing an algorithm that sucks every available cycle out of your control computer.

My personal preference for creating such controls is to design the software so that virtually every factor can be changed on the fly during operation. You can then run the system and observe its response to changes in various parameters. For example, the robot can be set to drive backward and forward between two points while various gains are tried. *In other words, let the system talk to you.*

Figure 9-5 shows typical code for a thermal PID control using many of the terms and tricks just discussed.

```
'Calculate the PID for a single control using error
'proportional, error integral, rabbit, and rabbit derivative
'terms. Error and rabbit derivative gains are non-symmetric.

'The process is controlled by an array of singles called
'"ControlSingles". Each term also has a limit band on its error.
'If the error is greater than the band, then the band value is
'substituted for the limit.

'Routine returns the power command as an integer between 0 and 9999.
Public Static Function DoPIDs(TempRabbit As Single,
Temp As Single) As Integer

Dim Error As Single      'Raw error
Dim LimError As Single  'Error or limit, whichever is smaller.
Dim OutputAccum As Long
Dim PGain As Long 'Proportional command Gain (0 to 9999)
Dim DGain As Long 'Derivative command Gain
Dim IGain As Long 'Integral command Gain
Dim RGain As Long 'Setpoint rabbit command Gain
Dim PTerm As Single
Dim ITerm As Single
```

Figure 9-5: A Band-Limited Hybrid Control with Integral Hold-Off.

```
Dim DTerm As Single
Dim RTerm As Single
Dim RabbitDeriv As Single
Dim LastTempRabbit As Single
Dim IntegralHold(MaxZones) As Integer '0=No Hold, 1=allow pos.
'only, -1=allow neg.

  On Error Resume Next

  'Calculate the error and rabbit derivative.
  Error = TempRabbit - Temp
  RabbitDeriv = TempRabbit - LastTempRabbit
  LastTempRabbit = TempRabbit

  'Get the Rabbit and error gains
  RGain = ControlSingles(RabbitGain)   'Rabbit gain is always
  positive.
  'Some gains depend on error polarity
  If Error >= 0 Then 'For positive errors use positive gains
     PGain = ControlSingles(PposGain)
     IGain = ControlSingles(IposGain)
  Else
     PGain = ControlSingles(PnegGain)
     IGain = ControlSingles(InegGain)
  End If
  '
  'Since there is no error derivative term, we will use
  'Dgain to mean the rabbit derivative gain. Its gain
  'depends on the polarity of the rabbit derivative.
  If RabbitDeriv > 0 Then
     DGain = ControlSingles(DposGain)
  Else
     DGain = ControlSingles(DnegGain)
  End If

  'Now do the calculation for each term.
  'First limit the error to the band for each gain
  If Error > ControlSingles(PBand) Then
     LimError = ControlSingles(PBand)
  ElseIf Error < -ControlSingles(PBand) Then
     LimError = -ControlSingles(PBand)
```

Figure 9-5: Continued

```
Else
   LimError = Error
End If

PTerm = CDbl((PGain * LimError)/100)

If Error > ControlSingles(IBand) Then
   LimError = ControlSingles(IBand)
ElseIf Error < -ControlSingles(IBand) Then
   LimError = -ControlSingles(IBand)
Else
   LimError = Error
End If

'The I term is cumulative, so it's gain range is 1/100th
'that of P. Integral is bled off while rabbit is moving, or
'if the output accumulator has gone below zero with a negative
'integral or over full range with a positive integral.

If Abs(RabbitDeriv) < ControlSingles(AttackRate)/30 Then
   If (LimError > 0 And IntegralHold >= 0) Or_
      (LimError < 0 And IntegralHold <= 0) Then
      ITerm = LimitTerm(ITerm + ((IGain * LimError)/10000))
   End If
Else 'Bleed off the i term.
   ITerm = 0.99 * ITerm
End If

DTerm = LimitTerm(RabbitDeriv * DGain * 10)
RTerm = LimitTerm(TempRabbit * (RGain) / 500)
OutputAccum = PTerm + ITerm + DTerm + RTerm
'Limit the output accumulator and flag the integrator
'if the output goes out of range. (In a properly tuned
'control, this should never happen).
If OutputAccum > 9999 Then
   OutputAccum = 9999
   IntegralHold = -1 'Allow only down integration
ElseIf OutputAccum < 0 Then
   OutputAccum = 0
   IntegralHold = 1 'Allow only upward integration
Else
   IntegralHold = 0 'Allow both directions.
End If
```

Figure 9-5: Continued

```
    DoPIDs = CInt(OutputAccum)

End Function


'---------------------------------------------------------------
'Prevent Terms from exceeding integer range for logging.
Private Function LimitTerm(Term As Single)
  If Term > 9999 Then
     LimitTerm = 9999
  ElseIf Term < -9999 Then
     LimitTerm = -9999
  Else
     LimitTerm = Term
  End If
End Function
```

Figure 9-5: Continued

9.5   Robot Drive Controls

The drive system of a mobile robot is subject to widely varying conditions, and thus it presents several challenges not found in the simpler control systems already discussed. These include:

1. Drive/brake asymmetry

2. Quadrant crossing nonlinearity due to gear backlash

3. Variable acceleration profiles due to unexpected obstacle detection

4. Variable drag due to surface variations

5. Variable inertia and drag due to payload variations.

9.5.1   Drive/Brake Asymmetry

There are two basic categories of drive gear reduction: back drivable and non-back drivable. The first category includes pinion, chain, and belt reducers, while the second includes worm and screw reducers. Servos using non-back drivable reducers are easier to control because variations in the load are largely isolated from the motor. Back-drivable reducers are much more efficient, but more difficult to control because load fluctuations and inertial forces are transmitted back to influence the motor's position.

The response of either type servo is usually very nonlinear with respect to driving and braking. To achieve a given acceleration may require very significant motor current, while achieving the same magnitude of deceleration may require only a minute reverse current. For this reason, asymmetric gains are usually required in both cases.

9.5.2 Quadrant Crossing Nonlinearity

Most drives are either two or four quadrant. A two-quadrant drive can drive in one direction and brake. A four-quadrant drive can drive and brake in both directions. The term quadrant thus comes from these four possible actions of the drive. A nasty phenomenon that appears in the back drivable servos is that of nonlinear regions in a servo's load. A drive motor using a pinion gear box is an excellent example. When the power transitions from forward to reverse (called *changing quadrants*), the servo will cross a zero load region caused by the gear backlash. If the control does not take this into account, it may become unstable after freewheeling across this region and slamming into the opposite gear face. One trick for handling this discontinuity is to sense that the control has crossed quadrants and reduce its total gain after the crossing. This gain can then be smoothly restored over several ticks if the control remains in the new quadrant.

9.5.3 Natural Deceleration

Any servo will have its own natural deceleration rate. This is the deceleration rate at which the servo will slow if power is removed. If the programmed deceleration is lower than the natural deceleration, the servo never goes into braking and the response is usually smooth.

Thus, during normal operation the ugly affects of quadrant crossing may not be noticed. If, however, the robot must decelerate unexpectedly to avoid a collision, the servo may react violently, and/or the robot may tip over forward. For this reason, a maximum deceleration is usually specified, beyond which the robot will not attempt to brake. In this way the robot will slow as much as possible before the impact. If a mechanical bumper is designed into the front of the robot, it can be made to absorb any minor impact that cannot be avoided by the control system.

9.5.4 Freewheeling

Perhaps the most difficult control problem for a drive servo is that of going down a ramp. Any back drivable drive servo will exhibit a freewheeling velocity on a given ramp. This is the speed at which the robot will roll down the ramp in an unpowered state. At this speed, the surface drag and internal drag of the servo are equal to the gravitational force multiplied by the sine of the slope. The freewheeling speed is thus load dependent.

If a robot attempts to go down a ramp at a speed that is greater than its natural free-wheeling speed for the given slope, then the servo will remain in the forward driving quadrant. If the robot attempts to go slower than the freewheeling speed, then the servo will remain in the braking quadrant. The problem comes when the speed goes between these two conditions. This condition usually occurs as soon as the robot moves over the crest of the ramp and needs to brake.

Under such transitions, both the quadrant discontinuity and drive/brake nonlinearity will act on the servo. This combination will make it very difficult to achieve smooth control, and the robot will lurch. Since lurching will throw the robot back and forth between driving and braking, the instability will often persist. The result roughly simulates an amphetamine junkie after enjoying a double espresso. If the gain ramping trick described above is not adequate, then it may be necessary to brake.

My dearly departed mother endeared herself to her auto mechanic by driving with one foot on the gas and the other on the brake. When she wished to go faster or slower she simply let up on the one while pushing harder on the other. This method of control, however ill-advised for an automobile, is one way of a robot maintaining smooth control while driving down ramps.

One sure-fire method of achieving smooth control on down ramps is to intentionally decrease the freewheeling velocity so that the servo remains in the drive quadrant. To accomplish this, one can use a mechanical brake or an electrical brake. The simplest electrical brake for a permanent magnet motor is to simply place a low value braking resistor across the armature. While a braking resistor or other form of braking will reduce the freewheeling speed of the robot, it will waste power. For this reason, brakes of any sort must be applied only when needed.

The ideal way to reduce the freewheeling velocity of a drive servo is through the use of circuitry that directs the back EMF of the motor into the battery. In this way, the battery recovers some energy while the robot is braking. The common way of doing this is through the use of a freewheeling diode in the motor control bridge.

I have not found simple freewheeling diodes to provide an adequate amount of braking in most downhill situations. This is because the back EMF must be greater than the battery voltage and the effective braking resistance includes both the battery internal resistance and the motor resistance. Thus, voltage-multiplying circuits are usually required if this type of braking is to actually be accomplished.

9.5.5 Drag Variations

A robot drive or steering servo is usually position commanded. The moment-to-moment position error of a cruising mobile robot is usually not critical. As long as the robot's odometry accurately registers the movements that are actually made, the position error can be corrected over the long term. For this reason, drag variations are not usually critical to normal driving.

However, sometimes a robot will be required to stop at a very accurate end position or heading. One such case might be when docking to a charger. In these cases, the closing error becomes critical.

Variations in drag caused by different surfaces and payloads can also cause another significant problem. If a robot's drive system is tuned for a hard surface, and it finds itself on a soft surface, then the servos will not have adequate gain to close as accurately.

If the robot stops on such a surface with its drive servo energized, then the position error may cause the servo to remain stopped but with a significant current being applied to the motor. This type of situation wastes battery power, and can overheat the motor. If the robot's gains are increased for such a surface, it may exhibit instability when it returns to a hard surface.

One solution to this situation is to ramp up the error proportional gain of the servo as its speed command reaches zero. When the P-gain has been at its upper limit (typically an order of magnitude higher than its base value) and long enough for the closing error to be minimized, the gain is smoothly reduced back to its base setting. I have found this technique quite effective.

9.6 Tuning Controls

The problem that now arises is one of tuning. There are many terms that all interplay. The first thing that is needed is a good diagnostic display upon which we can view the rabbit, reading, power output, and various term contributions. The graphs shown in this chapter are from an actual temperature control system written in Visual Basic. Readings are stored into a large array, and then displayed on the graphs in the scale and format desired.

Figure 9-6 shows the performance that can be obtained from a control like that shown in Figure 9-5. Notice this control cleanly tracks the rabbit by largely predicting the amount of power required. As the temperature rabbit is climbing linearly, the rabbit derivative term is constant, and as the rabbit begins to roll into the set point, this term backs off its contribution to the total power.

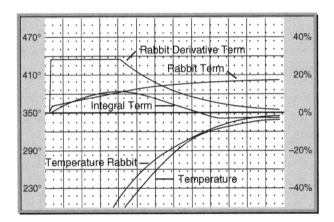

Figure 9-6: Combination Predictive and Reactive Control.

The rabbit term, on the other hand, represents the power required to remain at the present rabbit temperature. Thus, this term is proportional to the value of the rabbit. For clarity, the error proportional term is not shown, but it would have looked like that in Figure 9-3, reversing polarity as the temperature reading crossed the rabbit. The double effect of the rabbit derivative term and the error proportional term helped prevent overshooting of the set point.

Tuning such controls is an iterative process. Normally, the control is first run with all gains but one or two set to zero.

9.6.1 Learning Gains

For systems with an *error integral term*, it is a fairly simple matter to write code that can learn the rabbit gain just discussed. The process is one of finding reasonable parameters for the integral and proportional gains, and then allowing the system to reach a quiescent state in the region of typical operation.

The integral term should come to replace the proportional term, and at this point, it is divided by the set point to provide the *rabbit gain*. Since this will replace the power previously being supplied by the error integral term, the integral accumulator is dumped, and the new *rabbit term* generates the entire output required.

If the rabbit gain was already nonzero, then the old rabbit term must be added to the error integral term and then the result divided by the set point to calculate a corrected rabbit gain.

The rabbit derivative gain can then be determined by turning all gains to zero except the rabbit gain. The rabbit is then commanded to begin a smooth transition and the rabbit derivative gain is trimmed to make the slope of the reading match the slope of the rabbit.

9.6.2 Tuning Robot Position Controls

In the case of motor controls, the simple gain learning processes previously described may not be practical. It is theoretically possible to write adaptive controls that can adjust their own parameters dynamically, but such techniques are well beyond the scope of this chapter. The simplest method is the remote tuning process.

Motor controls may be velocity commanded, but in robots they are more commonly position commanded. The velocity desired is imparted through the motion of the rabbit. In most cases, absolute velocity accuracy is not as important as smoothness.

A position-seeking control will normally consist of a rabbit velocity term, a rabbit derivative (acceleration) term, an error proportional term, and a less significant error derivative term. Since no power is required to hold a robot at the same place (on a flat surface), there is no need for a rabbit term. On a slope, the error proportional term should be adequate to prevent the robot from rolling once it has stopped. This implies that the power will remain on to the control when the rabbit has stopped, and this brings about other issues.

9.7 Rabbits Chasing Rabbits

The smoother the rabbit's motion, the more likely the servo will be able to track it closely. For this reason, linear approximations (trapezoidal profiles) are not generally acceptable. Such a profile demands infinite acceleration at the knee points. Command curves with low second derivatives (jerk term) are also essential in some cases.

When we ascend to the level of making a robot navigate, there is much more involved than a simple single-axis servo chasing its rabbit. We must first generate a position rabbit that runs through imaginary space, and then determine the behavior we want from the steering and/or drive motors in order to follow it. If we are executing an arc, then the steering rabbit must be slaved to the drive rabbit, and both must be slaved to the position rabbit.

The important thing to remember is that the most elaborate building is made of lots of simple blocks. Likewise, all complex systems are constructed from simpler blocks. If these blocks are understood and perfected, they can be used to create systems that dependably generate very complex behavior.

9.8 Conclusions

The terms of the hybrid reactive and predictive control we have discussed are shown in Table 9-1 below, along with a matrix for the applications in which they may be useful.

Table 9-1

Term	Calculated From	Temperature Control	Position Control
Error proportional	Rabbit – reading (error)	✓	✓
Error derivative	Rate of change of error	✓	✓
Error integral	Accumulation of error	✓	
Rabbit	Rabbit value (temp or position)	✓	
Rabbit derivative	Rabbit rate of change (velocity)	✓	✓
Rabbit 2nd derivative	Rabbit velocity rate of change (acceleration)	✓	✓

The subject of controls is vast, and there are many techniques more advanced than the ones described here. Work has even been done with neural networks to solve the thornier problems. Even so, the simple architecture described here is one that I have found to be very useful, and it should serve as a starting point. More importantly, it demonstrates a way of thinking through such problems logically.

CHAPTER 10

Application Examples

David J. Katz and Rick Gentile

If there's any sure hit for a product, it's something that helps people communicate. Witness Apple's success with the iPod series, and the hysterical furor over the 2007 release of their iPhone. YouTube went from a startup to $1.6 billion buyout in merely two years. Infotainment, too, soars. Auto companies speculate that consumers assign some 70% of the value of a car not to stopping and going, but to the cool features delivered by a host of embedded processors.

In the embedded space much of the underlying technology behind communications and infotainment comes under the "media processing" rubric. Behind that nearly always stands one or more DSP processors. In this chapter, David Katz and Rick Gentile describe a number of DSP applications for the infotainment and automotive market. Though they base the text on Analog Device's Blackfin processor, the discussion scales equally well to any DSP device.

They start off with a description of the "car of the future," though that future may be only a few short years from this writing. This vehicle won't quite drive itself (darn!), but will automatically monitor blind spots, and give drivers a virtual slap across the face if they wander off the road or out of their lane.

David and Rick show how video cameras mounted at strategic points on the car will acquire data. That's the easy part. From there a very fast processor or network of CPUs (all DSPs) will have to extract collision and lane-wandering data from the video streams. That's a tough problem. But it's surprisingly analogous to what infotainment systems do as they process MPEG and JPEG files. For to create an MPEG file the computer must extract difference information from the video. What has changed between frames? It's not too far from that to "is that other vehicle now on a collision course?"

In this chapter the authors explain how convolutions are used in image processing. Many of us have used one-dimensional convolutions for signal smoothing (for instance, FIR and IIR filters). Those ideas extend to images to detect edges and to smooth data. The authors show how to efficiently implement a convolution inside a DSP processor.

From there they go on to the specifics behind JPEG and MPEG files. It's surprisingly complex, but a cleverly thought-out DSP solution makes the problem at least tractable. But it takes a lot of computation, so they also cover strategies for partitioning it across two or more DSP cores. With so many vendors offering multiple-core processing, this is an increasingly important (and difficult) problem.

—Jack Ganssle

10.1 Introduction

We'll first look at an automotive driver assistance application, seeing how a video-based lane departure warning system can utilize a media framework. Then, we'll turn our attention to implementation of a JPEG algorithm, which contains a lot of the key elements of more complex image and video codecs. Next, we'll extend this discussion to the MPEG realm, outlining how an MPEG-2 video encoder algorithm can be partitioned efficiently across a dual-core processor. Finally, we'll switch gears and talk about the porting of open-source C code to the Blackfin processor environment. In doing so, we'll undergo an enlightening analysis of the code optimization process—which optimization steps give you the most "bang for your buck."

So let's get started . . .

10.2 Automotive Driver Assistance

With the dozens of processors controlling every aspect of today's automobiles, not a single aspect of the "vehicle experience" remains untouched by technology. Whether it's climate control, engine control, or entertainment, there has been constant evolution in capabilities and manufacturer offerings over the last decade. One of the drivers for this evolution, the rapidly increasing performance trend of media processors, is set to have a profound impact on another critical automotive component—the safety subsystem.

While most current safety features utilize a wide array of sensors—most involving microwaves, infrared light, lasers, or acceleration and position detection—only recently have processors emerged that can meet the real-time video requirements that make image processing a viable safety technology. Let's look at the roles for embedded media processing in the emerging field of video-based automotive safety, using the Blackfin processor as a basis for discussion.

10.2.1 Automotive Safety Systems

The entire category of car safety demands a high-performance media processor for many reasons. For one, since the response times are so critical to saving lives, video filtering and image processing must be done in a real-time, deterministic manner. There is a natural desire to maximize video frame rate and resolution to the highest level that a processor can handle for a given application, since this provides the best data for decision making. Additionally, the processor needs to compare vehicle speeds and relative vehicle-object distances against desired conditions, again in real time. Furthermore, the processor must interact with many vehicle subsystems (such as the engine, braking, steering and airbag controllers), process sensor information from all these systems in real time, and provide appropriate audiovisual output to the driver interface environment. Finally, the processor should be able to interface to navigation and telecommunication systems to log accidents and call for assistance.

Smart Airbags

One emerging use of media processors in automotive safety is for an "intelligent airbag system," which bases its deployment decisions on who is sitting in the seat opposite the airbag. Presently, weight-based systems are most popular, but video sensing will soon become prevalent. Either thermal or regular cameras may be used, at rates up to 200 frames/sec, and more than one might be employed to provide a stereo image of the occupant. In any case, the goal is to characterize not only the size of the occupant, but also her position or posture. Among other things, image processing algorithms must account for the differentiation between a person's head and other body parts in determining body position. Ultimately, in the event of a collision the system may choose to restrict deployment entirely, deploy with a lower force, or fully deploy.

In this system, the media processor reads in multiple image streams at high frame rates, processes the images to profile a seat's occupant size and position under all types of lighting conditions, and constantly monitors all the crash sensors placed throughout the car in order to make the best deployment decision possible in a matter of milliseconds.

Collision Avoidance and Adaptive Cruise Control

Another high-profile safety application is Adaptive Cruise Control (ACC)/Collision Avoidance. ACC is a convenience feature that controls engine and braking systems to regulate distance and speed of the car relative to the vehicle ahead. The sensors involved employ a combination of microwave, radar, infrared, and video technology. A media processor might process between 17 and 30 frames/sec from a roadway-focused camera mounted near the rearview mirror of the car. The image processing algorithms involved include frame-to-frame image comparisons, object recognition, and contrast equalization for varying lighting scenarios. Goals of the video sensor input include providing information about lane boundaries, obstacle categorization, and road curvature profiling.

Whereas ACC systems are promoted as a convenience feature, collision avoidance systems aim to actively avoid accidents by coordinating the braking, steering, and engine controllers of the car. As such, they have been slower to evolve, given the legal ramifications of such an endeavor. The estimated widespread deployment of these systems is 2010. However, since the automotive design cycle is typically a five-year span, the design of such systems is ongoing.

Collision warning systems are a subset of the collision avoidance category. They provide a warning of an impending accident, but don't actively avoid it. There are two main subcategories within this niche:

- **Blind spot monitors**: Cameras are mounted strategically around the periphery of the car to provide visual display of the driver's blind spots, as well as to sound a warning if the processor senses another vehicle is present in a blind-spot zone. These systems also serve as back-up warnings, cautioning the driver if there is an obstruction in the rear of the car while it is shifted in reverse. The display might be integrated directly into the rear-view mirror, providing a full unobstructed view of the car's surroundings. Moreover, video of "blind spots" within the car cabin may also be included, to allow the driver to, say, monitor a rear-facing infant.

- **Lane departure monitors**: These systems can act as driver fatigue monitors, as well as notify drivers if it is unsafe to change lanes or if they are straying out of a lane or off the road. Cameras facing forward monitor the car's position relative to the center-line and side markers of the roadway up to 50–75 feet in front of the car. The system sounds an alarm if the car starts to leave the lane unintentionally.

Figure 10-1 gives an indication of where image sensors might be placed throughout a vehicle, including how a lane departure system might be integrated into the chassis. There are a few things to note here. First, multiple sensors can be shared across the different automotive safety functions. For example, the rear-facing sensors can be used when the vehicle is backing up, as well as to track lanes as the vehicle moves forward. In addition, the lane departure system might accept feeds from any number of camera sources, choosing the appropriate inputs for a given situation. At a minimum, a video stream feeds the embedded processor. In more advanced systems, other sensor information, such as position data from GPS receivers, also weighs in.

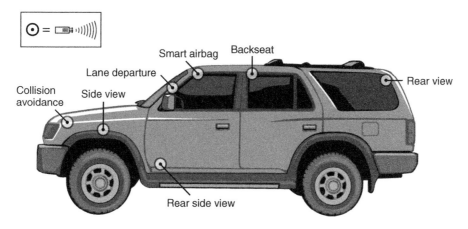

Figure 10-1: Basic Camera Placement Regions for Automotive Safety Applications.

10.2.2   Lane Departure: A System Example

Having discussed the role that a media processor can play in video-based automotive safety applications, it is instructive to analyze typical components of just such an application. To that end, let's probe further into a lane departure monitoring system. Figure 10-2a shows the basic operational steps in the system, while Figure 10-2b illustrates the processor's connection to external subsystems within the car.

The overall system diagram of Figure 10-2a is fairly straightforward, considering the complexity of the signal processing functions being performed. What is interesting about a video-based lane departure system is that, instead of having an analog signal chain, the bulk of the processing is image-based, carried out within the processor.

This is very advantageous from a system bill-of-materials standpoint. The outputs of the vehicle-based system consist of some indication to the driver to correct the car's path before the vehicle leaves the lane unintentionally. This may be in the form of an audible "rumble strip" warning, or perhaps just a cautionary chime or voice.

(a)

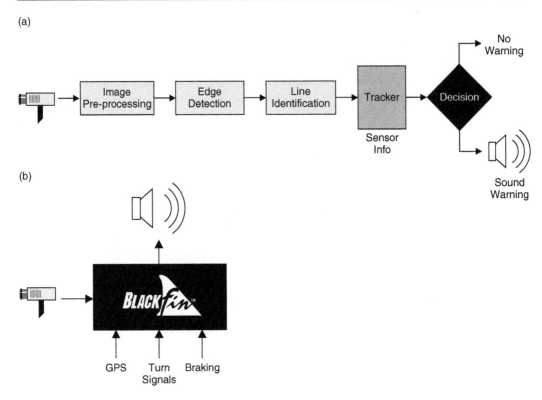

(b)

Figure 10-2: (a) Basic Steps in a Lane Departure Algorithm and (b) How the Processor Might Connect to the Outside World.

The video input system to the embedded processor must perform in harsh environments, including wide and drastic temperature shifts and changing road conditions. As the data stream enters the processor, it is transformed instantaneously into a form that can be processed to output a decision. At the simplest level, the lane departure system looks for the vehicle's position with respect to the lane markings in the road. To the processor, this means the incoming stream of road imagery must be transformed into a series of lines that delineate the road surface.

Lines can be found within a field of data by looking for edges. These edges form the boundaries that the vehicle should stay within while it is moving forward. The processor must track these line markers and determine whether or not to notify the driver of irregularities.

Keep in mind that several other automobile systems also feed the lane departure system. For example, the braking system and the turn signals will typically disable warnings during intentional lane changes and slow turns.

Let's now drill deeper into the basic components of our lane departure system example. Figure 10-3 follows the same basic operational flow as Figure 10-2a, but with more insight into the algorithms being performed. The video stream coming into the system needs to be filtered to reduce noise caused by temperature, motion, and electromagnetic interference. Without this step, it would be difficult to find clean lane markings. The next processing step involves edge detection, and if the system is set up properly, the edges found will represent the lane markings. These lines must then be matched to the direction and position of the vehicle. For this step, we will describe something called the Hough Transform. The output of this step will be tracked across frames of images, and a decision will be made based on all the compiled information. The final challenge is to send a warning in a timely manner without sounding false alarms.

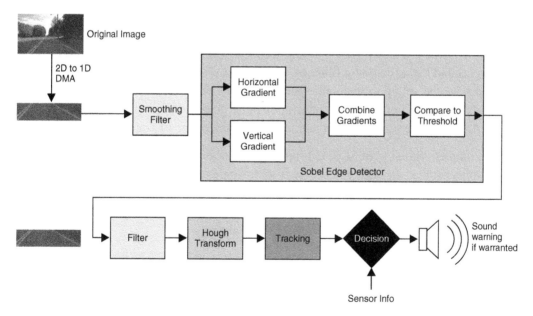

Figure 10-3: Algorithm Flow Showing Results of Intermediate Image Processing Steps.

Image Acquisition

For automotive safety applications, image resolutions typically range from VGA (640 × 480 pixels/image) down to QVGA (320 × 240 pixels/image). Regardless of the actual image size, the format of the data transferred remains the same, but because less data is transferred, lower clock speeds can be used. Moreover, for applications like lane departure warning systems, sometimes only grayscale images are required. Therefore, data bandwidth is halved (from 16 bits/pixel to 8 bits/pixel), because no chroma information is needed.

Memory and Data Movement: Efficient memory usage is an important consideration for system designers, because external memories are expensive and their access times are slow. Therefore, it is important to be judicious in transferring only the video data needed for the application. By intelligently decoding ITU-R BT.656 preamble codes, an interface like Blackfin's PPI can aid this "data filtering" operation. For example, in some applications, only the active video fields are required. In a full NTSC stream, this results in a 25% reduction in the amount of data brought into the system, because horizontal and vertical blanking data is ignored and not transferred into memory. What's more, this lower data rate helps conserve bandwidth on the internal and external data buses.

Because video data rates are so demanding, frame buffers must be set up in external memory, as shown in Figure 10-4. In this scenario, while the processor operates on one buffer, a second buffer is being filled by the PPI via a DMA transfer. A simple semaphore can be set up to maintain synchronization between the frames. With the Blackfin's 2D DMA controller, an interrupt can be generated virtually anywhere, but it is typically configured to occur at the end of each video line or frame.

Once a complete frame is in SDRAM, the data is normally transferred into internal L1 data memory so that the core can access it with single-cycle latency. To do this, the DMA controller can use two-dimensional transfers to bring in pixel blocks. Figure 10-5 shows how a 16×16 macroblock, a construct used in many compression algorithms, can be stored linearly in L1 memory via a 2D DMA engine.

From a performance standpoint, up to four unique internal SDRAM banks can be active at any time. This means that in the video framework, no additional bank activation latencies are observed when the 2D-to-1D DMA is pulling data from one bank while the PPI is feeding another.

Projection Correction

The camera used for the lane departure system typically sits in the center-top location of the front windshield, facing forward. In other cases, the camera is located in the rear

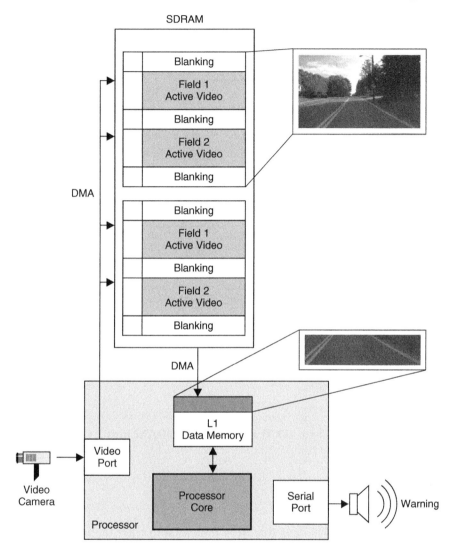

Figure 10-4: Use of Internal Memory for Frame Buffering.

windshield and faces the road already traveled. In some systems, a "bird's eye" camera is selected to give the broadest perspective of the upcoming roadway. This type of camera can be used in lieu of multiple line-of-sight cameras. In this case, the view is warped because of the wide-angle lens, so the output image must be remapped into a linear view before parsing the picture content.

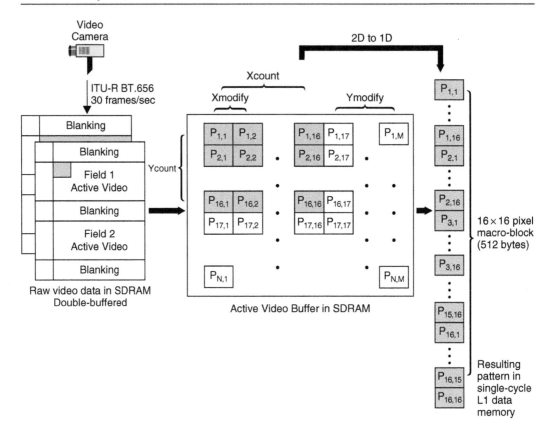

Figure 10-5: A 2D-to-1D Transfer from SDRAM into L1 Memory.

Image Filtering

Before doing any type of edge detection, it is important to filter the image to smooth out any noise picked up during image capture. This is essential because noise introduced into an edge detector can result in false edges output from the detector.

Obviously, an image filter needs to operate fast enough to keep up with the succession of input images. Thus, it is imperative that image filter kernels be optimized for execution in the fewest possible number of processor cycles. One effective means of filtering is accomplished with a basic 2D convolution operation. Let's look in some detail at how this computation can be performed efficiently on a Blackfin processor.

The high-level algorithm can be described in the following steps:

1. Place the center of the mask over an element of the input matrix.

2. Multiply each pixel in the mask neighborhood by the corresponding filter mask element.

3. Sum each of the multiplies into a single result.

4. Place each sum in a location corresponding to the center of the mask in the output matrix.

Figure 10-6 shows three matrices: an input matrix F, a 3×3 mask matrix H, and an output matrix G.

After each output point is computed, the mask is moved to the right by one element. When using a circular buffer structure, on the image edges the algorithm wraps around to the first element in the next row. For example, when the mask is centered on element F2m, the H23 element of the mask matrix is multiplied by element F31 of the input matrix. As a result, the usable section of the output matrix is reduced by one element along each edge of the image.

By aligning the input data properly, both Blackfin multiply-accumulate (MAC) units can be used in a single processor cycle to process two output points at a time. During this

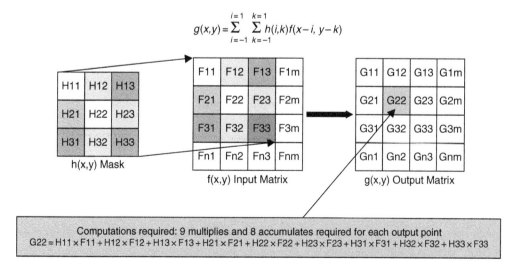

$$g(x,y) = \sum_{i=-1}^{i=1} \sum_{k=-1}^{k=1} h(i,k)f(x-i, y-k)$$

Computations required: 9 multiplies and 8 accumulates required for each output point
G22 = H11 × F11 + H12 × F12 + H13 × F13 + H21 × F21 + H22 × F22 + H23 × F23 + H31 × F31 + H32 × F32 + H33 × F33

Figure 10-6: Computation of 3 x 3 Convolution Output.

same cycle, multiple data fetches occur in parallel with the MAC operation. This method, shown in Figure 10-7, allows efficient computation of two output points for each loop iteration, or effectively 4.5 cycles per pixel instead of the 9 cycles per pixel of Figure 10-6.

G11	G12	G13	G1m
G21	G22	G23	G2m
G31	G32	G33	G3m
Gn1	Gn2	Gn3	Gnm

g(x,y) Output Matrix

▢ G22 = H11 × F11 + H12 × F12 + H13 × F13 + H21 × F21 + H22 × F22 + H23 × F23 + H31 × F31 + H32 × F32 + H33 × F33

▣ G23 = H11 × F12 + H12 × F13 + H13 × F14 + H21 × F22 + H22 × F23 + H23 × F24 + H31 × F32 + H32 × F33 + H33 × F34

Loop Cycle (F11 Loaded into R0.H, F12 Loaded into R0.L, H11 Loaded into R1.L)

1	A1 = R0.H * R1.L,	A0 = R0.L * R1.L	‖ R0.L = w[I0++] ‖ R2 = [I3++]; //I0-I3 are index regs
2	A1 += R0.L * R1.H,	A0 += R0.H * R1.H	‖ R0.H = w[I0--]; // w[ ] is 16-bit access
3	A1 += R0.H * R2.L,	A0 += R0.L * R2.L	‖ R0 = [I1++] ‖ R3 = [I3++];
4	A1 += R0.H * R2.H,	A0 += R0.L * R2.H	‖ R0.L = w[I1++];
5	A1 += R0.L * R3.L,	A0 += R0.H * R3.L	‖ R0.H = w[I1--] ‖ R1 = [I3++];
6	A1 += R0.H * R3.H,	A0 += R0.L * R3.H	‖ R0 = [I2++];
7	A1 += R0.H * R1.L,	A0 += R0.L * R1.L	‖ R0.L = w[I2++] ‖ R2 = [I3++];
8	A1 += R0.L * R1.H,	A0 += R0.H * R1.H	‖ R0.H = w[I2--] ‖ R1 = [I3++];
9	R6.H = (A1 += R0.H * R2.L),	R6.L = (A0 += R0.L * R2.L);	//Accumulate for 2 outputs

Each time through this loop yields two output points
Total Cycles = 9 for every 2 pixels => 4.5 cycles per pixel

Figure 10-7: Efficient Implementation of 3 x 3 Convolution.

Edge Detection

A wide variety of edge detection techniques are in common use. Before considering how an edge can be detected, we must first settle on a definition of what constitutes an edge. We can then find ways to enhance the edges we are seeking to improve the chances of detection. Because image sensors are non-ideal, two factors must be considered in any approach we take: ambient noise and the effects of quantization errors.

Noise in the image will almost guarantee that pixels with equal grayscale level in the original image will not have equal levels in the output image. Noise will be introduced by many factors that can't be easily controlled, such as ambient temperature, vehicular motion, and outside weather conditions. Moreover, quantization errors in the image will result in edge boundaries extending across a number of pixels. Because of these factors, any appropriate image-processing algorithm must keep noise immunity as a prime goal.

One popular edge detection method uses a set of common derivative-based operators to help locate edges within the image. Each of the derivative operators is designed to find places where there are changes in intensity. In this scheme, the edges can be modeled by a smaller image that contains the properties of an ideal edge.

We'll discuss the Sobel Edge Detector because it is easy to understand and illustrates principles that extend into more complex schemes. The Sobel Detector uses two convolution kernels to compute gradients for both horizontal and vertical edges. The first is designed to detect changes in vertical contrast (S_x). The second detects changes in horizontal contrast (S_y).

$$S_x = \begin{bmatrix} -1 & 0 & 1 \\ -2 & 0 & 2 \\ -1 & 0 & 1 \end{bmatrix} \quad S_y = \begin{bmatrix} -1 & -2 & -1 \\ 0 & 0 & 0 \\ 1 & 2 & 1 \end{bmatrix}$$

The output matrix holds an "edge likelihood" magnitude (based on horizontal and vertical convolutions) for each pixel in the image. This matrix is then thresholded in order to take advantage of the fact that large responses in magnitude correspond to edges within the image. Therefore, at the input of the Hough Transform stage, the image consists only of either "pure white" or "pure black" pixels, with no intermediate gradations.

If the true magnitude is not required for an application, this can save a costly square root operation. Other common techniques to build a threshold matrix include summing the gradients from each pixel or simply taking the larger of the two gradients.

Straight Line Detection: Hough Transform

The Hough Transform is a widely used method for finding global patterns such as lines, circles, and ellipses in an image by localizing them in a parameterized space. It is especially useful in lane detection, because *lines* can be easily detected as *points* in Hough Transform space, based on the polar representation of Equation 10-1:

$$\rho = x \cos \Phi + y \sin \Phi$$ **Equation 10-1**

The meaning of Equation 10-1 can be visualized by extending a perpendicular from the given line to the origin, such that Φ is the angle that the perpendicular makes with the abscissa, and ρ is the length of the perpendicular. Thus, one pair of coordinates (ρ, Φ) can fully describe the line. To demonstrate this concept, we will look at the lines L1 and L2 in Figure 10-8. As shown in Figure 10-8a, L1's location is defined by Φ_1 and the length of the perpendicular extending from the X-Y origin to L1, while L2's position is defined by Φ_2 and the length of the perpendicular extending from the X-Y origin to L2.

Another way to look at the Hough Transform is to consider a possible way that the algorithm could be implemented intuitively:

1.   Visit only white pixels in the binary image.

2.   For each pixel and every Φ value being considered, draw a line through the pixel at angle Φ from the origin. Then calculate ρ, which is the length of the perpendicular between the origin and the line under consideration.

3.   Record this (ρ, Φ) pair in an accumulation table.

4.   Repeat steps 1–3 for every white pixel in the image.

5.   Search the accumulation table for the (ρ, Φ) pairs encountered most often. These pairs describe the most probable "lines" in the input image, because in order to register a high accumulation value, there had to be many white pixels that existed along the line described by the (ρ, Φ) pair.

The Hough Transform is computationally intensive, because a sinusoidal curve is calculated for each pixel in the input image. However, certain techniques can speed up the computation considerably. First, some of the computation terms can be computed ahead of time, so that they can be referenced quickly through a lookup table. In a fixed-point architecture like Blackfin's, it is very useful to store the lookup table only for the cosine function. Since the sine values are 90 degrees out of phase with the cosines, the same table can be used for both, with an offset as appropriate. Given the lookup tables, the computation of Equation 10-1 can be represented as two fixed-point multiplications and one addition.

Another factor that can improve Hough performance is a set of assumptions about the nature and location of lane markings within the input image. By considering only those input points that could potentially be lane markings, a large number of unnecessary calculations can be avoided, since only a narrow range of Φ values need be considered for each white pixel.

(a)

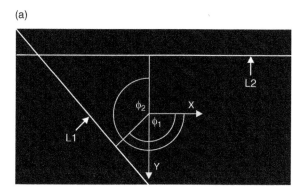

Figure 10-8(a): Line Segments L1 and L2 Can Be Described by the Lengths and Angles of Perpendicular Lines Extending from the Origin.

(b)

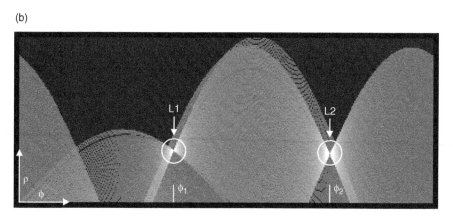

Figure 10-8(b): The Hough Transform of the Two Line Segments in Figure 10-8a. The two bright regions correspond to local maxima, which can be used to reconstruct the two segments L1 and L2 of Figure 10-8a.

Among the parameters useful for further analysis are the offset from the camera's center axis, the widths of the detected lines, and the angles with respect to the position of the camera. Since lane markings in many highway systems are standardized, a set of rules can eliminate some lines from the list of lane-marking candidates. The set of possible lane-marking variables can then be used to derive the position of the car.

Lane Tracking

Lane information can be determined from a variety of possible sources within an automobile. This information can be combined with measurements of vehicle-related parameters (e.g., velocity, acceleration, etc.) to assist in lane tracking. Based on the results of these measurements, the lane departure system can make an intelligent decision as to whether an unintentional departure is in progress.

The problem of estimating lane geometry is a challenge that often calls for using a Kalman filter to track the road curvature. Specifically, the Kalman filter can predict future road information, which can then be used in the next frame to reduce the computational load presented by the Hough Transform.

As described earlier, the Hough Transform is used to find lines in each image. But we need to track these lines over a series of images. In general, a Kalman filter can be described as a recursive filter that estimates the future state of an object. In this case, the object is a line. The state of the line is based on its location and its motion path across several frames.

Along with the road state itself, the Kalman filter provides a variance for each state. The predicted state and the variance can be used to narrow the search space of the Hough Transform in future frames, which saves processing cycles.

Decision Making

From experience, we know that false positives are always undesirable. There is no quicker way to have a consumer disable an optional safety feature than to have it indicate a problem that does not really exist.

With a processing framework in place, system designers can add their own intellectual property to the decision phase of each of the processing threads. The simplest approach might take into account other vehicle attributes in the decision process. For example, a lane-change warning could be suppressed when a lane change is perceived to be intentional—as when a blinker is used or when the brake is applied. More complex systems may factor in GPS coordinate data, occupant driving profile, time of day, weather and other parameters.

Clearly, we have only described an example framework for how an image-based lane departure system might be structured. The point is, with a flexible media processor in the design, there is plenty of room for feature additions and algorithm optimizations.

10.3 Baseline JPEG Compression Overview

Image compression, once the domain of the PC, is now pervasive in embedded environments. This trend is largely a result of the increased processing power and multimedia capabilities of new embedded processors. Consequently, it now becomes advantageous for embedded designers to gain a better grasp of image algorithms in an effort to implement them efficiently. This section examines the Baseline JPEG compression standard, one of the most popular image compression algorithms. While not covered explicitly here, the JPEG decode process is a straightforward inversion of the encode process.

Although there are many specified versions of JPEG, Baseline JPEG compression (referred to herein simply as "JPEG") embodies a minimum set of requirements. It is lossy, such that the original image cannot be exactly reconstructed (although a "Lossless JPEG" specification exists as well). JPEG exploits the characteristics of human vision, eliminating or reducing data to which the eye is less sensitive. JPEG works well on grayscale and color images, especially on photographs, but it is not intended for two-tone images. Figure 10-9 shows the basic encode process, which we'll examine below in some detail.

10.3.1 Preprocessing

Color Space

The incoming uncompressed image might be stored in one of several formats. One popular format is 24 bit/pixel RGB—that is, 8 bits each of red, green, and blue subpixels. However, in viewing these separate R, G, and B subchannels for a given image, there is usually a clear visual correlation between the three pictures. Therefore, in order to achieve better compression ratios, it is common to decorrelate the RGB fields

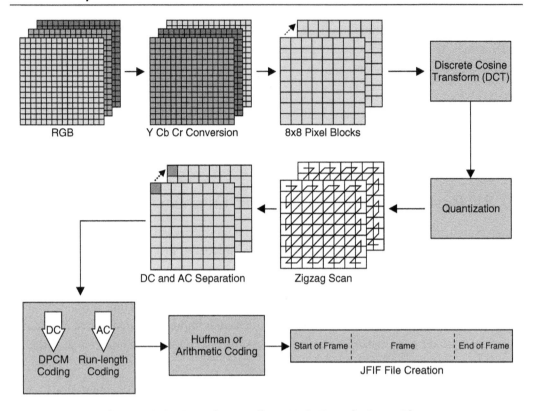

Figure 10-9: Sample Baseline JPEG Encode Data Flow.

into separate luma (*Y*) and chroma (*Cb*, *Cr*) components. The equations to do this are:

$$Y' = (0.299)\,R' + (0.587)\,G' + (0.114)\,B'$$

$$Cb = -(0.168)\,R' - (0.330)\,G' + (0.498)\,B' + 128$$

$$Cr = (0.498)\,R' - (0.417)\,G' - (0.081)\,B' + 128$$

10.3.2 Spatial Filtering

Whichever format is chosen, the image needs to be separated into *Y*, *Cb*, and *Cr* buffers, because the JPEG algorithm acts on each component individually and in identical fashion. If the chroma are subsampled, this is simply akin to running the JPEG algorithm on an image of smaller size.

JPEG operates on 8 × 8-byte blocks of data, as shown in Figure 10-9. Therefore, in each image buffer the data is partitioned into these blocks, from left to right and top to bottom. These blocks do not overlap, and if the image dimensions are not multiples of 8, the last row and/or column of the image is duplicated as needed.

10.3.3 *Discrete Cosine Transform (DCT)*

The DCT stage of JPEG exploits the fact that the human eye favors low-frequency image information over high-frequency details. The 8 × 8 DCT transforms the image from the spatial domain into the frequency domain. Although other frequency transforms can be effective as well, the DCT was chosen because of its decorrelation features, image independence, efficiency of compacting image energy, and orthogonality (which makes the inverse DCT very straightforward). Also, the separable nature of the 2D DCT allows the computation of a 1D DCT on the eight column vectors, followed by a 1D DCT on the eight row vectors of the resulting 8 × 8 matrix. The 8 × 8 DCT can be written as follows:

$$Y_{mn} = \tfrac{1}{4} K_m K_n \sum_{i=0}^{i=7} \sum_{j=0}^{j=7} x_{ij} \cos\left((2i+1)\,m\pi/16\right) \cos\left((2j+1)\,n\pi/16\right), \text{ where}$$

Y_{mn} = the output DCT coefficient at row m, column n of the 8 × 8 output block

x_{ij} = the input spatial image coordinate at row i, column j of the 8 × 8 input block

$K_m = \frac{1}{\sqrt{2}}$ for $m = 0$, or 1 otherwise

$K_n = \frac{1}{\sqrt{2}}$ for $n = 0$, or 1 otherwise

For uniform handling of different image components, the DCT coder usually requires that the expected average value for all pixels is zero. Therefore, before the DCT is performed, a value of 128 may be subtracted from each pixel (normally ranging from 0 to 255) to shift it to a range of −127 to 127. This offset has no effect on the ac characteristics of the image block.

It is instructive to take a visual view of the DCT transform. Refer to the DCT basis functions shown in Figure 10-10. In performing a DCT on an 8 × 8 image block, what we are essentially doing is correlating the input image with each of the 64 DCT basis functions and recording the relative strength of correlation as coefficients in the output DCT matrix.

For example, the coefficient in the output DCT matrix at (2,1) corresponds to the strength of the correlation between the basis function at (2,1) and the entire 8×8 input image block. The coefficients corresponding to high-frequency details are located to the right and bottom of the output DCT block, and it is precisely these weights which we try to nullify—the more zeros in the 8×8 DCT block, the higher the compression achieved. In the quantization step below, we'll discuss how to maximize the number of zeros in the matrix.

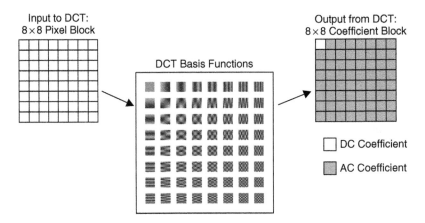

Input to DCT:
8×8 Pixel Block

DCT Basis Functions

Output from DCT:
8×8 Coefficient Block

☐ DC Coefficient

▨ AC Coefficient

Figure 10-10: DCT Basis Functions.

10.3.4 Quantization

After the DCT has been performed on the 8×8 image block, the results are quantized in order to achieve large gains in compression ratio. Quantization refers to the process of representing the actual coefficient values as one of a set of predetermined allowable values, so that the overall data can be encoded in fewer bits (because the allowable values are a small fraction of all possible values).

Remember that the human eye is much more attuned to low-frequency information than high-frequency details. Therefore, small errors in high-frequency representation are not easily noticed, and eliminating some high-frequency components entirely is often visually acceptable. The JPEG quantization process takes advantage of this to reduce the amount of DCT information that needs to be coded for a given 8×8 block.

Quantization is the key irreversible step in the JPEG process. Although performing an inverse DCT will not exactly reproduce the original input 8×8 input matrix because of rounding error, the result is usually quite visually acceptable. However, after quantization, the precision of the original unquantized coefficients is lost forever. Thus, quantization only occurs in lossy compression algorithms, such as the Baseline JPEG implementation we're discussing here.

The quantization process is straightforward once the quantization table is assembled, but the table itself can be quite complex, often with a separate quantization coefficient for each element of the 8 × 8 DCT output block. Because this output matrix corresponds to how strongly the input image block exhibits each of the 64 DCT basis functions, it is possible to experimentally determine how important each DCT frequency is to the human eye and quantize accordingly. In other words, frequencies towards the bottom and right of the basis functions in Figure 10-10 will tend to have large values in their quantization table entries, because this will tend to zero out the higher-frequency elements of the DCT output block. The actual process of quantization is a simple element-wise division between the DCT output coefficient and the quantization coefficient for a given row and column.

A "Quality Scaling Factor" can be applied to the quantization matrix to balance between image quality and compressed image size. For the sample tables mentioned in the JPEG standard, typical quality values range from 0.5 (high recovered image quality) to 5 (high compression ratio).

10.3.5 Zigzag Sorting

Next, we prepare the quantized data in an efficient format for encoding. As we have seen from the DCT output, the quantized coefficients have a greater chance of being zero as the horizontal and vertical frequency values increase. To exploit this behavior, we can rearrange the coefficients into a one-dimensional array sorted from the dc value to the highest-order spatial frequency coefficient, as shown in Figure 10-11. This is accomplished using *zigzag sorting,* a process which traverses the 8 × 8 block in a back-and-forth direction of increasing spatial frequency. On the left side of the diagram, we see the index numbers of the quantized output DCT matrix. Using the scan pattern shown in the middle of the figure, we can produce a new matrix as shown in the right side of the figure.

Each quantized DCT output matrix is run through the zigzag scan process. The first element in each 64 × 1 array represents the dc coefficient from the DCT matrix, and the remaining 63 coefficients represent the ac components. These two types of information

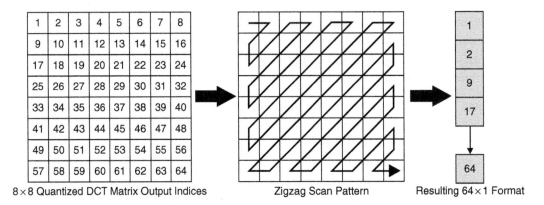

8×8 Quantized DCT Matrix Output Indices   Zigzag Scan Pattern   Resulting 64×1 Format

Figure 10-11: Illustration of Zigzag Scan for Efficient Coefficient Encoding.

are different enough to warrant separating them and applying different methods of coding to achieve optimal compression efficiency.

All of the dc coefficients (one from each DCT output block) must be grouped together in a separate list. At this point, the dc coefficients will be encoded as a group, and each set of ac values will be encoded separately.

10.3.6   Coding the DC Coefficients

The dc components represent the intensity of each 8×8 pixel block. Because of this, significant correlation exists between adjacent blocks. So, while the dc coefficient of any given input array is fairly unpredictable by itself, real images usually do not vary widely in a localized area. As such, the previous dc value can be used to predict the current dc coefficient value. By using a differential prediction model (DPCM), we can increase the probability that the value we encode will be small, and thus reduce the number of bits in the compressed image.

To obtain the coding value we simply subtract the dc coefficient of the previously processed 8×8 pixel block from the dc coefficient of the current block. This value is called the "DPCM difference." Once this value is calculated, it is compared to a table to identify the symbol group to which it belongs (based on its magnitude), and it is then encoded appropriately using an entropy encoding scheme such as Huffman coding.

10.3.7   Coding the AC Coefficients (Run-Length Coding)

Because the values of the ac coefficients tend towards zero after the quantization step, these coefficients are run-length encoded. The concept of run-length encoding is a straightforward principle. In real image sequences, pixels of the same value can always be represented as individual bytes, but it doesn't make sense to send the same value over and over again. For cxample, we have seen that the quantized output of the DCT blocks produces many zero-valued bytes. The zigzag ordering helps produce these zeros in groups at the end of each sequence.

Instead of coding each zero individually, we simply encode the number of zeros in a given "run." This run-length coded information is then variable-length coded (VLC), usually using Huffman codes.

10.3.8   Entropy Encoding

The next processing block in the JPEG encode sequence is known as the *entropy encoder*. In this stage, a final lossless compression is performed on the quantized DCT coefficients to increase the overall compression ratio achieved.

Entropy encoding is a compression technique that uses a series of bit codes to represent a set of possible symbols. Table 10-1a shows an obvious 2-bit encode sequence with four possible output symbols (A, B, C, and D).

In this example, we can uniquely describe each symbol with two bits of information. Because each symbol is represented as a 2-bit quantity, we refer to the code as "fixed length." Fixed length codes are most often applied in systems where each of the symbols occurs with equal probability.

Table 10-1 (a) Example of entropy encoding with equal symbol probabilities
(b) Example of entropy encoding with weighted symbol probabilities

(a)

Symbol	Bit Code
A	00
B	01
C	10
D	11

(b)

Symbol	Probability	Bit Code
A	0.50	0
B	0.30	10
C	0.10	110
D	0.10	111

In reality, most symbols do not occur with equal probability. In these cases, we can take advantage of this fact and reduce the average number of bits used to compress the sequence. The length of the code used for each symbol can be varied based on the probability of the symbol's occurrence. By encoding the most common symbols with shorter bit sequences and the less frequently used symbols with longer bit sequences, we can easily improve on the average number of bits used to encode a sequence.

Huffman Coding

Huffman coding is a variable-length encoding technique that is used to compress a stream of symbols with a known probability distribution. Huffman code sequences are built with a code tree structure by pairing symbols with the two least probabilities of occurrence in a sequence. Each time a pair is combined, the new symbol takes on the sum of the two separate probabilities. This process continues with each node having its two probabilities combined and then being paired with the next smallest probable symbol until all the symbols are represented in the coded tree.

Once the code tree is constructed, code sequences can be assigned within the tree structure by assigning a 0 or a 1 bit to each branch. The code for each symbol is then read by concatenating each bit from the branches, starting at the center of the structure and extending to the branch for which the relevant symbol is defined. This procedure produces a unique code for each symbol data size that is guaranteed to be optimal.

Even though the codes can have different lengths, each code can be unambiguously decoded if the symbols are read beginning with the most significant bit. The JPEG standard provides a number of Huffman code tables to describe the different ways of encoding the input quantized data—for instance, depending on whether luminance or chrominance information is being processed.

Table 10-1b uses the same example as in Table 10-1a, but with differently weighted symbol probabilities. The Bit Code column shows the modified encode sequence. As we can see, the most likely symbol is encoded with a single bit, while the least likely symbols require 3 bits each.

With this table, we can determine the average bit length for a code per symbol.

$$(0.5 \times 1\,\text{bit}) + (0.3 \times 2\,\text{bits}) + (0.1 \times 3\,\text{bits}) + (0.1 \times 3\,\text{bits}) = 1.7\,\text{bits/symbol}$$

In this example, the variable length encode scheme produces an average of 1.7 bits/symbol, which is more efficient than the 2 bits/symbol that the fixed-length code of Table 10-1a produced.

Another entropy encoding scheme used in JPEG is Arithmetic Coding. While this algorithm provides for better compression than Huffman coding because it uses adaptive techniques that make it easier to achieve the entropy rate, the additional processing required may not justify the fairly small increase in compression. In addition, there are patent restrictions on Arithmetic Coding.

10.3.9 JPEG File Interchange Format (JFIF)

The encoded data is written into the JPEG File Interchange Format (JFIF), which, as the name suggests, is a simplified format allowing JPEG-compressed images to be shared across multiple platforms and applications. JFIF includes embedded image and coding parameters, framed by appropriate header information. Specifically, aside from the encoded data, a JFIF file must store all coding and quantization tables that are necessary for the JPEG decoder to do its job properly.

10.4 MPEG-2 Encoding

As a natural outgrowth of JPEG's popularity for still image compression, Motion JPEG (M-JPEG) gained acceptance as a "barebones" method of video compression. In M-JPEG, each frame of video is encoded as a JPEG image, and frames are transmitted or stored sequentially.

MPEG-1 soon emerged to satisfy the requirement to transmit motion video over a T1 data line, as well as to match the access times of CD-ROMs. MPEG-2 soon followed, and it further improved attainable compression ratios and provided more flexibility on image quality and bandwidth tradeoffs. MPEG-1 and MPEG-2 differ from M-JPEG in that they allow temporal, in addition to spatial, compression. Temporal compression adds another dimension to the encode challenge, a realm that is addressed using motion estimation techniques.

Both MPEG-1 and MPEG-2 have asymmetric encode and decode operations. That is, the encoding process is much more computationally intensive than the decoding process. This is why we'll choose a dual-core processor as the basis for describing the framework for an MPEG-2 encoder implementation.

One point to note about MPEG algorithms—the encoder is not specifically defined by the MPEG organization, but rather the encoded bit stream is specified. This means that the encoder can be implemented in a variety of ways, as long as it creates a compliant bit stream. The decoder, however, must be able to handle any MPEG-compliant bit stream.

Figure 10-12 provides a block diagram for a sample MPEG-2 encoder. Some of the steps it shows are very similar to the ones we described earlier in this chapter for JPEG. The biggest difference is in how motion within frames is handled, so let's start our discussion there. Once you understand these basics, we'll proceed to examine some possible algorithm partitions on a dual-core processor.

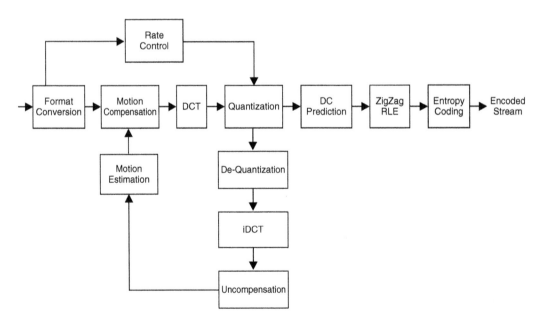

Figure 10-12: Block Diagram for Sample MPEG-2 Encoder.

10.4.1 Motion Estimation/Compensation

M-JPEG encodes only the information contained in one image frame at a time. As a result, the algorithm does not take advantage of any correlation between frames. In reality, there can be a high degree of correlation between frames, and if we're able to take this fact into account, we can greatly reduce the rate of the encoded bit stream.

You can imagine that if you look at consecutive frames of a video stream and only encode (and thus transmit) the "differences," you can conserve a lot of bandwidth over an approach like M-JPEG, which has no frame-to-frame "memory." A simple example illustrates this point. Consider a video sequence with a constant background but a small moving object in the foreground. Substantially, the frames are identical except with regard to the position of this foreground object. Therefore, if the background is encoded only once, and subsequent frames only encode the new position of this object, the output bit rate needed to faithfully reproduce the original sequence is markedly reduced.

However, now consider the case where the inter-frame data is largely identical, but the whole image might shift slightly between frames. Since this amounts to a slowly moving background image with no foreground objects, it seems that we'd have to revert to M-JPEG-style encoding, because for all intents and purposes, one frame is different from the next on a pixel-for-pixel basis.

But, if we could instead calculate the movement of a macroblock between frames, we can transmit information that tells the decoder to use the macroblock from the previous frame, along with a motion vector, to derive the proper macroblock location for the current frame. The motion vector informs the decoder where the macroblock was located in the previous frame. To obtain the vector, we employ an algorithm called "motion estimation." When the decoder uses this vector to account for motion, it employs "motion compensation."

The challenge is detecting motion between frames on a macroblock basis. In an ideal world, the macroblock would be identical between frames, but shifted in the image. In reality, things like shading and lighting may alter the macroblock contents, in which case the search must be adjusted from an exact match to something that is "close enough." This is where a "Subtract/Absolute/Accumulate (SAA)" instruction can come in handy. Using SAA, the encoder can compare macroblocks against their positions in a reference frame, created using the feedback loop shown in Figure 10-12.

You might be saying, "Well, that sounds fine, but how do you go about searching for the match?" The more you search the reference frame, the more cycles are required to perform the encoding function. The challenge is determining where to search within the next frame. This is a system consideration and really depends on how many cycles you have available to allocate to this function. If an encoder is used in a real-time application such as a videoconferencing system or a video phone, the processor performing the encode function must be sized assuming the maximum amount of processing each frame. Some common methods include searching only a subset of the frame, and searching through various levels of pixel-reduced image frames.

10.4.2 Frame Types

MPEG-encoded frames are classified into I, P, and B frames.

I frames represent *intra-coded* frames. Each of these macroblocks are encoded using only information contained within that same frame. The I frame is similar in this respect to a JPEG frame. I frames are required to provide the decoder with a place to start for prediction, and they also provide a baseline for error recovery.

There are two types of *inter-coded* frames, P frames and B frames. Inter-coded frames use information outside the current frame in the encode process.

P frames represent *predicted* frames. These macroblocks may be coded with forward prediction from references made from previous I and P frames, or they may be intra-coded.

B frames represent *bidirectional* predicted frames. The macroblocks within these frames may be coded with forward prediction based on data from previous I or P references. Or, they may be coded with backward prediction from the most recent I or P reference frame.

Note that in P and B pictures, macroblocks may be skipped and not sent at all. The decoder then uses the anchor reference pictures (I frames) for prediction with no error. B pictures are never used as prediction references.

There are many techniques that can be used to space out I, B, and P frames. Our purpose here is only to provide some background on the general concepts. As an example, one technique involves using something similar to a histogram to help determine when a new

I frame needs to be generated. If the difference in histogram results between frames exceeds a certain threshold, it may make sense to generate another I frame.

10.4.3 Format Conversion

All profiles of MPEG-2 support 4:2:0 coding. In this system, data comes from an ITU-R BT.656 source in 4:2:2 interleaved format. The first step in the encoding process involves converting the data to a 4:2:0 format.

The first frame in every group of pictures (GOP) in the sequence is an intra frame (I frame), which is independently coded. As such, the macroblocks from these frames go through the stages of a discrete cosine transform (DCT), quantization, zigzag scan, and run-length encoding (RLE). These algorithms all follow the descriptions we provided in our JPEG discussion earlier in this chapter.

The DCT converts the input so it can be processed in the frequency domain.

The encoder quantizes the DCT coefficients to reduce the number of bits required to represent them. Higher-frequency coefficients have larger quantized step sizes than lower frequency coefficients, for the reasons we described in the JPEG section.

A reference frame is created by "de-quantizing" the output of the quantization step.

Quantization scales down the input, and the zigzag scan and RLE both take advantage of spatial correlation in order to compress the input stream into the final bit stream.

During the encoding process, frames are reconstructed for reference. These reference frames are used during the predictive coding of future frames. To reconstruct a reference frame, the input results of the quantization stage are de-quantized. Next, an inverse DCT is performed. If a predictive frame is currently being encoded, the final stage of reconstructing the reference is uncompensation.

The encoding of predictive frames calls for the extra stages of motion estimation and motion compensation, which compute temporal correlation and produce a motion vector. Information is gathered at different stages so that the rate control task can update the quantization stage, which will in turn stabilize the output bit rate.

The table and pie chart of Figure 10-13 provide a representative cycle distribution of the algorithm on a Blackfin processor.

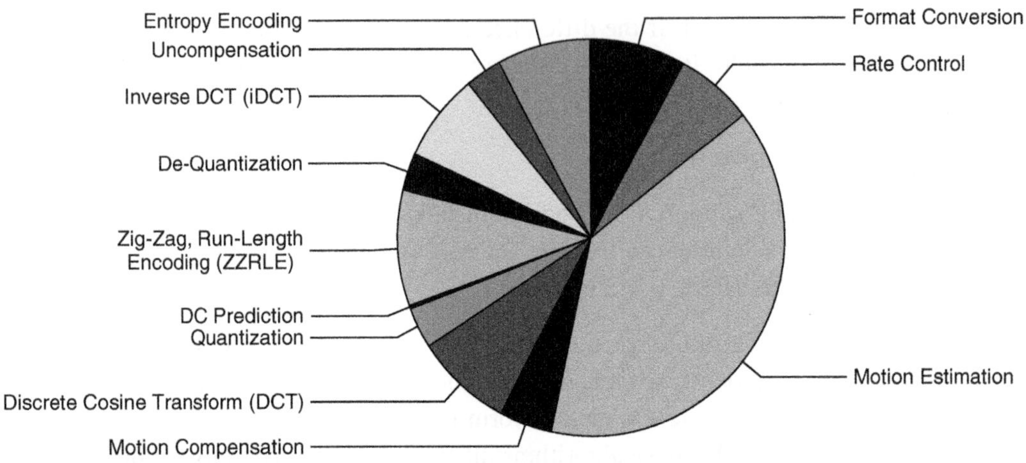

Task Name	Percentage of total algorithm	Parallelism possible?
Format Conversion	8%	Yes
Rate Control	7%	No
Motion Estimation	39%	Dependent on Algorithm
Motion Compensation	4%	Yes
Discrete Cosine Transform (DCT)	8%	Yes
Quantization	3%	Yes
DC Prediction	1%	No
Zig-Zag, Run-Length Encoding (ZZRLE)	9%	Yes
De-Quantization	3%	Yes
Inverse DCT (iDCT)	7%	Yes
Uncompensation	3%	Yes
Entropy Encoding	8%	No

Figure 10-13: MPEG-2 Encoder Task Workload (Cycles per Pixel).

10.4.4 MPEG-2 Encoder Frameworks

Now that you have a basic understanding of the MPEG-2 Encoder implementation on a single-core processor, let's add a twist. How can this algorithm be partitioned across a dual-core symmetric multiprocessor like the ADSP-BF561?

To achieve maximum processing speed for the encoder, we want to utilize both cores efficiently by partitioning code and data across each core. There are many ways to split the algorithm in a parallel pattern, taking into account task dependency and data flow. An efficient parallel system will have multiple execution units with balanced workloads. We can partition the tasks so that both cores are in an execution phase most of the time.

The first model we will demonstrate is the master-slave model. It treats both cores as thread execution units. The program flow is almost the same as for a single-core processor implementation. One of the cores is designated the "master," and it handles the host processing and signal processing. The program running on the master is similar to the code base for a single-core program, except that it parallelizes tasks wherever possible and assigns a portion of the parallel workload to the "slave." After it completes a given task, it waits for the slave to finish the same task, and then it resynchronizes the system.

For example, the DCT stage in the MPEG-2 encoder needs to perform the discrete cosine transform on N 8×8 input blocks. These DCTs can be executed independently. The master can create a thread request to the slave, assigning it $N/2$ of the blocks, so that each core ends up only doing $N/2$ DCTs. After both of them finish executing, they will resynchronize with each other, and the master will continue processing the next task.

For those tasks that have to execute sequentially, the master core has two options. It either continues executing instructions itself while the slave is left idle, or it assigns the task(s) to the slave core, during which time the master can handle I/O or host data transfer. The master-slave model achieves speedup from those tasks that can be executed in parallel. In the MPEG-2 encoder, 75% of the tasks can be executed in parallel. Figure 10-14 indicates this graphically. The tasks in the left column execute on the master, while the tasks in the right column execute on the slave. The speedup of this model is approximately 1.75.

While the master-slave model is easy to implement and is scalable, it does not fully utilize processor resources. When the task is not able to run in parallel, one core has to idle and wait for the other core to send another request. When the slave processor is not finished with a task, the master core may have to wait to resynchronize.

Let's now consider how things change if we use the pipelined model of Figure 10-15 instead of the master-slave configuration. Here, the task flow is separated vertically by the two cores. The first portion of the algorithm runs on the first processor core, and the

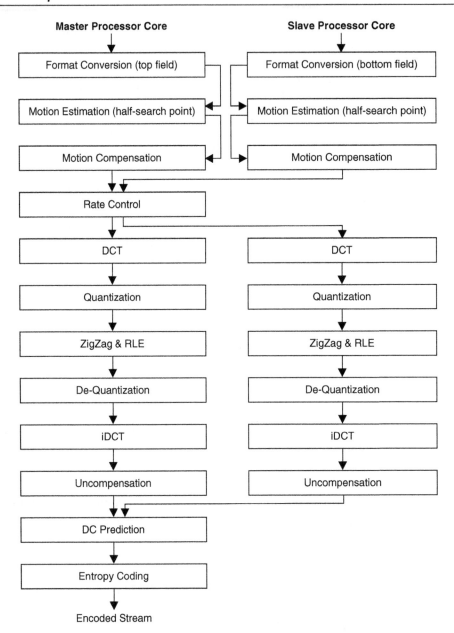

Figure 10-14: Master-Slave MPEG-2 Encoder Implementation on Dual-Core Processor.

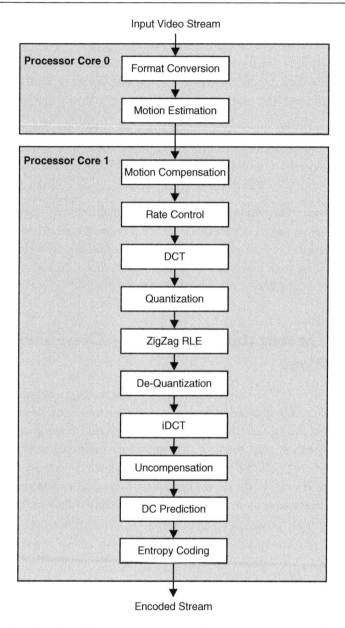

Figure 10-15: Pipelined MPEG-2 Encoder Implementation on Dual-Core Processor.

remainder executes on the second core. Because the two processor cores are "peers" to each other, there is no concept of master-slave in this model. The division of the tasks needs to be balanced between the two cores such that both have almost the same workload. Due to the data dependency between the two cores, typically one core will process the data that feeds into the other core—hence the term "pipelined."

In the task flow diagram, we see that the first processor will perform format conversion and motion estimation. This is about 46% of the overall workload. After motion estimation, the first processor will save the motion vectors in a shared memory region. The second processor will then use these motion vectors to complete the remaining 54% of the algorithm.

In this pipelined model, the two processor cores are much more loosely synchronized than in the master-slave model. Consequently, the pipelined model has lower communication overhead, which makes it much more efficient. However, when more than two processor cores are involved, the master-slave model has an advantage, because the pipelined model is not very scaleable, due to the nature of the code partitioning.

10.5   Code Optimization Study Using Open-Source Algorithms

Although "open source" C/C++ code is becoming an increasingly popular alternative to royalty-based algorithms in embedded processing applications, it carries with it some challenges. Foremost among these is how to optimize the code to work well on the chosen processor. This is a crucial issue, because a compiler for a given processor family will cater to that processor's strengths, at the possible expense of inefficiencies in other areas. This leads to uneven performance of the same algorithm when run out-of-the-box on different platforms. This section will explore the porting of opens source algorithms to the Blackfin processor, outlining in the process a "plan of attack" leading to code optimization.

10.5.1   What Is Open Source?

The generally understood definition of "open source" refers to any project with source code that is made available to other programmers. Open source software typically is developed collaboratively among a community of software programmers and distributed freely. The Linux operating system, for example, was developed this way. If all goes well, the resulting effort provides a continuously evolving, robust application. The application is well-tested because so many different applications take advantage of the code. Programmers do not have to pay for the code or develop it themselves, and they can therefore accelerate their project schedule.

The certification stamp of "Open Source" is owned by the Open Source Initiative (OSI). Code that is developed to be freely shared and evolved can use the Open Source trademark if the distribution terms conform to the OSI's Open Source Definition. This requires that the software be redistributed to others under certain guidelines. For example, under the General Purpose License (GPL), source code must be made available so that other developers will be able to improve or evolve it.

What Is Ogg?

There is a whole community of developers who devote their time to the cause of creating open standards and applications for digital media. One such group is the Xiph.org Foundation, a nonprofit corporation whose purpose is to support and develop free, open protocols and software to serve the public, developer, and business markets. This umbrella organization (see Figure 10-16) oversees the administration of such technologies as video (Theora), music (the lossy Vorbis and lossless Flac), and speech (Speex) codecs.

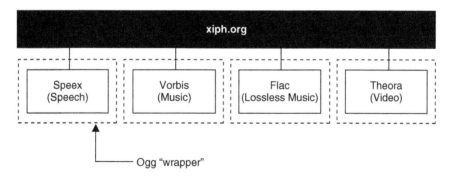

Figure 10-16: Xiph.org Open-Source "Umbrella."

The term *Ogg* denotes the container format that holds multimedia data. It generally serves as a prefix to the specific codec that generates the data. One audio codec we'll discuss here is called Vorbis and uses Ogg to store its bit streams as files, so it is usually called "Ogg Vorbis." In fact, some portable media players are advertised as supporting OGG files, where the "Vorbis" part is implicit. Speex, a speech codec discussed below, also uses the Ogg format to store its bit streams as files on a computer. However, VoIP networks and other real-time communications systems do not require file storage capability, and a network layer like the Real-time Transfer Protocol (RTP) is used to encapsulate these streams. As a result, even Vorbis can lose its Ogg shell when it is transported across a network via a multicast distribution server.

What Is Vorbis?

Vorbis is a fully open, patent-free, and royalty-free audio compression format. In many respects, it is very similar in function to the ubiquitous MPEG-1/2 layer 3 (MP3) format and the newer MPEG-4 (AAC) formats. This codec was designed for mid- to high-quality (8 kHz to 48 kHz, >16 bit, polyphonic) audio at variable bit rates from 16 to 128 kbps/channel, so it is an ideal format for music.

The original Vorbis implementation was developed using floating-point arithmetic, mainly because of programming ease that led to faster release. Since most battery-powered embedded systems (like portable MP3 players) utilize less expensive and more battery-efficient fixed-point processors, the open-source community of developers created a fixed-point implementation of the Vorbis decoder. Dubbed *Tremor*, the source code to this fixed-point Vorbis decoder was released under a license that allows it to be incorporated into open-source and commercial systems.

Before choosing a specific fixed-point architecture for porting the Vorbis decoder, it is important to analyze the types of processing involved in recovering audio from a compressed bit stream. A generalized processor flow for the Vorbis decode process (and other similar algorithms) is shown in Figure 10-17. Like many other decode algorithms, there are two main stages: front-end and back-end.

During the front-end stage, the main activities are header and packet unpacking, table lookups, Huffman decoding, etc. These types of operations involve a lot of conditional code and tend to take up a relatively large amount of program space. Therefore, embedded developers commonly use microcontrollers (MCUs) for this stage.

Back-end processing is defined by filtering functions, inverse transforms, and general vector operations. In contrast to the front-end phase, the back-end stage involves more loop constructs and memory accesses, and it is quite often smaller in code size. For these reasons, back-end processing in embedded systems has historically been dominated by full-fledged DSPs.

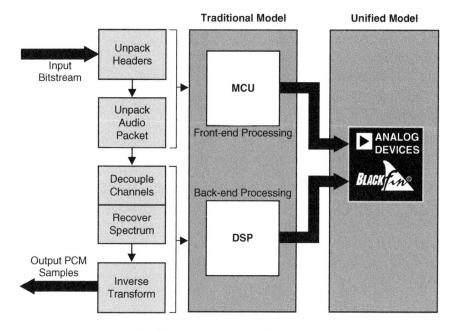

Figure 10-17: Generalized Processor Flow for the Vorbis Decode Process.

The Blackfin processor architecture unifies MCU and DSP functionality, so there is no longer a need for two separate devices. The architecture allows for an efficient implementation of both front-end and back-end processing on a single chip.

What Is Speex?

Speex is an open-source, patent-free audio compression format designed for speech. While Vorbis is mostly aimed at compressing music and audio in general, Speex targets speech only. For that reason, Speex can achieve much better results than Vorbis on speech at the same quality level.

Just as Vorbis competes with already existing royalty-based algorithms like MP3 and AAC, Speex shares space in the speech codec market with GSM-EFR and the G.72x algorithms, such as G.729 and G.722. Speex also has many features that are not present in most other codecs. These include variable bit rate (VBR), integration of multiple sampling rates in the same bit stream (8 kHz, 16 kHz, and 32 kHz), and stereo encoding support. Also, the original design goal for Speex was to facilitate incorporation into Internet applications, so it is a very capable component of VoIP phone systems.

Besides its unique technical features, the biggest advantages of using Speex are its (lack of) cost and the fact that it can be distributed and modified to conform to a specific application. The source code is distributed under a license similar to that of Vorbis. Because the maintainers of the project realized the importance of embedding Speex into small fixed-point processors, a fixed-point implementation has been incorporated into the main code branch.

10.5.2   Optimizing Vorbis and Speex on Blackfin

"Out-of-the-box" code performance is paramount when an existing application, such as Vorbis or Speex, is ported to a new processor. However, there are many techniques available for optimizing overall performance, some requiring only minimal extra effort. Software engineers can reap big payback by familiarizing themselves with these procedures.

The first step in porting any piece of software to an embedded processor is to customize the low-level I/O routines. As an example, the reference code for both Vorbis and Speex assumes the data originates from a file and the processed output is stored into a file. This is mainly because both implementations were first developed to run and be easily tested on a Unix/Linux system where file I/O routines are available in the operating system. On an embedded media system, however, the input and/or output are often data converters that translate between the digital and real-world analog domains. Figure 10-18 shows a conceptual overview of a possible Vorbis-based media player implementation. The input bit stream is transferred from a flash memory, and the decoder output is driven to an audio DAC. While some media applications like the portable music player still use files to store data, many systems replace storage with a network connection.

When it comes to actually optimizing a system like the Vorbis decoder to run efficiently, it is a good idea to have an organized plan of attack. One possibility is to focus first on optimizing the algorithm from within C, then move on to streamlining system data flows, and finally tweak individual pieces of the code at an assembly level. To show the efficacy of this method, Figure 10-19 illustrates a representative drop in processor load through successive optimization steps.

Compiler Optimization

Probably the most useful tool for code optimization is a good profiler. Using the Statistical Profiler in Visual DSP++ for Blackfin allows a programmer to quickly focus

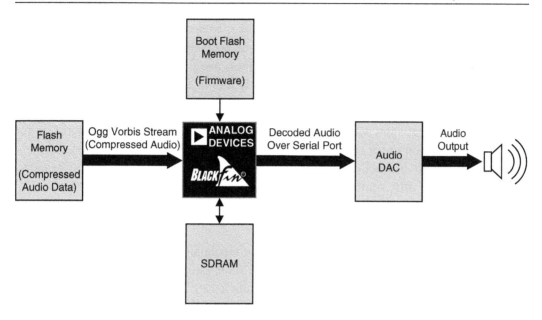

Figure 10-18: Example Vorbis Media Player Implementation.

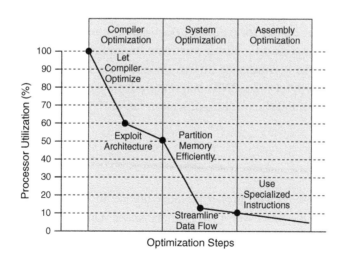

Figure 10-19: Steps in Optimizing Vorbis Source Code on Blackfin, Leading to Significantly Reduced Processor Utilization.

on hotspots that become apparent as the processor is executing code. A generally accepted rule of thumb states that 20% of the code takes 80% of the processing time. Focusing on these critical sections yields the highest marginal returns. It turns out loops are prime candidates for optimization in media algorithms like Vorbis. This makes sense, given that DSP-centric number-crunching usually occurs inside loops.

There are also global approaches to code optimization. First, a compiler can optimize for memory conservation or for speed. Functions can also be considered for automatic inlining of assembly instructions into the C code. This, again, creates a tradeoff between space and speed. Lastly, some compilers can use a two-phase process to derive relationships between various source files within a single project to further speed up code execution (inter-procedural analysis).

As mentioned above, most reference software for media algorithms uses floating-point arithmetic. Those that are written with fractional fixed-point machines in mind are actually still missing a critical component. The language of choice for the majority of codec algorithms is C, but the C language doesn't natively support the use of fractional fixed-point data. For this reason, many fractional fixed-point algorithms are emulated with integer math. While this makes the code highly portable, it doesn't approach the performance attainable by rewriting some math functions with machine-specific compiler constructs for highest computational efficiency.

A specific example of this is shown in Figure 10-20. The left column shows the C code and Blackfin compiler output for emulated fractional arithmetic that works on all integer machines. One call to perform a 32-bit fractional multiplication takes 49 cycles. The right column shows the performance improvement by utilizing a Blackfin compiler intrinsic function (mult_fr1x32x32NS) that takes advantage of the underlying fractional hardware. With this straightforward modification, an 88% speedup is achieved.

System Optimization

System optimization starts with proper memory layout. In the best case, all code and data would fit inside the processor's L1 memory space. Unfortunately, this is not always possible, especially when large C-based applications are implemented within a networked application.

The real dilemma is that processors are optimized to move data independently of the core via DMA, but MCU programmers typically run using a cache model instead. While

Original	Improved

union magic {
 struct {
  int lo;
  int hi;
 } halves;
 long long whole;
};
 	
int MULT31_original(int x, int y) {
  union magic magic;
  magic.whole = (long long)x * y;
  magic.whole = magic.whole << 1;
  return magic.halves.hi;
 	
R1 = R0 >>> 31;
R3 = R2 >>> 31;
[SP + 0xc] = R3;
CALL ___mulli3 ; // 43 cycles
R1 <<= 0x1;
R0 >>= 0x1f;
R0 = R1 | R0;
 	
49 cycles
	int MULT31_improved(int x, int y) {
	 return mult_fr1x32x32NS(x,y);
	}
	
	A1 = R0.L * R1.L (FU);
	A1 = A1 >> 16;
	A1 += R0.H * R1.L (M),A0 = R0.H * R1.H;
	A1 += R1.H * R0.L (M);
	A1 = A1 >>> 15;
	R0 = (A0 += A1);
	
	6 cycles (12% of original)

Figure 10-20: Compiler Intrinsic Functions Are an Important Optimization Tool.

core fetches are an inescapable reality, using DMA or cache for large transfers is mandatory to preserve performance.

One important memory attribute is the ability to arbitrate requests without core intervention. Because internal memory is typically constructed in sub-banks, simultaneous access by the DMA controller and the core can be accomplished in a single cycle by placing data in separate sub-banks. For example, the core can be operating on data in one sub-bank while the DMA is filling a new buffer in a second sub-bank. Under certain conditions, simultaneous access to the same sub-bank is also possible.

When access is made to external memory, there is usually only one physical bus available. As a result, the arbitration function becomes more critical. When you consider that, on any given cycle, external memory may be accessed to fill an instruction cache-line at the same time it serves as a source and destination for incoming and outgoing data, the challenge becomes clear.

Instruction Execution

As you know by now, SDRAM is slower than L1 SRAM, but it's necessary for storing large programs and data buffers. However, there are several ways for programmers to take advantage of the fast L1 memory. If the target application fits directly into L1 memory, no special action is required, other than for the programmer to map the application code directly to this memory space. In the Vorbis example described above, this is the case.

If the application code is too large for internal memory, as is the case when adding, say, a networking component to a Vorbis codec, a caching mechanism can be used to allow single-cycle access to larger, less expensive external memories. The key advantage of this process is that the programmer does not have to manage the movement of code into and out of the cache.

Using cache is best when the code being executed is somewhat linear in nature. The instruction cache really performs two roles. First, it helps pre-fetch instructions from external memory in a more efficient manner. Also, since caches usually operate with some type of "least recently used" algorithm, instructions that run the most often tend to be retained in cache. Therefore, if the code has already been fetched once, and if it hasn't yet been replaced, the code will be ready for execution the next time through the loop.

Data Management

Now that we have discussed how code is best managed to improve performance, let's review the options for data movement. As an alternative to cache, data can be moved in and out of L1 memory using DMA. While the core is operating on one section of memory, the DMA is bringing in the next data buffer to be processed.

Wherever possible, DMA should always be employed for moving data. As an example, our Vorbis implementation uses DMA to transfer audio buffers to the audio converter.

For this audio application, a double-buffer scheme is used to accommodate the DMA engine. As one half of the circular double buffer is emptied by the serial port DMA, the other half is filled with decoded audio data. To throttle the rate at which the compressed data is decoded, the DMA interrupt service routine (ISR) modifies a semaphore that the decoder can read in order to make sure that it is safe to write to a specific half of the double buffer. On a system with no operating system (OS), polling a semaphore equates to wasted CPU cycles; however, under an OS, the scheduler can switch to another task (like a user interface) to keep the processor busy with real work.

Using DMA can, however, lead to incorrect results if data coherency is not considered. For this reason, the audio buffer bound for the audio DAC should be placed in a noncacheable memory space, since the cache might otherwise hold a newer version of the data than the buffer the DMA would be transferring.

Assembly Optimization

The final phase of optimization involves rewriting isolated segments of the open-source C code in assembly language. The best candidates for an assembly rewrite are normally interrupt service routines and reusable signal processing modules.

The motivation for writing interrupt handlers in assembly is that an inefficient ISR will slow the responses of other interrupt handlers. As an example, some schemes must format in the audio ISR any AC97 data bound for the audio DAC. Because this happens on a periodic basis, a long audio ISR can slow down responses of other events. Rewriting this interrupt handler in assembly is the best way to decrease its cycle count.

A good example of a reusable signal processing module is the modified discrete cosine transform (MDCT) used in the back-end Vorbis processing to transform a time-domain signal into a frequency domain representation. The compiler cannot produce code as tightly as a skilled assembly programmer can, so the C version of the MDCT won't be as efficient. An assembly version of the same function can exploit the hardware features of the processor's architecture, such as single-cycle butterfly add/subtract and hardware bit-reversal.

CHAPTER 11

Analog I/Os

Jean LaBrosse

At age 12 or 13 I remember asking my dad, a mechanical engineer, about designing houses. "Does an architect compute floor loads for every beam in every home he creates?" For it seemed to me inconceivable that one would do so much work, especially in that era long before computer aided designing did so much computation for us. He laughed, and introduced me to the world of engineering handbooks.

Architects and civil engineers do do a lot of computation. Finite element analysis, once the province of the most exotic applications, is today available on any desktop PC. But they live and breathe by their handbooks. How wide can a floor span be in an average home, given certain joist sizes? Look it up.

Jean LaBrosse's Embedded Systems Building Blocks *is that sort of handbook for firmware engineers. It's a compendium of canned solutions, complete handlers for a wide variety of I/O devices. This chapter is from that book's section about analog inputs and outputs. Unfortunately, the world is not digital so we're forced to convert raw analog signals to and from the binary representations our computers can handle. That's not always an easy job. Here, Jean not only shows how to do this, but he gives canned code that does all the heavy lifting.*

Jean describes the scaling math that's behind any A/D conversion clearly and in sufficient detail so that you can incorporate any A/D in your design. He shows the problems with, and solutions to, the offset and gain problems that plague every analog front end.

Then there's the code.

I've long admired Jean's coding style, which first came to my attention when reading his uC/OS operating system. It's clear. It's simple. Everything is formatted the same way; he codes to a standard, and that standardization is reflected in the clarity of expression that I wish we would all emulate.

Then there's the documentation.

For each function Jean documents, in prose that's not part of the listing, the input and output arguments, in every case gives the legal range of each variable. Every function has a notes/warnings paragraph that, when appropriate, lists assumptions the user needs to success-fully use the code in any application. Every program has many underlying assumptions; reuse has largely failed because these are never documented. Here they are.

This chapter is akin to the handbooks routinely employed by other engineering disciplines. It's the "software IC" that the industry has long dreamed of. The code works, and has been tested in hundreds of applications. Read the listings and learn how beautiful code is structured. Use the code and get a better product to market faster.

—Jack Ganssle

Natural parameters such as temperature, pressure, displacement, altitude, humidity, flow, etc., are *analog*. In other words, the value taken by these parameters can change continuously instead of in discrete steps. To be manipulated by a computer, these analog parameters must be converted to digital. This is called *analog-to-digital conversion*.

Certain analog parameters can also be controlled. For example, the speed of an automobile is adjusted by changing the position of the *throttle*. The exact position of the throttle depends on many factors, such as wind resistance, whether you are going uphill or downhill, etc. You can control the flow of liquids or gases by adjusting the opening of a valve. (Flow, in this case, is not necessarily proportional to the opening of the valve, but this is a different issue.) The position of the heads in some hard disk drives is controlled by *voice coil* type *actuators*. An actuator is a device that converts electrical or pneumatic signals into linear motion. To be controlled by a computer, analog parameters must be converted from their digital form to analog. This is called *digital-to-analog conversion*.

This chapter discusses software issues relating to analog-to-digital conversions and digital-to-analog conversions. I will also describe how I implemented an analog I/O module. The analog I/O module offers the following features:

- Reads and scales from 1 to 250 analog inputs.

- Updates and scales from 1 to 250 analog outputs.

- Each analog I/O channel can define its own scaling function.

- Your application obtains *Engineering Units* from analog input channels instead of ADC counts.

- Your application provides *Engineering Units* to analog output channels instead of DAC counts.

This chapter assumes you understand the concept of fixed-point math.

11.1 Analog Inputs

A typical analog-to-digital system generally consists of the following circuit elements:

- transducer

- amplifier

- filter

- multiplexer

- analog-to-digital converter (ADC)

The interconnection of these components is shown in Figure 11-1. The inputs to the system are the physical parameters to measure (pressure, temperature, flow, position, etc.).

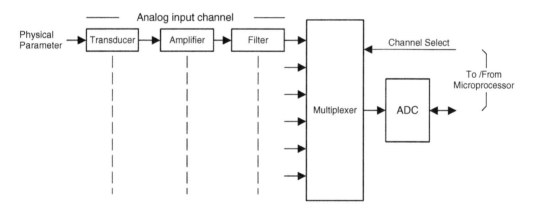

Figure 11-1: Analog-to-Digital Conversion.

The physical parameter is first converted into an electrical signal by a *transducer*. Transducers are available to convert temperature, pressure, humidity, position, etc., to electrical signals. An *amplifier* is generally used to increase the amplitude of the transducer output to a more usable level for further processing (typically between 1 and 10 volts); the output of a transducer may produce a signal in the microvolt to millivolt range. The amplifier is frequently followed by a *low pass filter*, which is used to reduce unwanted high-frequency electrical noise. The process described previously is usually called *input conditioning* and each conditioned input is also referred to as an *analog input channel*. Analog input channels are *multiplexed* into an *analog-to-digital converter* (ADC) because ADCs are often expensive devices. The ADC converts each analog input signal to digital form. The microprocessor is responsible for selecting which analog input it wants to convert and also for initiating the conversion process for the selected channel. The block diagram of Figure 11-1 can be augmented by adding a *sample-and-hold* stage between the multiplexer and the ADC which would be used to ensure that the level of the signal is constant while a conversion is taking place.

The process of converting analog signals to digital is a complex topic and is covered in great details in many books (see the Bibliography). In this chapter, I will concentrate mostly on some of the software aspects. Analog-to-digital conversion basically consists of transforming a continuous analog signal into a set of digital codes. This is called *quantizing*. Figure 11-2 shows how a 0-to-10 volt signal is quantized into a 3-bit code.

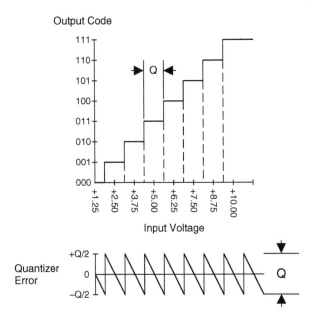

Figure 11-2: Quantizing an Analog Signal.

There are several important points to note about Figure 11-2. First, the *resolution* of the quantizer is defined by the number of bits it uses. An 8-bit quantizer will divide the input level into 256 steps. A 12-bit quantizer will divide the input level into 4,096 steps. Thus, a 12-bit quantizer has a higher resolution than an 8-bit quantizer. The number of steps for the quantizer is 2^n where n corresponds to the number of bits used. Quantizers (or ADCs) are commercially available from 4 to 24 bits. The required resolution is dictated by the application. There are literally hundreds of ADCs to choose from, and generally cost increases with resolution.

An important point to make is that the maximum value of the digital code of an ADC, namely all 1s (ones), does not correspond with the analog Full Scale (FS) but rather, one Least Significant Bit (LSB) less than full scale or:

$$Maximum\_value\_of\_digital\_code = FS \times (1 - 2^{-n})$$ **Equation 11-1**

For example, a 12-bit ADC with a 0 to +10 V analog range has a maximum digital code of 0x0FFF (4095) and a maximum analog value of $+10\,V \times (1 - 2^{-12})$ or +9.99756 V. In other words, the maximum analog value of the converter never quite reaches the point defined as full scale. At any part of the input range of the ADC, there is a small range of analog values within which the same code is produced. This small range in values is known as the *quantization size*, or *quantum*, Q.

The quantum in Figure 11-2 is 1.25 V and is found by dividing the full scale analog range by the number of steps of the quantizer. Q is thus given by the following equation:

$$Q = \frac{FSV}{2^n}$$ **Equation 11-2**

Q is the smallest analog difference that can be distinguished by the quantizer.

FSV is the full scale voltage range.

n corresponds to the number of bits used by the quantizer (i.e., ADC).

As shown in Figure 11-2 (Quantizer Error), a saw tooth error function is obtained if the ADC input is moved through its range of analog values and the difference between output and input is taken. For example, any voltage between 1.875 V and 3.125 V will produce the binary code 010.

All ADCs require a small but significant amount of time to quantize an analog signal. The time it takes to make the conversion depends on several factors: the converter resolution, the conversion technique, and the technology used to manufacture the ADC. The *conversion speed* (how fast an analog voltage is converted to digital) required for a particular application depends on how fast the signal to be converted is changing and on the desired accuracy. The *conversion time* (inverse of conversion speed) is frequently called *aperture time*. If the analog signal to measure varies by more than the resolution of the quantizer during the conversion time, then a *sample-and-hold* circuit should be used. ADCs are available with conversion speeds ranging from about three conversions per second to well over 100 million conversions per second.

11.2 Reading an ADC

The method used to read the ADC depends on how fast the ADC converts an analog voltage to a binary code. In most cases, however, the ADC must be explicitly triggered to perform a conversion. In other words, you must issue a command to the ADC to start the conversion process. Very fast ADCs, those that can convert an analog signal in less than 1 μS, generally have dedicated hardware to handle the fast conversion rate and will typically buffer the *samples*. When the buffer is full, the analog samples are processed *offline*. This is basically how a digital storage oscilloscope works. At the other end of the spectrum, ADCs used in voltmeters are generally slow (about 200 mS) but accurate ($4\frac{1}{2}$ digits, or 0.005 percent).

The actual method used to read an ADC depends on many factors: the conversion time of the ADC, how often you need the analog value converted, how many channels you have to read, etc. The next three sections describe some possible methods of reading an ADC.

11.2.1 Reading an ADC, Method #1

The scheme shown in Figure 11-3 assumes that the ADC conversion time is relatively slow (greater than about 5 mS). Here a driver (a function) reads an analog input channel and returns the result of the conversion to your application. Your application calls the driver in Figure 11-3 and passes it the desired channel to read. The driver starts by selecting (through the multiplexer) the desired analog channel (①) to read. Before starting the conversion, you may want to wait a few microseconds to allow for the signal to propagate through the multiplexer and stabilize. If you don't wait for the multiplexer's output to stabilize, your readings may be unstable. Next, the ADC is triggered to start the conversion (②). The driver then delays to allow for the conversion to complete (③). Note that the delay time must be longer than the conversion time of the ADC. After the delay, the driver assumes that the conversion is complete and reads the ADC (④). The binary result is then returned to your application (⑤).

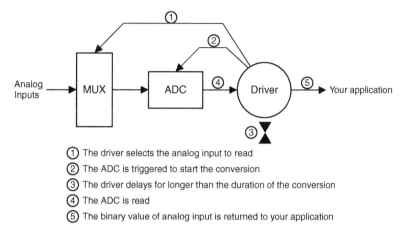

① The driver selects the analog input to read
② The ADC is triggered to start the conversion
③ The driver delays for longer than the duration of the conversion
④ The ADC is read
⑤ The binary value of analog input is returned to your application

Figure 11-3: Reading an ADC (Method #1).

This method is simple and can be used with slow-changing analog signals. For example, you can use this method when measuring the temperature of a room (which doesn't change very quickly).

11.2.2 Reading an ADC, Method #2

You can actually use a signal provided by most ADCs (i.e., the *End Of Conversion* (EOC) signal) to tell your driver when the ADC has completed its conversion. The code and your hardware in this case will be a little more complicated, but this method is more efficient.

Again, your application calls the driver by passing it the analog input channel to read. The driver shown in Figure 11-4 starts by selecting (through the multiplexer) the desired analog channel (①). At this point, you should again wait a few microseconds to allow for the signal to propagate through the multiplexer and stabilize. The ADC is then triggered to start the conversion (②). The driver then waits for a semaphore (③) with a timeout. A timeout is used to detect a hardware malfunction. In other words, you don't want the driver to wait forever if the ADC fails (i.e., never finishes the conversion). When the analog conversion completes, the ADC generates an interrupt (④). The ADC conversion-complete ISR signals the semaphore (⑤), which notifies the driver that the ADC has completed its conversion. When the driver gets to execute, it reads the ADC (⑥) and returns the binary result to your application (⑦). The pseudocode for both the driver and the ISR follows.

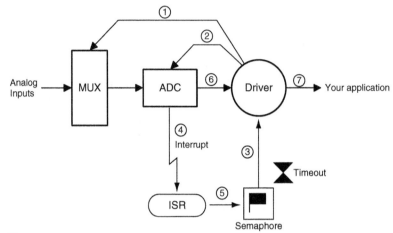

① The driver selects the desired analog input to read

② The ADC is triggered to start the conversion

③ The driver waits for the semaphore to be signalled (with timeout)

④ The end of conversion generates an interrupt

⑤ The end of conversion ISR signals the semaphore

⑥ The driver reads the ADC

⑦ The binary value of the analog input is returned to your application

Figure 11-4: Reading an ADC (Method #2).

```
ReadAnalogInputChannel (Channel#)
{
      Select the desired analog input channel;
      Wait for MUX output to stabilize;
      Start ADC conversion;
      Wait for signal from ADC ISR (with timeout);
      if (Timed out) {
         Signal error;
      } else {
         Read ADC and return result to the caller;
      }
}
Conversion complete ISR
{
      Signal conversion complete semaphore;
}
```

You would use this method if the conversion time of the ADC is greater than the execution time of the ISR and the call to wait for the semaphore. For example, your ADC takes 1 mS to perform a conversion, and the total execution time of the ISR and the call to wait for the semaphore requires only about 50 μS. If the execution time of the ISR and the call to wait for the semaphore is greater than the conversion time of the ADC, you might as well wait in a software loop (polling the ADC's EOC line) until the ADC completes its conversion. This method will be discussed next.

11.2.3 Reading an ADC, Method #3

The third method can be used if the conversion time of the ADC is less than the time needed to process the interrupt and wait for the semaphore, as described in the previous method. For example, depending the microprocessor, an ADC with a conversion time less than 25 μS cannot afford the overhead of an interrupt and a semaphore which could take over 50 μS. In other words, the execution time to handle the interrupt overhead and the time to signal and wait for the semaphore can take more than 25 μS. This is true of most 8-bit and some 16-bit microprocessors.

Your application calls the driver shown in Figure 11-5 by passing it the desired analog input channel to read. The driver starts by selecting (through the multiplexer) the channel to read (①). Again, before starting the conversion, you may want to wait a few microseconds to allow for the signal to propagate through the multiplexer and stabilize. The ADC is then triggered to start the conversion (②). The driver then waits (③) in a software loop for the ADC to complete its conversion. While waiting in the loop, the driver monitors the status (the EOC) or the *BUSY* signal of the ADC. You need to ensure that you have a way to prevent an infinite loop if your hardware becomes defective. An infinite loop is avoided by using a software counter which is decremented every time through the polling loop (see the pseudocode following). The initial value of the counter is determined from the execution time of each iteration of the polling loop. For example, if you have an ADC that should perform a conversion in 50 μS and each iteration through the polling loop takes 5 μS, you will need to load the counter with a value of at least 10. You want to use the loop counter as an indication of a hardware malfunction and not to indicate when the ADC is done converting. Based on experience, you should load the loop counter so that a timeout occurs when the polling time exceeds the ADC conversion by about 25 to 50 percent. In other words, you would load the counter with a value between 13 and 15 in my example. When the ADC finally signals an end of conversion, the driver reads the ADC (④) and returns the binary result to your application (⑤).

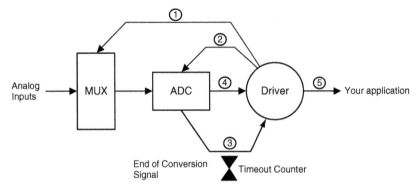

① The driver selects the desired analog input to read
② The ADC is triggered to start the conversion
③ The driver waits for the ADC to complete its conversion (with timeout)
④ The driver reads the ADC
⑤ The binary value of analog input is returned to your application

Figure 11-5: Reading an ADC (Method #3).

The pseudocode for the driver is:

```
ReadAnalogInputChannel (Channel#)
{
    Select the desired analog input channel (i.e. MUX);
    Wait for MUX output to stabilize;
    Start ADC conversion;
    Load timeout counter;
    while (ADC Busy && Counter-- > 0)  /* Polling Loop        */
        ;
    if (Counter == 0) {    /* Check for hardware malfunction */
        Signal error;
    } else {
        Read ADC and return result to the caller;
    }
}
```

Actually, I prefer this method because:

- You can get fairly inexpensive fast ADCs ($\sim$25 μS conversion time).

- You don't have the added complexity of an ISR.

- Your signal has less time to change during a conversion.

- This method imposes very little overhead on your CPU.

- The polling loop can be interrupted to service interrupts.

11.2.4   Reading an ADC, Miscellaneous

The nice thing about reading analog input channels through drivers is that the implementation details are hidden from your application. You can use any of the three drivers shown without changing your application code.

By always returning the same number of bits to your application, you can make your application insensitive to the actual number of bits of the ADC. In other words, if the ADC driver always returned a signed 16-bit number irrespective of the actual number of bits for the ADC, your application would not have to be adjusted every time you changed the word size of your ADC. This is actually quite easy to accomplish, as shown in Figure 11-6. All you need to do is to shift left the binary value of the ADC until the most significant bit of the ADC value is in bit position number 14 of the result. I use a 16-bit signed result because the computations required to scale the result of the ADC need to be signed. This will be described in the next section. If you deal with higher

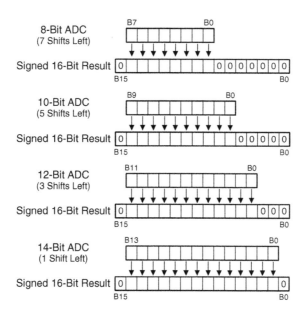

Figure 11-6: ADC Driver Always Returning a Signed 16-bit Result.

resolution ADCs, you may want to write your drivers and application code to assume signed 32-bit values.

For example, an 8-bit ADC can measure a voltage between 0 and 0.996094 (255/256) of the full scale voltage. This is the same as (255 << 7)/32768, or 0.996094. Similarly, a 12-bit ADC can measure a voltage between 0 and 4095/4096 or 0.999756, which is the same as (4095 << 3)/32768 (i.e., 0.999756). You can thus hide the details about how many bits each ADC has with respect to your application without losing any accuracy.

11.3   Temperature Measurement Example

As we have seen, an ADC produces a binary code based on a full scale voltage. If you are measuring a temperature, for example, this information means very little to you. What you really want to know is the temperature of what you are measuring. The circuit in Figure 11-7 shows a commonly used temperature sensor *Integrated Circuit* (IC), the National Semiconductor LM34A.

The LM34A produces a voltage that is directly proportional to the temperature surrounding it, specifically, 10 mV/°F. Note that you can also obtain the temperature in degrees Celsius by using an LM35A. The amplifier is designed to have a gain of 2.5, and thus –50 to 300 °F will produce a voltage of –1.25 to 7.50 volts. By using a 10-bit ADC, you can obtain a resolution of about 0.342 °F (350 °F/1024). Note that the ADC can only convert positive voltages, and thus a bias of 1.25 volts is introduced following the amplification stage to ensure that a positive voltage is present at the input of the ADC for the complete temperature range. With this bias, –50 °F will appear as 0 V, 0 °F will be 1.25 V and 300 °F will be 8.75 V. The value obtained at the ADC is given by:

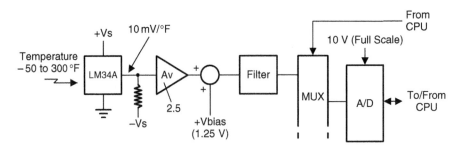

Figure 11-7: Temperature Measurement Using an LM34A.

$$ADC_{counts} = \frac{(Temperature_{(°F)} \times 0.01_{V/(°F)} \times 2.5_{A_V} + 1.25_{V_{bias}}) \times 1023_{counts}}{10_{V_{FullScale}}}$$

Equation 11-3

counts is an industry standard convention that means the *binary* value of the ADC.

$0.01_{V/(°F)}$ corresponds to the transducer transfer function—10 mV/°F—specified by National Semiconductor.

2.5 is the gain of the amplifier stage and is established by the hardware designer.

1.25 is the bias voltage to ensure that the ADC always reads a positive voltage.

1023 is the maximum binary value taken by a 10-bit converter.

$10_{V_{FullScale}}$ is the full scale voltage.

For example, a temperature of 100 °F would have a value of 383 counts (actually, 383.625). Note that the ADC can produce only integer values, and thus the actual value of 383.625 is truncated to 383. To obtain the temperature read at the sensor, you need to rearrange Equation 11-3 so that temperature is given as a function of ADC counts, as shown in Equation 11-4. This process is often called converting ADC counts to *engineering units* (E.U.):

$$Temperature_{(°F)} = \frac{\dfrac{ADC_{counts} \times 10_{V_{FullScale}}}{1023_{counts}} - V_{bias}}{0.01_{V/(°F)} \times 2.5_{A_V}}$$

Equation 11-4

The general form for this equation is:

$$Temperature_{(°F)} = \frac{\dfrac{ADC_{counts} \times FSV}{2^n - 1} - V_{bias}}{Transducer_{V/(EU)} \times A_V}$$

Equation 11-5

E.U. is the engineering unit of the transducer (°F, PSI, Feet, etc.).

V_{bias} is the bias voltage added to the output of the amplifier stage to allow the ADC to read negative values.

FSV is the full scale voltage of the ADC.

$Transducer_{V/(EU)}$ corresponds to the number of volts produced by the transducer per engineering unit.

A_V is the gain of the amplifier stage.

n is the resolution of the ADC (in number of bits).

You can also write Equation 11-5 as follows:

$$E.U. = \frac{(ADC_{counts} - Bias_{counts}) \times FSV}{Transducer_{V/(EU)} \times A_V \times (2^n - 1)}$$

Equation 11-6

In this case, $Bias_{counts}$ corresponds to the ADC counts of the bias voltage as is given by the following equation:

$$Bias_{counts} = \frac{V_{bias} \times (2^n - 1)}{FSV}$$

Equation 11-7

Note that most of the terms in Equation 11-6 are known when the system is designed, and thus, to save processing time, they should not be evaluated at run time. In other words, you could rewrite the equation as follows:

$$E.U. = (ADC_{counts} - ConvOffset_{counts}) \times ConvGain_{(EU)/(count)}$$

Equation 11-8

where:

$$ConvGain_{(EU)/(count)} = \frac{FSV}{Transducer_{V/(EU)} \times A_V \times (2^n - 1)}$$

Equation 11-9

Note that the units of the conversion gain (*ConvGain*) are E.U. per ADC count.

$$ConvOffset_{counts} = -\left(\frac{V_{bias} \times (2^n - 1)}{FSV}\right)$$

Equation 11-10

In the temperature measurement example, the conversion gain would be 0.391007 and the conversion offset would be 127.875. You can apply fixed-point arithmetic and scale factors to the temperature measurement example. The temperature of the LM34A sensor is given by:

$$Temperature_{(\circ F)} = (ADC_{counts} + ConvOffset_{counts}) \times ConvGain_{(\circ F)/(count)}$$

Equation 11-11

Remember that you have a 10-bit ADC, and thus the range of counts is from 0 to 1023. You can scale this number by multiplying the ADC counts by 32 (shifting left five places). To perform the subtraction with the bias, you need to scale the bias (i.e., conversion offset) by the same value, or $127.875 \times 32 = 4092S - 5$. The gain (0.391007) can be scaled by multiplying by 65536, and thus the conversion gain is 25625S – 16. The temperature is thus given by:

$$\text{Temperature}(°F)\,S\text{--}21 = ((\text{ADCcounts} << 5)S\text{--}5 - \mathbf{4092}S - 5) \times \mathbf{25625}S - 16$$

Equation 11-12

or

$$\text{Temperature}(°F)\,S\text{--}6 = (((\text{ADCcounts} << 5)S\text{--}5 - \mathbf{4092}S - 5) \times \mathbf{25625}S - 16) >> 15$$

Equation 11-13

From Equation 11-3, 150 °F would produce 511 ADC counts. Substituting 511 counts in Equation 11-12 produces the following:

$$\text{Temperature}(°F)\,S\text{--}21 = (\mathbf{165352}S\text{--}5 - \mathbf{4092}S\text{--}5) \times \mathbf{25625}S\text{--}16,$$

or

$$\text{Temperature}(°F)\,S\text{--}21 = \mathbf{314162500}S\text{--}21 \text{ (i.e., 149.80)}$$

or using Equation 11-13:

$$\text{Temperature}(°F)\,S\text{--}6 = \mathbf{9587}S\text{--}6 \text{ (i.e., 149.80)}$$

The C code to convert the ADC counts to temperature is:

```
INT16S RdTemp(INT16S raw)
{
      INT16S cnts;
      INT16S temp;


      cnts = (raw << 5) -  4092;
      temp = (INT16S)(((INT32S)cnts * (INT32S)25625) >> 15L);
      return (temp);        /* Result is scaled S- 6 */
}
```

Note that `raw` corresponds to the ADC counts (10 bits). The total counts (`cnts`) number is computed separately because a good compiler should perform this operation using 16-bit arithmetic instead of 32-bit (which would be faster). Counts and gain are then converted to INT32S because the multiplication needs 30-bit precision. The result is divided by 32768 so that it fits back into a 16-bit signed variable. Finally, the temperature is returned in °F scaled S–6. You could obtain the temperature to the nearest degree by first adding 32 (0.5) and then dividing the result by 64. In other words, by rounding the result.

The electronic components used to provide the amplification and the bias voltage are generally inaccurate. Oddly enough, extra components can be added to allow the amplification stage and bias voltage to be precisely adjusted (that is, *calibrated*). Adding such components, however, adds recurring cost to your system. Component inaccuracies easily can be compensated in software by modifying Equation 11-8 as:

$$EU = (ADC_{counts} + ConvOffset_{counts} + CalOffset_{counts}) \qquad \textbf{Equation 11-14}$$
$$\times ConvGain_{(EU)/(count)} \times CalGain$$

The calibration gain (*CalGain*) and calibration offset (*CalOffset*) would be entered by a calibration technician using a keyboard/display or through a communications port. Both calibration parameters could then be stored in a non-volatile memory device such as battery backed-up RAM, EEPROM, or even a floppy disk. The adjustment range of the calibration parameters is based on the accuracy of the electronic components used. A 10 percent adjustment range should be sufficient for most situations. For the calibration gain, all we need is an adjustment range between 0.90 (**14745**S–14) and 1.10 (**18022**S–14). In our example, all we need is an adjustment range between –100 (**–3200**S–5) and +100 (**3200**S–5) for the calibration offset when using a 10-bit ADC. The new C code to convert raw ADC counts to a temperature is:

```
INT16S RdTemp(INT16S raw)
{
        INT16S cnts;
        INT16S temp;


        cnts  = (raw << 5) -  4092 + CalOffset;
        temp  = (INT16S)(((INT32S)cnts * (INT32S)25625) >> 15L);
        temp  = (INT16S)(((INT32S)temp * (INT32S)CalGain) >> 14L);
        return (temp);          /* Result is scaled S- 6 */
}
```

For example, if the actual gain of the amplification stage of our temperature measurement example was 2.45 instead of 2.50, then CalGain would be set to

1.020408 (**16718**S–14). Similarly, if the bias voltage was 1.27 V instead of 1.25 V, then you would have to subtract 0.02 V, or 65 counts (see Equation 11-10). In other words, `CalOffset` would be set to –**65**S–5.

11.4 Analog Outputs

A typical digital to analog system generally consists of the following circuit elements:

- digital to analog converter (DAC)
- filter
- amplifier
- transducer

Digital-to-analog converters (DACs) are generally inexpensive devices, and thus each analog output channel can have its own DAC, as shown in Figure 11-8. The DAC

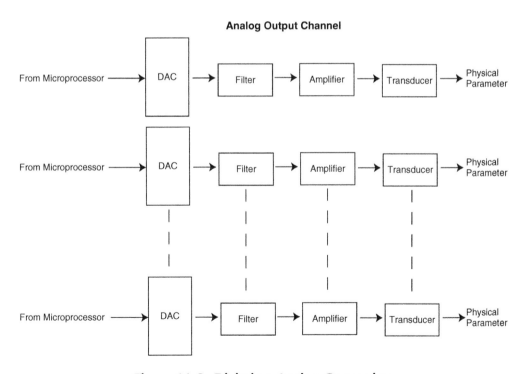

Figure 11-8: Digital-to-Analog Conversion.

converts a binary value provided by a microprocessor to either a current or a voltage (depending on the DAC). The voltage or current is *filtered* to smooth out the step changes. An *amplifier* stage is sometimes used to increase the amplitude or power drive capability of the analog output channel in order to properly interface with the transducer. The *transducer* is used to convert the electrical signal to a physical quantity. For example, transducers are available to convert electrical signals to pressures (known as *current-to-pressure transducers*, or I to P). These pressures can be—and often are—used to control other physical devices.

DACs are commercially available with resolutions from 4 to 16 bits. The resolution to choose from is application specific. There are literally hundreds of DACs to choose from. Generally, the cost of DACs increases with resolution and conversion speed. DACs are much faster than ADCs. Conversion time (also called *settling time*) is always less than a few microseconds and can be as fast as 5 nS (nanosecond). Very fast DACs are used in video applications, and because of their higher cost and lower resolution (8 bits), very fast DACs are seldom used in industrial applications.

A digital-to-analog conversion is handled exclusively in hardware. From a software standpoint, updating a DAC is as simple as writing the binary value to one or more (if more than 8 bits) I/O port locations or memory locations (when DACs are memory mapped).

11.5   Temperature Display Example

Suppose you wanted to display the temperature read by our LM34A on a meter, as shown in Figure 11-9.

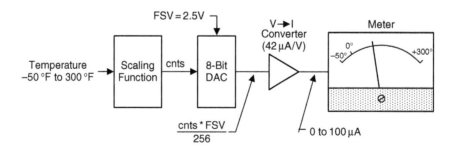

Figure 11-9: Temperature Display.

An 8-bit DAC is deemed sufficient considering the accuracy of these types of meters. The DAC is followed by a circuit that converts the voltage output of the DAC to a current (a V → *I Converter*). The Full Scale Voltage (FSV) of the DAC is set to 2.5 volts. The current converter is designed to produce about 42 μA/V, and the meter requires 100 μA for full scale. Your task is to write a function that takes the temperature (−50 °F to +300 °F) as an input and produces the proper output current (0 to 100 μA) to drive the meter.

The relationship between the temperature and the meter current is shown in Figure 11-10.

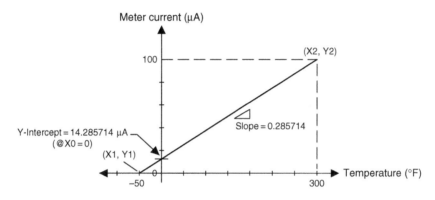

Figure 11-10: Temperature to DAC Counts Scaling.

The graph can also be represented by the following linear equation:

$$y = m \times x + b$$

Equation 11-15

where *m* is the *slope* and *b* is the *Y-intercept* (the value on the y-axis when *x* is 0). The slope gives us the current per degree of temperature and is given by:

$$m = \frac{(Y_2 - Y_1)}{(X_2 - X_1)}$$

Equation 11-16

In this case, the slope is 100 μA/350 °F , or 0.285714 μA/°F. The Y-intercept (i.e., Y_0) is given by:

$$Y_0 = m \times (X_0 - X_1) + Y_1$$

Equation 11-17

By substituting the values of m, Y_1, X_1, and X_0 (i.e., 0) in Equation 11-17, you obtain a Y-intercept of $14.285714\,\mu A$. The meter current thus is given by:

$$Meter_{\mu A} = 0.285714_{(\mu A)/(°F)} \times Temperature_{°F} + 14.285714_{\mu A} \qquad \textbf{Equation 11-18}$$

The meter current is also given by:

$$Meter_{\mu A} = \frac{DAC_{counts} \times FSV}{256} \times 42_{(\mu A)/V} \qquad \textbf{Equation 11-19}$$

Combining Equations 11-18 and 11-19, I obtain:

$$0.285714_{(\mu A)/(°F)} \times Temperature_{°F} + 14.285714_{\mu A} = \frac{DAC_{counts} \times 2.5}{256} \times 42_{(\mu A)/V}$$

$$\textbf{Equation 11-20}$$

Solving for DAC_{counts}, I obtain:

$$DAC_{counts} = INT\left(\frac{0.285714 \times 256}{2.5 \times 42_{(\mu A)/V}} \times Temperature_{°F} \times \frac{14.285714 \times 256}{2.5 \times 42_{(\mu A)/V}}\right)$$

$$\textbf{Equation 11-21}$$

Note that $INT()$ means that only the integer portion of the result is retained. As you can see, Equation 11-22 is also a linear equation, where m is 0.696598 and b is 34.829931. DAC_{counts} thus are given by:

$$DAC_{counts} = INT(0.696598_{(counts)/(°F)} \times Temperature_{°F} + 34.829931_{counts})$$

$$\textbf{Equation 11-22}$$

Substituting $-50°F$ in Equation 11-22, I obtain 0 counts (as I should). Similarly, substituting $300°F$ in Equation 11-22, I obtain 243 counts, which should produce $100\,\mu A$.

As with analog inputs, the electronic components used in circuits such as the voltage-to-current converter are generally inaccurate. You can compensate for component inaccuracies in software by modifying Equation 11-22 as:

$$DAC_{counts} = INT(0.696598_{(counts)/(°F)} \times Temperature_{°F}$$
$$\times CalGain + 34.829931_{counts} + CalOffset) \qquad \textbf{Equation 11-23}$$

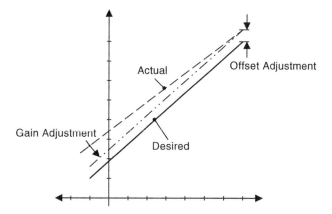

Figure 11-11: Calibration Gain and Offset Adjustments (Exaggerated).

The effect of the calibration gain and offset is shown in Figure 11-11, which has been exaggerated for sake of discussion. The actual curve that you get from an incorrect gain and offset needs to be adjusted, as shown in Figure 11-11.

The adjustment range of the calibration parameters is based on the accuracy of the electronic components. Based on experience, a 10 percent adjustment range should be sufficient for most situations. For the calibration gain, you only need an adjustment range between 0.90 and 1.10. For the calibration offset, you need an adjustment range between −25 and +25 for an 8-bit ADC. What would happen if the voltage-to-current converter was actually putting out $40\,\mu A/V$ instead of 42 (a 5 percent error)? In this case, the slope in Figure 11-11 (see Equation 11-21, substituting 40 instead of 42) would need to be adjusted to 0.731428 and the intercept would need to be 36.571428. This can be accomplished by setting `CalGain` and `CalOffset` to 1.05 and 1.741497 respectively.

The general form for Equation 11-23 is:

$$DAC_{counts} = INT(ConvGain_{(counts)/(EU)} \times CalGain \times Input_{EU}$$
$$+ ConvOffset_{counts} + CalOffset_{counts}) \qquad \textbf{Equation 11-24}$$

11.6 Analog I/O Module

In this chapter, I provide you with a complete analog I/O module that will allow you to read and scale up to 250 analog inputs and scale and update up to 250 analog output channels. Each analog input channel is scanned at a regular interval and the scan rate for each channel can be programmed individually. This allows you to determine whether some analog inputs are scanned more often than others. Similarly, each analog output channel is updated at a regular interval and the update rate for each channel can also be programmed individually. This allows you to establish which analog outputs are to be updated more often.

The source code for the analog I/O module is found in the \SOFTWARE\BLOCKS\ AIO\SOURCE directory. The source code is found in the files AIO.C (Listing 11-1) and AIO.H (Listing 11-2) at the end of the chapter. As a convention, all functions and variables related to the analog I/O module start with either AIO (functions and variables common to both analog inputs and outputs), AI (analog input functions and variables), or AO (analog output functions and variables). Similarly, #defines constants will either start with AIO_, AI_, or AO_.

11.7 Internals

The analog I/O module makes extensive use of floating-point arithmetic (additions, multiplications, and divisions). The reason I chose to use floating-point instead of integer arithmetic is that it is very difficult to make a general purpose analog I/O module using integer arithmetic. The analog I/O module can become CPU-intensive unless you have hardware-assisted floating-point (i.e., a math coprocessor). The analog I/O module, however, can be easily modified to make use of integer arithmetic if you have a dedicated application.

Figure 11-12 shows a block diagram of the analog I/O module. You should also refer to Listings 11-1 and 11-2 for the following description. As shown, the analog I/O module consists of a single task (AIOTask()) that executes at a regular interval (set by AIO_TASK_DLY). AIOTask() can manage as many analog inputs and outputs as your application requires (up to 250 each). The analog I/O module must be initialized by calling AIOInit(). AIOInit() initializes all analog input channels, all analog output channels, the hardware (ADCs and DACs), a semaphore used to ensure exclusive access to the internal data structures used by the analog I/O module, and finally, AIOInit() creates AIOTask().

AITbl[] is a table that contains configuration and run-time information for each analog input channel. An entry in AITbl[] is a structure defined in AIO.H and is called AIO. AIUpdate() is charged with reading all of the analog input channels on a regular basis. AIUpdate() calls AIRd()and passes it a logical channel number (0..AIO_MAX_AI – 1). AIRd() is responsible for selecting the proper analog input through one or more multiplexers (based on the logical channel number), starting and waiting for the proper ADC to convert (if more than one is used), and for returning raw counts to AIUpdate(). AIRd() is the only function that knows about your hardware, and thus AIRd() can easily be adapted to your environment.

AOTbl[] is a table that contains configuration and run-time information for each analog output channel. An entry in AOTbl[] also uses the AIO structure. AOUpdate() is responsible for updating all of the analog output channels on a regular basis. AOUpdate() calls AOWr() and passes it a logical channel number (0..AIO_MAX_AO − 1) and the raw DAC counts. AOWr() is responsible for outputting the raw counts to the proper DAC based on the logical channel. AOWr() is the only function that knows about your hardware, and thus AOWr() can easily be adapted to your environment.

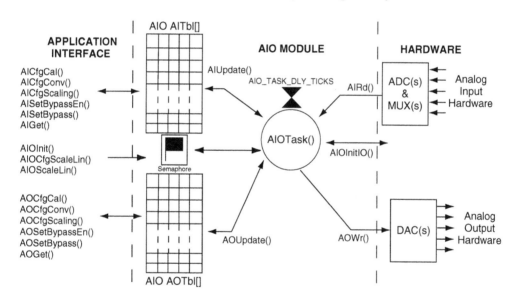

Figure 11-12: AIO Module Flow Diagram.

Figure 11-13 shows a flow diagram of a single analog input channel. Note that I used electrical symbols to represent functions performed in software. .AIO??? are all members of the AIO structure. AIUpdate() updates each channel as described in the following paragraphs.

The raw counts obtained from AIRd() are placed in the channel's .AIORaw variable. The raw counts are then added to .AIOCalOffset and .AIOConvOffset. The result of this operation is then multiplied by .AIOCalGain and .AIOConvGain.

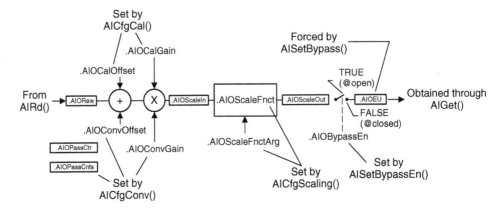

Figure 11-13: Analog Input Channel Flow Diagram.

These mathematical operations are basically used to implement Equation 11-14:

$$.\texttt{AIOScaleIN} = (.\texttt{AIORaw} + .\texttt{AIOConvOffset} + \\ .\texttt{AIOCalOffset}) \times .\texttt{AIOConvGain} \times .\texttt{AIOCalGain}$$

Equation 11-25

.AIScaleFnct is a pointer to a function that is executed when the channel is updated. The function allows you to apply further processing when reading an analog input. For example, a *Resistance Temperature Detector* (RTD) is a device that requires special processing. The temperature at the RTD is proportional to the resistance of the RTD (but is nonlinear). A scaling function can thus be written to convert .AIOScaleIn (the resistance of the RTD) to a temperature in degrees Fahrenheit (placed in.AIOScaleOut). There are many types of RTDs, and thus you need to be able to specify the actual type used. This is where .AIOScaleFnctArg comes in. .AIOScaleFnctArg is a pointer to any arguments that your scaling function requires. In the case of an RTD, this argument can specify the type of RTD used. The scaling function that you write must be declared as:

```
void AIOScale???(AIO *paio);
```

When called, your scaling function will receive a pointer to the AIO channel to scale (or linearize). The input to your function is available in paio->AIOScaleIn, and your function must place the result in paio->AIOScaleOut. Any arguments to the scaling function are found through paio->AIOScaleFnctArg. If you do not have any linearization function, the value of .AIOScaleIn is simply copied to .AIOScaleOut by AIUpdate().

`.AIOBypassEn` is a software switch that is used to prevent the analog input from being updated. This feature allows your application code to "bypass" the channel and force a value into `.AIOEU`. When another part of your application code tries to read the analog input channel, it will actually be getting the forced value instead of what the sensor is measuring. I have found this feature to be invaluable.

`.AIOEU` is the value that your application code will obtain when it needs the latest value read by the analog input channel (by calling `AIGet()`). `.AIOEU` contains engineering units. This means that if the analog input channel monitors a pressure, your application code will obtain a value in either PSI, KPa, InHgg, etc.

`.AIOPassCnts` allows your application code to specify how often the analog input channel is to be updated. In fact, `.AIOPassCnts` specifies how many analog input scans are needed before the channel is updated. In other words, if analog inputs are read every 50 mS and you specify a pass count of 20, then the analog input channel will be read every 1000 mS (i.e., 1 second).

Figure 11-14 shows a flow diagram of a single analog output channel. Note that I used electrical symbols to represent functions performed in software. As with analog input channels, `.AIO???` are all members of the `AIO` structure. `AOUpdate()` updates each channel as described in the following paragraphs.

Your application deposits the value for the analog output channel by calling `AOSet()`. This value is passed in engineering units. This means that if the analog output channel controls a meter that displays the RPM of a rotating device, you call `AOSet()` by specifying an RPM and the analog output channels take care of figuring out how much voltage or current is needed to display the RPM.

`.AIOBypassEn` is a software switch used to override the value that your application code is trying to put out on the analog output channel. Another function provided by the analog I/O module is used to load `.AIOScaleIn`. This feature is very useful for debugging purposes because it allows you to test your output independently of the application code.

`.AIScaleFnct` is a pointer to a function that is executed when the analog output channel is updated. The function allows you to apply further processing prior to updating an analog output. For example, a 0 to 100 mA output may be controlling a valve. If the flow through the valve is proportional to the output—but nonlinear, the function can make the valve action look linear with respect to your application. If your software needs to support different types of valves, you can specify which valve is being used through

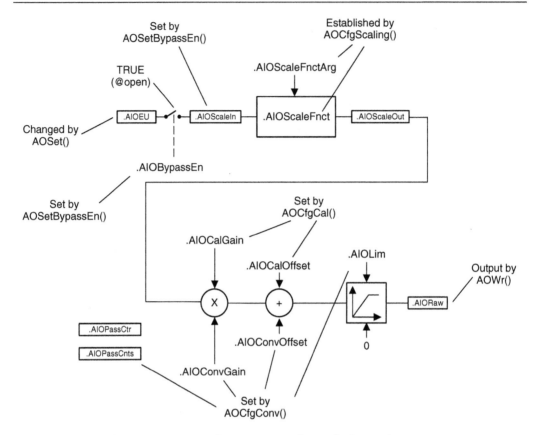

Figure 11-14: Analog Output Channel Flow Diagram.

`.AIOScaleFnctArg`. `.AIOScaleFnctArg` is a pointer to any arguments that your scaling function requires. The scaling function that you write must be declared as follows:

```
void AIOScale???(AIO *paio);
```

When called, your scaling function will receive a pointer to the `AIO` channel to scale (or linearize). The input to your function is available in `paio->AIOScaleIn`, and your function must place the result in `paio->AIOScaleOut`. Any arguments to your function are found through `paio->AIOScaleFnctArg`. If you do not have any linearization function, the value of `.AIOScaleIn` is simply copied to `.AIOScaleOut` by `AOUpdate()`.

`.AIOScaleOut` is then multiplied by `.AIOCalGain` and `.AIOConvGain`. The result of the multiplication is the added to `.AIOCalOffset` and `.AIOConvOffset`.

The result of this operation is deposited in .AIORaw so that it can be sent to the proper DAC by AOWr().

$$.AIORaw = .AIOScaleOut \times .AIOConvGain$$
$$\times .AIOCalGain + .AIOConvOffset + .AIOCalOffset$$

Equation 11-26

.AIOLim is used to ensure that .AIORaw does not exceed the maximum counts allowed by the DAC. For example, an 8-bit DAC has a range of 0 to 255 counts. An output of 256 counts to a DAC would appear to the DAC as 0 (the lower eight bits of 100000000_2). .AIOLim contains the maximum count that can be sent to the DAC (255 for an 8-bit DAC).

.AIOPassCnts allows your application code to specify how often the analog output channel is to be updated. In fact, .AIOPassCnts specifies how many analog output scans are needed before the channel is updated. In other words, if analog outputs are updated every 50 mS and you specify a pass count of 5, the analog output channel will only be updated every 250 mS.

11.8 Interface Functions

Your application software knows about the analog I/O module through the interface functions shown in Figure 11-15.

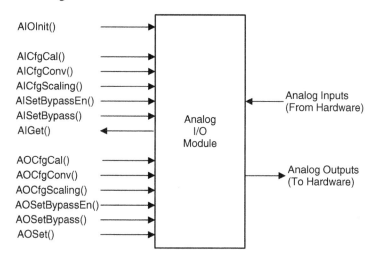

Figure 11-15: Analog I/O Module Interface Functions.

11.8.1 AICfgCal()

INT8U AICfgCal(INT8U n, FP32 gain, FP32 offset);

AICfgCal() is used to set the calibration gain and offset of an analog input channel. The analog I/O module implements Equation 11-14, and this function is used to set the value of CalGain and CalOffset.

Arguments

n is the desired analog input channel to configure. Analog input channels are numbered from 0 to AIO_MAX_AI − 1.

gain is a multiplying factor that is used to compensate for component inaccuracies and doesn't have any units. The gain would be entered by a calibration technician and stored in some form of non-volatile memory device such as an EEPROM or battery-backed-up RAM.

offset is a value that is added to the raw counts of the ADC to compensate for offset type errors caused by component inaccuracies. The offset would also be entered by a calibration technician and stored in some form of non-volatile memory device such as an EEPROM or battery-backed-up RAM.

Return Value

AICfgCal() returns 0 upon success and 1 if the analog input channel you specified is not within 0 and AIO_MAX_AI − 1.

Notes/Warnings

None

Example

```
void main (void)
{
    .
    .
    .
    /* Calibration gain and offset obtained by technician */
    AICfgCal(0, (FP32)1.09, (FP32)10.0);
    .
    .
    .
}
```

11.8.2 AICfgConv()

INT8U AICfgConv(INT8U n, FP32 gain, FP32 offset, INT8U pass);

`AICfgConv()` is used to set the conversion gain, offset, and the value of the pass counter for an analog input channel. The analog I/O module implements Equation 11-14, and this function is used to set the value of `ConvGain` and `ConvOffset`.

Arguments

n is the desired analog input channel to configure. Analog input channels are numbered from 0 to `AIO_MAX_AI - 1`.

gain is the conversion gain of the ADC channel in engineering units per count (E.U./count). `gain` is given by Equation 11-9 which is repeated in Equation 11-27 for your convenience:

$$gain_{(EU)/(count)} = \frac{FSV}{Transducer_{V/(EU)} \times A_V \times (2^{bits} - 1)} \qquad \textbf{Equation 11-27}$$

FSV is the Full Scale Voltage of the ADC and typically is the reference voltage used with the ADC.

Transducer$_{(V/EU)}$ corresponds to the number of volts produced by the transducer per engineering unit. For example, the LM34A produces 0.01 volt per degree Fahrenheit.

A$_V$ is the gain of the amplifier stage of an analog input channel (see Figure 11-1).

bits is the number of bits of the ADC.

offset is used to bias the ADC counts. `offset` is given by Equation 11-10 which is repeated in Equation 11-28 for your convenience.

$$offset_{counts} = \frac{V_{bias} \times (2^{bits} - 1)}{FSV} \qquad \textbf{Equation 11-28}$$

V$_{bias}$ is the bias voltage added to the output of the amplifier stage to allow the ADC to read negative values (see Figure 11-7 for an example on how to use the bias).

pass is used to specify a *pass count*. The pass count specifies to the module how often the analog channel will be read. The analog I/O module reads all analog input channels on a regular basis every so many clock ticks. This is called *scanning*. `pass` specifies how many scans are needed to read the analog input channel. For example, suppose the analog I/O module's scan rate is 10 Hz and you specify a *pass count* of 5 for analog input channel #0. Analog input channel #0 will be read every half second. I included a pass count because some analog input channels may not need to be read as often as others. For example, if you wanted the program to read the temperature of a room, you could tell it to read the temperature every 250 scans (or every 25 seconds, as in my example).

Return Value

AICfgConv() returns 0 upon success and 1 if the analog input channel you specified is not within 0 and AIO_MAX_AI − 1.

Notes/Warnings

None

Example

```
void main (void)
{
    .
    .
    /* Conversion gain and offset obtained by hardware
        engineer */
    AICfgConv(0, (FP32)1.987, (FP32)123.0, 1);
    .
    .
}
```

11.8.3 AICfgScaling()

INT8U AICfgScaling(INT8U n, void (fnct)(AIO *paio), void *arg);

AICfgScaling() is used to specify a scaling function to be executed when the analog input channel is read. The scaling function allows you to apply further processing when reading an analog input. There is no need to call AICfgScaling() if the analog input channel does not need a scaling function. In fact, if you don't define a scaling function the member .AIOScalingIn will simply be copied to .AIOScalingOut by AIUpdate() (see code).

Arguments

n is the desired analog input channel to configure. Analog input channels are numbered from 0 to AIO_MAX_AI − 1.

fnct is a pointer to the scaling function that will be executed when the analog input channel is read. You must write fnct to expect an argument. Specifically, fnct must be written to receive a pointer to the analog I/O data structure called AIO as shown in the code fragment following this paragraph. You specify a NULL pointer to prevent a previously configured channel from using a scaling function:

```
void fnct (AIO *paio);
```

arg is a pointer to any arguments or parameters needed for the scaling function. This argument can be used to specify specific options about the scaling being performed.

Return Value

AICfgScaling() returns 0 upon success and 1 if the analog input channel you specified is not within 0 and AIO_MAX_AI − 1.

Notes/Warnings

The scaling function is assumed to take its input from paio->AIOScaleIn and produce its result in paio->AIOScaleOut.

Example

```
INT8U ThermoType = THERMO_TYPE_J;



void main (void)
{
    .
    .
    AICfgScaling(0, ThermoLin, (void *)&ThermoType);
    .
    .
}
```

```
void ThermoLin (AIO *paio)
{
    /* Function to linearize a thermocouple */
    paio->AIOScaleIn is assumed to contain the number
        of millivolts for the thermocouple.
    paio->AIOScaleOut is where the temperature of
        the thermocouple is assumed to be saved to.
    paio->AIOScaleFnctArg could have also indicated
        the type of thermocouple used as well as whether
        the temperature is in degrees F or C.
}
```

11.8.4   AIGet()

INT8U AIGet(INT8U n, FP32 *pval);

The current value of the analog input channel can be obtained by calling AIGet(). The value obtained is in engineering units or physical units. For example, if the analog input channel is measuring a temperature from a thermocouple then the value returned is the number of degrees at the thermocouple.

Arguments

n is the desired analog input channel. Analog input channels are numbered from 0 to AIO_MAX_AI − 1.

pval is a pointer to where the value of the analog input channel will be stored.

Return Value

AIGet() returns 0 upon success and 1 if the analog input channel you specified is not within 0 and AIO_MAX_AI − 1.

Notes/Warnings

The value returned is the last 'scanned' value. In other words, an ADC conversion is not performed when you call this function—AIOTask() is responsible for 'scanning' the analog input on a continuous basis.

Example

```
void Task (void *pdata)
{
    INT8U err;
    FP32  eu;


    for (;;) {
        .

        .
        err = AIGet(0, &eu);   /* Get current value of
                                    analog input #0 */

        .
    }
}
```

11.8.5 AIOInit()

void AIOInit(void);

AIOInit() is the initialization code for the analog I/O module. AIOInit() must be called before you use any of the other analog I/O module functions. AIOInit() is responsible for initializing the internal variables used by the module and for creating the task that will update the analog inputs and outputs.

Arguments

None

Return Value

None

Notes/Warnings

You are expected to provide the value of the following compile-time configuration constants (see "Analog I/O Module, Configuration"):

AIO_TASK_STK_SIZE

AIO_TASK_PRIO

AIO_MAX_AI

AIO_MAX_AO

Example

```
void main (void
{
    .
    .
    AIOInit();
    .
    .
}
```

11.8.6 AISetBypass()

INT8U AISetBypass(INT8U n, FP32 val);

Your application software can bypass or override the analog input channel value by using this function. `AISetBypass()` doesn't do anything unless you *open* the bypass *switch* by calling `AISetBypassEn()`.

Arguments

n is the desired analog input channel to override. Analog input channels are numbered from 0 to AIO_MAX_AI – 1.

val is the value you want `AIGet()` to return to your application. The value you pass to `AISetBypass()` is in engineering units.

Return Value

`AISetBypass()` returns 0 upon success and 1 if the analog input channel you specified is not within 0 and AIO_MAX_AI – 1.

Notes/Warnings

`AISetBypass()` forces the value of `.AIOEU` in Figure 11-13 when `.AIOBypassEn` is set to TRUE.

Example

```
void Task (void *pdata)
{
    FP32 val;


    for (;;) {
        .
        .
        val = Get value from keyboard;
        AISetBypass(0, (FP32)val);
        .
    }
}
```

11.8.7 AISetBypassEn()

INT8U AISetBypassEn(INT8U n, BOOLEAN state);

AISetBypassEn() allows your application code to prevent the analog input channel from being updated. This permits another part of your application to set the value returned by AIGet(). In other words, you can "fool" the application code that monitors the analog input channel into thinking that the value is coming from a sensor, when in fact, the value returned by the analog input channel can come from another source. The value of the analog input channel is set by AISetBypass(). AISetBypassEn() and AISetBypass() are very useful functions for debugging.

Arguments

n is the desired analog input channel to bypass. Analog input channels are numbered from 0 to AIO_MAX_AI − 1.

state is the state of the bypass switch. When TRUE, the bypass switch is open (i.e., the analog input channel is bypassed). When FALSE, the bypass switch is closed (i.e., the analog input channel is not bypassed).

Return Value

AISetBypassEn() returns 0 upon success and 1 if the analog input channel you specified is not within 0 and AIO_MAX_AI − 1.

Notes/Warnings

`AISetBypassEn()` forces the value of `.AIOBypassEn` in Figure 11-13.

Example

```
void main (void)
{
    .
    .
    .
    AISetBypassEn(0, TRUE);
    .
    .
    .
}
```

11.8.8 AOCfgCal()

INT8U AOCfgCal(INT8U n, FP32 gain, FP32 offset);

`AOCfgCal()` is used to set the calibration gain and offset of an analog output channel. An analog output channel basically implements a generalization of Equation 11-23, as shown in Equation 11-29:

$$DAC_{counts} = INT(.AIOConvGain_{(counts/EU)} \times .AIOCalGain$$
$$\times .AIOScaleOut_{(EU)} + .AIOConvOffset_{counts} + .AIOCalOffset_{(counts)})$$

Equation 11-29

You can specify a calibration gain (`.AIOCalGain`) and offset (`.AIOCalOffset`) to compensate for component inaccuracies.

Arguments

n is the desired analog output channel. Analog output channels are numbered from 0 to `AIO_MAX_AO - 1`.

gain is a multiplying factor that is used to compensate for component inaccuracies and doesn't have any units. `gain` sets the value of `.AIOCalGain` in Figure 11-13. The `gain` would be entered by a calibration technician and stored in some form of non-volatile memory device such as an EEPROM or battery-backed-up RAM.

offset is a value that is added to the raw counts before outputing to a DAC to compensate for offset-type errors caused by component inaccuracies. `offset` sets the value of `.AIOCalOffset` in Figure 11-13. The `offset` would also be entered by a calibration technician and stored in some form of non-volatile memory device such as an EEPROM or battery-backed-up RAM.

Return Value

`AOCfgCal()` returns 0 upon success and 1 if the analog output channel you specified is not within 0 and `AIO_MAX_AO – 1`.

Notes/Warnings

None

Example

```
void main (void)
{
    .
    .
    AOCfgCal(0, (FP32)1.05, (FP32)10.6);
    .
    .
}
```

11.8.9 AOCfgConv()

INT8U AOCfgConv(INT8U n, FP32 gain, FP32 offset, INT16S lim,
INT8U pass);

`AOCfgConv()` is used to set the conversion gain, conversion offset, and the value of the pass counter for an analog output channel. An analog output channel basically implements a generalization of Equation 11-23, as shown in Equation 11-29. `AOCfgConv()` is used to set the value of `.AIOConvGain` and `.AIOConvOffset`.

Arguments

n is the desired analog output channel to configure. Analog output channels are numbered from 0 to `AIO_MAX_AO – 1`.

gain is the conversion gain for the analog output channel in counts per engineering unit (counts/E.U.). gain sets the .AIOConvGain field of Figure 11-14.

offset is used to bias the DAC counts and sets the .AIOConvOffset field of Figure 11-14.

lim is used to specify the maximum count that can be sent to the DAC. This argument ensures that the DAC will never be written with a count larger than lim. For example, an 8-bit DAC has a maximum count of 255 ($2^n- 1$). lim sets the .AIOLim field of Figure 11-14.

pass is used to specify a *pass count*. The pass count is used to specify to the module how often the analog channel will be updated. The analog I/O module updates all analog output channel on a regular basis every so many clock ticks. This is called *scanning*. pass specifies how many scans are needed to update the specific analog output channel. For example, suppose the analog I/O module scan rate is 10 Hz and you specify a *pass count* of 2 for analog output channel #4. In this case, analog output channel #4 will be updated five times per second. I included a pass count because some analog output channels may not need to be updated as often as others. pass sets the .AIOPassCnts field of Figure 11-14.

Return Value

AOCfgConv() returns 0 upon success and 1 if the analog output channel you specified is not within 0 and AIO_MAX_AO − 1.

Notes/Warnings

None

Example

```
void main (void)
{
    .
    .
    AOCfgConv(0, (FP32)1.05, (FP32)10.6, 0x0FFF, 1);
    .
    .
}
```

11.8.10 AOCfgScaling()

INT8U AOCfgScaling(INT8U n, void (*fnct)(AIO *paio), void *arg);

AOCfgScaling() is used to specify a scaling function to be executed when the analog output channel is updated. The scaling function allows you to apply further processing before updating an analog output. You don't need to call this function if your analog output channel doesn't need a scaling function. In this case, the .AIOScaleIn field will simply be copied to the .AIOScalingOut field by AOUpdate()(see code).

Arguments

n is the desired analog output channel. Analog output channels are numbered from 0 to AIO_MAX_AO − 1.

fnct is a pointer to the scaling function that will be executed when the analog output channel is updated. fnct sets the value of .AIOScaleFnct in Figure 11-14. fnct must be written to receive a pointer to the analog I/O data structure called AIO as follows:

void fnct (AIO *paio);

arg is a pointer to any arguments or parameters needed for the scaling function. arg sets the value of .AIOScaleFnctArg in Figure 11-14. This argument can be used to specify specific options about the scaling being performed.

Return Value

AOCfgScaling() returns 0 upon success and 1 if the analog output channel you specified is not within 0 and AIO_MAX_AO − 1.

Notes/Warnings

The scaling function is assumed to take its input from paio->AIOScaleIn and produce its result in paio->AIOScaleOut.

Example

```
void main (void)
{
    .
    .
    AOCfgScaling(0, ActLin, (void *)0);
    .
    .
.
}



void ActLin (AIO *paio)
{
    /* Linearize actuator function */
    paio->AIOScaleIn is the input value to the scaling function.
    paio->AIOScaleOut is where the scaling function will
        place the result.
    paio->AIOScaleFnctArg in this case is not used but could
        be made to tell ActLin() the type of actuator
        to linearize.
}
```

11.8.11   AOSet()

INT8U AOSet(INT8U n, FP32 val);

This function is used by your application software to set the value of the analog output channel. The value you set the channel to is specified in engineering units. In other words, if your analog output channel has been configured to control the position of a valve in percent, then you would pass the desired percentage of position you desire (a number between 0.0 and 100.0).

Arguments

n is the desired analog output channel. Analog output channels are numbered from 0 to AIO_MAX_AO − 1.

val is the desired value for the analog output channel and is specified in engineering units.

Return Value

AOSet() returns 0 upon success and 1 if the analog output channel you specified is not within 0 and AIO_MAX_AO − 1.

Notes/Warnings

None

Example

```
void Task (void *pdata)
{
    FP32 valve;


    for (;;) {
        .
        .
        valve = Get desired value position from user;
        AOSet(0, (FP32)valve);
        .
    }
}
```

11.8.12 AOSetBypass()

INT8U AOSetBypass(INT8U n, FP32 val);

Your application software can bypass or override the analog output channel value by using this function. AOSetBypass() doesn't do anything unless you open the bypass switch by calling AOSetBypassEn(), as described previously. As with AOSet(), the value you set the channel to is specified in engineering units.

Arguments

n is the desired analog output channel. Analog output channels are numbered from 0 to AIO_MAX_AO − 1.

val is the value that you want to force into the analog output channel (in engineering units).

Return Value

AOSetBypass() returns 0 upon success and 1 if the analog output channel you specified is not within 0 and AIO_MAX_AO − 1.

Notes/Warnings

None

Example

```
void Task (void *pdata)
{
     FP32 val;

     for (;;) {
          .

          .
          val = Get value from keyboard;
          AOSetBypass(0, (FP32)val);
          .
     }
}
```

11.8.13   AOSetBypassEn()

INT8U AOSetBypassEn(INT8U n, BOOLEAN state);

AOSetBypassEn() allows you to prevent your application from changing the value of an analog output channel. This allows you to gain control of the analog output channel from elsewhere in your application code. This is a quite useful feature because it allows you to test your analog output channels one by one. In other words, you can set an analog output to any desired value even though your application software is trying to control the output. The value of the analog output channel is set by AOSetBypass(). AOSetBypassEn() and AOSetBypass() are very useful for debugging.

Arguments

n is the desired analog output channel. Analog output channels are numbered from 0 to AIO_MAX_AO – 1.

state is the state of the bypass *switch*. When TRUE, the bypass switch is opened (i.e., the analog output channel is bypassed). When FALSE, the bypass switch is closed (i.e., the analog output channel is not bypassed).

Return Value

AOSetBypassEn() returns 0 upon success and 1 if the analog output channel you specified is not within 0 and AIO_MAX_AO – 1.

Notes/Warnings

None

Example

```
void main (void)
{
    .
    .
    AOSetBypassEn(0, TRUE);
    .
    .
}
```

11.9 Analog I/O Module, Configuration

Configuration of the analog I/O module is quite simple.

1. You need to define the value of five #defines. The #defines are found in AIO.H (or CFG.H).

    AIO_TASK_PRIO is used to set the priority of the analog I/O module task.

    AIO_TASK_DLY is used to establish how often the analog I/O module will be executed.

AIO_TASK_DLY determines the number of milliseconds to delay between execution of the analog I/O task.

Warning

Because μC/OS-II provides a more convenient function (i.e., OSTimeDlyHMSM()) to specify the task execution period in hours, minutes, seconds and milliseconds, AIO_TASK_DLY_TICKS is no longer used and AIO_TASK_DLY now specifies the scan period in milliseconds instead of ticks.

AIO_TASK_STK_SIZE specifies the size of the stack (in bus width units) allocated to the analog I/O task. The number of bytes allocated for the stack is thus given by: AIO_TASK_STK_SIZE times size of(OS_STK).

Warning

μC/OS-II assumes the stack is specified in stack width elements.

AIO_MAX_AI determines the number of analog input channels that will be handled by the analog I/O task.

AIO_MAX_AO determines the number of analog output channels handled by the analog I/O task.

2. You will need to define how analog inputs are read (i.e., how to read your ADC(s). ADCs must all be handled through AIRd(). The function prototype for AIRd() is:

INT16S AIRd (INT8U ch);

AIRd() is called by AIUpdate() (see code) and is passed the logical channel number (0 to AIO_MAX_AI − 1). You must translate this logical channel into code that selects the proper multiplexer for the desired channel, start the ADC, wait for the conversion to complete, read the ADC, and finally, return the ADC's counts.

3. You will need to provide the code for the function that writes to all DACs (i.e., AOWr()). The function prototype for AOWr() is:

void AOWr (INT8U ch, INT16S raw);

AOWr() is called by AOUpdate() (see code) and is passed the logical channel number (0 to AIO_MAX_AO − 1). You must translate this logical channel into code that selects the proper DAC for the desired channel. `AOWr()` is also passed the counts to send to the DAC. Your code must thus write the counts to the proper DAC.

4. You will need to provide the hardware initialization function (`AIOInitIO()`), which is called by `AIOInit()`. The function prototype for `AIOInit()` is:

`void AIOInit (void);`

11.10   *How to Use the Analog I/O Module*

Let's assume that you need to read the analog inputs and control the analog outputs shown in Figure 11-16.

The analog I/O module has to read six analog inputs, and thus you will configure `AIO_MAX_AI` to 6. Similarly, to update three analog outputs, you need to set `AIO_MAX_AO` to 3. We can set `AIO_TASK_DLY` to 100 (i.e., milliseconds) because all analog I/Os need to be read or updated in multiples of 100 mS. Obviously, you need to

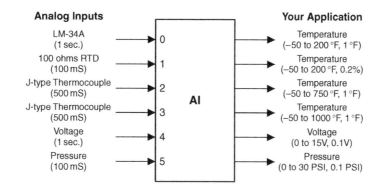

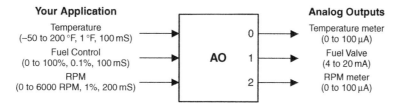

Figure 11-16: Using the Analog I/O Module.

allocate sufficient stack space (i.e., `AIO_TASK_STK_SIZE`) for `AIOTask()` as well as determine what priority (i.e., `AIO_TASK_PRIO`) you want to give to that task.

To initialize the analog I/O module, you need to call `AIOInit()` prior to using any of the analog I/O module functions. You would typically do this in `main()`:

```
void main (void)
{
    .
    OSInit();      /* Initialize the O.S. (mC/OS-II)    */
    .
    .
    AIOInit();     /* Initialize the analog I/O module */
    .
    .
    OSStart();     /* Start multitasking  (mC/OS-II)    */
}
```

You would initialize each one of the analog I/O channels from an application task, as shown in the code fragment following this paragraph. It is important that you do this at the task level because some of the analog I/O module services assume that the operating system is running in order to access the mutual exclusion semaphore (`AIOSem`).

```
void AppTask (void *data)
{
    data = data;
    /* Initialize analog I/O channels here ...*/
    .
    .
    for (;;) {
        /* Application task code ... */
    }
}
```

Let's assume the hardware designer came up with the circuit shown in Figure 11-17 to read the analog inputs. As you can see, each input has signal conditioning circuitry which feeds into a multiplexer. The multiplexer selects one of the analog inputs to be converted by a 12-bit analog-to-digital converter (ADC).

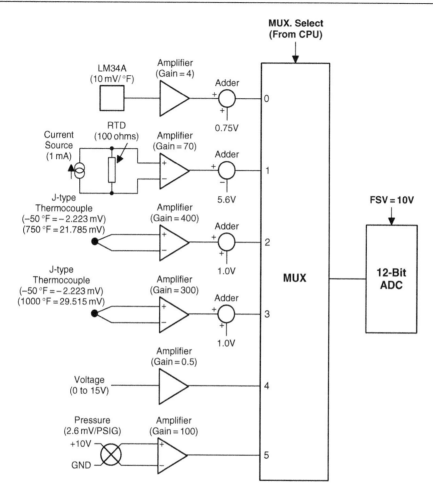

Figure 11-17: Analog Inputs.

11.11   How to Use the Analog I/O Module, AI #0

Analog input channel #0 is an LM–34A temperature sensor used to read temperatures from –50 to 200 °F. Using Equation 11-9, the conversion gain is:

$$ConvGain_{(EU)/(count)} = \frac{FSV}{Transducer_{V/(EU)} \times A_V \times (2^n - 1)}$$

$$ConvGain_{(°F)/(count)} = \frac{10}{0.01_{V/(°F)} \times 4 \times (2^{12} - 1)}$$

$$ConvGain_{(°F)/(count)} = 0.061050$$

Equation 11-30

From Equation 11-10 the conversion offset is:

$$ConvOffset_{counts} = -\left(\frac{V_{bias} \times (2^n - 1)}{FSV}\right)$$

$$ConvOffset_{counts} = -\left(\frac{0.75 \times (2^{12} - 1)}{10}\right) \qquad \textbf{Equation 11-31}$$

$$ConvOffset_{counts} = -307.125$$

The temperature at the LM34A is given by Equation 11-11 and is:

$$Temperature_{\circ F} = (ADC_{counts} + ConvOffset_{counts}) \times ConvGain_{(EU)/(count)}$$

$$Temperature_{\circ F} = (ADC_{counts} - 307.125) \times 0.061050 \qquad \textbf{Equation 11-32}$$

Because the LM-34A only needs to be read once per second, the pass counter for the channel will be set to 10 (i.e., $10 \times 100\,mS$ scan period).

11.11.1   How to Use the Analog I/O Module, **AI #1**

Analog input channel #1 is a 100-ohm Resistance Temperature Device (RTD). The RTD has about 80 ohms of resistance when the temperature at the RTD is –50 °F and 139 ohms when the temperature at the RTD is 200 °F. Unfortunately, the temperature at the RTD is not a linear function of resistance, and thus you will have to write a linearization function (beyond the scope of this chapter). The current source is used to develop a voltage across the RTD so that the resistance of the RTD can be measured. The circuit produces 1 mV per ohm (which is before the amplifier). By using Equations 11-9, 11-10, and 11-11, the resistance of the RTD is given by:

$$ConvGain_{(ohms)/(count)} = 0.034886$$

$$ConvOffset_{counts} = -2293.2 \qquad \textbf{Equation 11-33}$$

$$Resistance_{ohms} = (ADC_{counts} - 2293.2) \times 0.034886$$

The pass counter for analog input channel #1 will be set to 1 in order to read the RTD every 100 mS.

11.11.2   How to Use the Analog I/O Module, **AI #2**

Analog input channel #2 is a J-Type thermocouple (another temperature measurement device). If you want to get the official reference on thermocouples, you should get the

NIST Monograph 175 (see "Bibliography"). A thermocouple produces a small voltage (called the *Seebeck voltage*) that varies as a function of temperature. The temperature at the thermocouple is not a linear function of the voltage produced. To further complicate things, the temperature at the thermocouple is also a function of a reference temperature called the *Cold Junction*. Determining the temperature at the thermocouple is beyond the scope of this chapter. Let's say for now that all you need to do is to measure the voltage (actually millivolts) produced by the thermocouple. It is thus up to you to write a linearization function (also called *thermocouple compensation* function). A J-Type thermocouple produces −2.223 mV at −50 °F and 21.785 mV at 750 °F. This voltage is amplified by 400 so that it can be read by the ADC. A bias voltage is introduced to ensure that the ADC only sees positive voltages. From Equations 11-9, 11-10, and 11-11, the number of millivolts at the thermocouple is given by:

$$ConvGain_{(mV)/(count)} = 0.006105$$
$$ConvOffset_{counts} = -409.5 \qquad \textbf{Equation 11-34}$$
$$Thermocouple_{mV} = (ADC_{counts} - 409.5) \times 0.006105$$

All you have to do is linearize the thermocouple based on the number of millivolts read from the thermocouple. The pass counter for analog input channel #2 will be set to 5 in order to read the thermocouple every 500 mS.

11.11.3 How to Use the Analog I/O Module, AI #3

Analog input channel #3 is also a J-Type thermocouple. A J-Type thermocouple produces −2.223 mV at −50 °F and 29.515 mV at 1000 °F. This voltage is amplified by 300 so that it can be read by the ADC. The bias voltage is also introduced to ensure that the ADC only sees positive voltages. From Equations 11-9, 11-10, and 11-11, the number of millivolts at the thermocouple is given by:

$$ConvGain_{(mV)/(count)} = 0.008140$$
$$ConvOffset_{counts} = -409.5 \qquad \textbf{Equation 11-35}$$
$$Thermocouple_{mV} = (ADC_{counts} - 409.5) \times 0.008140$$

Again, all you have to do is linearize the thermocouple based on the number of millivolts read from the thermocouple. The pass counter for analog input channel #3 will also be set to 5 in order to read the thermocouple every 500 mS.

11.11.4 How to Use the Analog I/O Module, AI #4

Analog input channel #4 reads a voltage directly (maybe a battery). Because the voltage to read exceeds the FSV of the ADC, the hardware designer decided to simply divide the voltage in half. From Equations 11-9, 11-10, and 11-11, the voltage at the input is given by:

$$ConvGain_{(V)/(count)} = 0.004884$$
$$ConvOffset_{counts} = -0$$
$$Voltage_V = (ADC_{counts} - 0) \times 0.004884$$

Equation 11-36

The pass counter for analog input channel #4 will also be set to 10 in order to read the thermocouple every second.

11.11.5 How to Use the Analog I/O Module, AI #5

Analog input channel #5 reads a pressure from a pressure transducer which produces 2.6 mV/PSIG (pounds per square inch gauge). From Equations 11-9, 11-10, and 11-11, the pressure read by the transducer is given by:

$$ConvGain_{(PSIG)/(count)} = 0.009392$$
$$ConvOffset_{counts} = -0$$
$$Pressure_{PSIG} = (ADC_{counts} - 0) \times 0.009392$$

Equation 11-37

The pass counter for analog input channel #5 will be set to 1 in order to read the pressure every 100 mS.

Let's assume that the hardware designer came up with the circuit shown in Figure 11-18 to update the analog outputs.

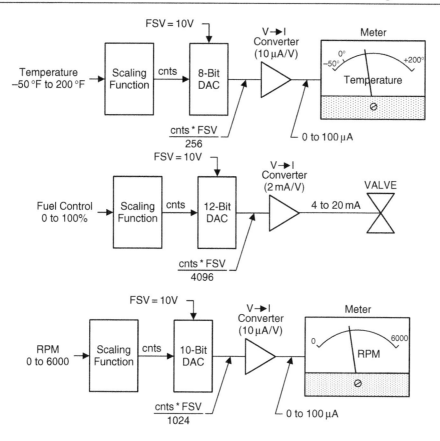

Figure 11-18: Analog Inputs.

11.11.6 How to Use the Analog I/O Module, AO #0

Analog output channel #0 is used to display temperatures from $-50\,°F$ to $200\,°F$ on a 0 to $100\,\mu A$ meter. A display of $-50\,\mu F$ is obtained with 0 DAC counts ($0\,\mu A$) while $200\,°F$ is obtained with 255 DAC counts ($99.609\,\mu A$). The DAC counts are given by:

$$ConvGain_{(counts)/(°F)} = 1.02$$
$$ConvOffset_{counts} = 51 \qquad \textbf{Equation 11-38}$$
$$DAC_{counts} = 1.02 \times Temperature_{°F} + 51$$

The pass counter for analog output channel #0 will be set to 1 in order to update the meter every $100\,mS$.

11.11.7 How to Use the Analog I/O Module, AO #1

Analog output channel #1 is used to control the opening of a valve. The valve is closed when the control current is 4 mA and wide open when the control current is 20 mA. The counts vs. output current is given by:

$$DAC_{counts} = \frac{2^{n-1} \times Out_{mA}}{FSV \times 2_{(mA)/V}}$$

Equation 11-39

A 12-bit DAC is used because a 10-bit DAC would not have the required resolution. Using a 10-bit DAC, 4 mA would require 205 counts (Equation 11-31), while 20 mA would require 1023 counts, a range of 818 counts, or 0.122 percent. Note that 11-bit DACs are not commercially available. A 12-bit DAC requires 819.2 counts for a 4 mA output and 4095 counts for 20 mA (actually 19.995 mA). The DAC counts required to control the DAC are given by:

$$ConvGain_{(counts)/\%} = \frac{4095 - 819.2}{100\% - 0\%} = 32.758$$

$$ConvOffset_{counts} = 819.2$$

Equation 11-40

$$DAC_{counts} = 32.758 \times Input_{\%} + 819.2$$

The pass counter for analog output channel #1 will be set to 1 in order to update the valve every 100 mS.

11.11.8 How to Use the Analog I/O Module, AO #2

Analog output channel #2 is used to display the RPM of a rotating device on a 0 to 100 μA meter. A display of 0 RPM is obtained with 0 DAC counts (0 μA), while 6000 RPM is obtained with 1023 DAC counts (99.902 μA). The DAC counts are given by:

$$ConvGain_{(counts)/(RPM)} = 0.1705$$

$$ConvOffset_{counts} = 0$$

Equation 11-41

$$DAC_{counts} = 0.1705 \times RPM + 0$$

The pass counter for analog output channel #2 will be set to 2 in order to update the meter every 200 mS.

The code to initialize the analog I/O channels is:

```
void AppInitAIO (void)
{
    AICfgConv(0, 0.061050,  307.125,   10);    /* Analog Inputs      */
    AICfgConv(1, 0.034886, 2293.2,      1);
    AICfgConv(2, 0.006105,  409.5,      5);
    ATCfgConv(3, 0.008140,  409.5,      5);
    AICfgConv(4, 0.004884,    0.0,     10);
    AICfgConv(5, 0.009392,    0.0,      1);

    AICfgScaling(1, /* Pointer to RTD code */, /* Pointer to args */);
    AICfgScaling(2, /* Pointer to TC  code */, /* Pointer to args */);
    AICfgScaling(3, /* Pointer to TC  code */, /* Pointer to args */);

    AOCfgConv(0,  1.02,      51.0,  255, 1);    /* Analog Outputs     */
    AOCfgConv(1, 32.758,    819.2, 4095, 2);
    AOCfgConv(2,  0.1705,     0.0, 1023, 2);
}
```

You can now obtain the value read by any analog input channels by using `AIGet()` and set any analog output channel by calling `AOSet()`.

References

Burns, G.W., Scroger, M.G., Strouse, G.F., Croarkin, M.C. and Guthrie, W.F. *Temperature-Electromotive Force Reference Functions and Tables for the Letter-Designated Thermocouple Types Based on the ITS-90 (NIST Monograph 175).* Gaithersburg, MD: United States Department of Commerce, National Institute of Standards and Technology.

Morgan, Don. *Numerical Methods, Real-Time and Embedded Systems Programming.* San Mateo, CA: M&T Publishing, Inc. ISBN 1-55851-232-2

U.S. Software
14215 NW Science Park Dr
Portland, OR 97229
(503) 641-8446

Zuch, Eugene L. *Data Acquisition and Conversion Handbook.* Mansfield, MA: Datel/Intersil, 1979.

Listing 11.1 AIO.C

```c
/*
*********************************************************************************
*                                Analog I/O Module
*
*                   (c) Copyright 1999, Jean J. Labrosse, Weston, FL
*                                  All Rights Reserved
*
* Filename   : AIO.C
* Programmer : Jean J. Labrosse
*********************************************************************************
*/

/*
*********************************************************************************
*                                INCLUDE FILES
*********************************************************************************
*/

#define    AIO_GLOBALS
#include "includes.h"

/*
*********************************************************************************
*                                LOCAL VARIABLES
*********************************************************************************
*/

static OS_STK    AIOTaskStk[AIO_TASK_STK_SIZE];
static OS_EVENT *AIOSem;

/*
*********************************************************************************
*                           LOCAL FUNCTION PROTOTYPES
*********************************************************************************
*/

        void    AIOTask(void *data);


static  void    AIInit(void);
static  void    AIUpdate(void);

static  void    AOInit(void);
static  void    AOUpdate(void);

/*$PAGE*/
```

Listing 11.1 (Continued) AIO.C

```
/*
*********************************************************************************
*                 CONFIGURE THE CALIBRATION PARAMETERS OF AN ANALOG INPUT CHANNEL
*
* Description : This function is used to configure an analog input channel.
* Arguments   : n      is the analog input channel to configure:
*                gain   is the calibration gain
*                offset is the calibration offset
* Returns     : 0      if successfull.
*                1      if you specified an invalid analog input channel number.
*********************************************************************************
*/

INT8U AICfgCal (INT8U n, FP32 gain, FP32 offset)
{
    INT8U err;
    AIO *paio;


    if (n < AIO_MAX_AI) {
        paio          = &AITbl[n]; /* Point to Analog Input structure            */
        OSSemPend(AIOSem, 0, &err);     /* Obtain exclusive access to AI channel      */
        paio->AIOCalGain   = gain;        /* Store new cal. gain and offset into struct */
        paio->AIOCalOffset = offset;
        paio->AIOGain      = paio->AIOCalGain * paio->AIOConvGain;   /* Compute overall
                                                                         gain       */
        paio->AIOOffset    = paio->AIOCalOffset + paio->AIOConvOffset;/* Compute overall
                                                                         offset     */
        OSSemPost(AIOSem);                                           /* Release AI
                                                                         channel    */

        return (0);
    } else {
        return (1);
    }
}

/*$PAGE*/
```

Listing 11.1 (Continued) AIO.C

```
/*
*********************************************************************************
*              CONFIGURE THE CONVERSION PARAMETERS OF AN ANALOG INPUT CHANNEL
*
* Description : This function is used to configure an analog input channel.
* Arguments   : n      is the analog channel to configure (0..AIO_MAX_AI-1).
*               gain   is the conversion gain
*               offset is the conversion offset
*               pass   is the value for the pass counts
* Returns     : 0      if successfull.
*               1      if you specified an invalid analog input channel number.
*********************************************************************************
*/

INT8U AICfgConv (INT8U n, FP32 gain, FP32 offset, INT8U pass)
{
    INT8U err;
    AIO *paio;


    if (n < AIO_MAX_AI) {
        paio               = &AITbl[n];  /* Point to Analog Input structure          */
        OSSemPend(AIOSem, 0, &err);       /* Obtain exclusive access to AI channel    */
        paio->AIOConvGain  = gain;        /* Store new conv. gain and offset into struct */
        paio->AIOConvOffset = offset;
        paio->AIOGain      = paio->AIOCalGain * paio->AIOConvGain;    /* Compute overall
                                                                        gain      */
        paio->AIOOffset    = paio->AIOCalOffset + paio->AIOConvOffset; /* Compute overall
                                                                        offset     */
        paio->AIOPassCnts  = pass;
        OSSemPost(AIOSem);                                           /* Release AI
        return (0);                                                    channel     */
    } else {
        return (1);
    }
}

/*$PAGE*/
```

Listing 11.1 (Continued) AIO.C

```
/*
*********************************************************************************
*                  CONFIGURE THE SCALING PARAMETERS OF AN ANALOG INPUT CHANNEL
*
* Description : This function is used to configure the scaling parameters
*               associated with an analog input channel.
* Arguments   : n    is the analog input channel to configure (0..AIO_MAX_AI-1).
*               arg  is a pointer to arguments needed by the scaling function
*               fnct is a pointer to a scaling function
* Returns     : 0    if successfull.
*               1    if you specified an invalid analog input channel number.
*********************************************************************************
*/

INT8U AICfgScaling (INT8U n, void (*fnct)(AIO *paio), void *arg)
{
    AIO *paio;


    if (n < AIO_MAX_AI) {
        paio                   = &AITbl[n];/* Faster to use a pointer to the structure */
        OS_ENTER_CRITICAL();
        paio->AIOScaleFnct     = (void (*)())fnct;
        paio->AIOScaleFnctArg = arg;
        OS_EXIT_CRITICAL();
        return (0);
    } else {
        return (1);
    }
}

/*$PAGE*/
```

Listing 11.1 (Continued) AIO.C

```
/*
*************************************************************************************
*                       GET THE VALUE OF AN ANALOG INPUT CHANNEL
*
* Description : This function is used to get the currect value of an analog input
*               (channel in engineering units).
* Arguments   : n    is the analog input channel (0..AIO_MAX_AI-1).
*               pval is a pointer to the destination engineering units
*                    of the analog input channel
* Returns     : 0       if successfull.
*               1       if you specified an invalid analog input channel number.
*                       In this case, the destination is not changed.
*************************************************************************************
*/

INT8U AIGet (INT8U n, FP32 *pval)
{
    AIO *paio;


    if (n < AIO_MAX_AI) {
        paio = &AITbl[n];
        OS_ENTER_CRITICAL(); /* Obtain exclusive access to AI channel                */
        *pval = paio->AIOEU; /* Get the engineering units of the analog input channel */
        OS_EXIT_CRITICAL();  /* Release AI channel                                    */
        return (0);
    } else {
        return (1);
    }
}

/*$PAGE*/
```

Listing 11.1 (Continued) AIO.C

```
/*
*********************************************************************************
*                         ANALOG INPUTS INITIALIZATION
*
* Description : This function initializes the analog input channels.
* Arguments   : None
* Returns     : None.
*********************************************************************************
*/

static void AIInit (void)
{
    INT8U i;
    AIO   *paio;


    paio = &AITbl[0];
    for (i = 0; i < AIO_MAX_AI; i++) {
        paio->AIOBypassEn    = FALSE;       /* Analog channel is not bypassed   */
        paio->AIORaw         = 0x0000;      /* Raw counts of ADC or DAC         */
        paio->AIOEU          = (FP32)0.0;   /* Engineering units of AI channel  */
        paio->AIOGain        = (FP32)1.0;   /* Total gain                       */
        paio->AIOOffset      = (FP32)0.0;   /* Total offset                     */
        paio->AIOLim         =      0;
        paio->AIOPassCnts    =      1;      /* Pass counts                      */
        paio->AIOPassCtr     =      1;      /* Pass counter                     */
        paio->AIOCalGain     = (FP32)1.0;   /* Calibration gain                 */
        paio->AIOCalOffset   = (FP32)0.0;   /* Calibration offset               */
        paio->AIOConvGain    = (FP32)1.0;   /* Conversion gain                  */
        paio->AIOConvOffset  = (FP32)0.0;   /* Conversion offset                */
        paio->AIOScaleIn     = (FP32)0.0;   /* Input to scaling function        */
        paio->AIOScaleOut    = (FP32)0.0;   /* Output of scaling function       */
        paio->AIOScaleFnct   = (void *)0;   /* No function to execute           */
        paio->AIOScaleFnctArg = (void *)0;  /* No arguments to scale function   */
        paio++;
    }
}

/*$PAGE*/
```

Listing 11.1 (Continued) AIO.C

```
/*
*********************************************************************************
*                       ANALOG I/O MANAGER INITIALIZATION
*
* Description : This function initializes the analog I/O manager module.
* Arguments   : None
* Returns     : None.
*********************************************************************************
*/
void AIOInit (void)
{
    INT8U err;

    AIInit();
    AOInit();
    AIOInitIO();
    AIOSem = OSSemCreate(1); /* Create a mutual exclusion semaphore for AIOs */
    OSTaskCreate(AIOTask, (void *)0, &AIOTaskStk[AIO_TASK_STK_SIZE], AIO_TASK_PRIO);
}

/*$PAGE*/

/*
*********************************************************************************
*                          ANALOG I/O MANAGER TASK
*
* Description : This task is created by AIOInit() and is responsible for updating
*               the analog inputs and analog outputs.
*               AIOTask() executes every AIO_TASK_DLY milliseconds.
* Arguments   : None.
* Returns     : None.
*********************************************************************************
*/
void AIOTask (void *data)
{
    INT8U err;

    data = data;                          /* Avoid compiler warning              */
    for (;;) {
        OSTimeDlyHMSM(0, 0, 0, AIO_TASK_DLY); /* Delay between execution of
                                               AIO manager                       */
        OSSemPend(AIOSem, 0, &err);       /* Obtain exclusive access to AI
                                               channels                          */
        AIUpdate();                       /* Update all AI channels              */
        OSSemPost(AIOSem);                /* Release AI channels (Allow high prio.
                                               task to run)                      */
        OSSemPend(AIOSem, 0, &err);       /* Obtain exclusive access to AO channels */
        AOUpdate();                       /* Update all AO channels              */
        OSSemPost(AIOSem);                /* Release AO channels (Allow high prio.
                                               task to run)                      */
    }
}

/*$PAGE*/
```

Listing 11.1 (Continued) AIO.C

```
/*
*********************************************************************************
*                   SET THE STATE OF THE BYPASSED ANALOG INPUT CHANNEL
*
* Description : This function is used to set the engineering units of
*               a bypassed analog input channel. This function is used
*               to simulate the presense of the sensor. This function is
*               Only valid if the bypass 'switch' is open.
* Arguments   : n   is the analog input channel (0..AIO_MAX_AI-1).
*               val is the value of the bypassed analog input channel:
* Returns     : 0       if successfull.
*               1       if you specified an invalid analog input channel number.
*               2       if AIOBypassEn was not set to TRUE
*********************************************************************************
*/

INT8U AISetBypass (INT8U n, FP32 val)
{
    AIO *paio;


    if (n < AIO_MAX_AI) {
        paio = &AITbl[n];                  /* Faster to use a pointer to the
                                              structure                       */
        if (paio->AIOBypassEn == TRUE) {   /* See if the analog input channel
                                              is bypassed                     */
            OS_ENTER_CRITICAL();
            paio->AIOEU = val;             /* Yes, then set the new value of
                                              the channel                     */
            OS_EXIT_CRITICAL();
            return (0);
        } else {
            return (2);
        }
    } else {
        return (1);
    }
}

/*$PAGE*/
```

Listing 11.1 (Continued) AIO.C

```
/*
********************************************************************************
*                        SET THE STATE OF THE BYPASS SWITCH
*
* Description : This function is used to set the state of the bypass switch.
*               The analog input channel is bypassed when the 'switch' is
*               open (i.e. AIOBypassEn is set to TRUE).
* Arguments   : n     is the analog input channel (0..AIO_MAX_AI-1).
*               state is the state of the bypass switch:
*                       FALSE disables the bypass (i.e. the bypass 'switch' is closed)
*                       TRUE enables the bypass (i.e. the bypass 'switch' is open)
* Returns     : 0     if successfull.
*               1     if you specified an invalid analog input channel number.
********************************************************************************
*/

INT8U AISetBypassEn (INT8U n, BOOLEAN state)
{
    if (n < AIO_MAX_AI) {
        AITbl[n].AIOBypassEn = state;
        return (0);
    } else {
        return (1);
    }
}

/*$PAGE*/
```

Listing 11.1 (Continued) AIO.C

```c
/*
*********************************************************************************
*                         UPDATE ALL ANALOG INPUT CHANNELS
*
* Description : This function processes all of the analog input channels.
* Arguments   : None.
* Returns     : None.
*********************************************************************************
*/

static void AIUpdate (void)
{
    INT8U i;
    AIO   *paio;


    paio = &AITbl[0];                       /* Point at first analog input channel    */
    for (i = 0; i < AIO_MAX_AI; i++) {      /* Process all analog input channels      */
        if (paio->AIOBypassEn == FALSE) {   /* See if analog input channel is bypassed */
            paio->AIOPassCtr--;             /* Decrement pass counter                 */
            if (paio->AIOPassCtr == 0) {    /* When pass counter reaches 0, read and
                                               scale AI                               */
                paio->AIOPassCtr = paio->AIOPassCnts; /* Reload pass counter           */
                paio->AIORaw = AIRd(i);                /* Read ADC for this channel    */
                paio->AIOScaleIn = ((FP32)paio->AIORaw + paio->AIOOffset)             */
                                   paio->AIOGain;
                if ((void *)paio->AIOScaleFnct != (void *)0) { /* See if function
                                                        defined                        */
                    (*paio->AIOScaleFnct)(paio);               /* Yes, execute function */
                } else {
                    paio->AIOScaleOut = paio->AIOScaleIn;   /* No, just copy data      */
                }
                paio->AIOEU = paio->AIOScaleOut;    /* Output of scaling fnct to E.U. */
            }
        }
        paio++;                                      /* Point at next AI channel        */
    }
}

/*$PAGE*/
```

Listing 11.1 (Continued) AIO.C

```
/*
*********************************************************************************
*             CONFIGURE THE CALIBRATION PARAMETERS OF AN ANALOG OUTPUT CHANNEL
*
* Description : This function is used to configure an analog output channel.
* Arguments   : n      is the analog output channel to configure (0..AIO_MAX_AO-1)
*               gain   is the calibration gain
*               offset is the calibration offset
* Returns     : 0      if successfull.
*               1      if you specified an invalid analog output channel number.
*********************************************************************************
*/

INT8U AOCfgCal (INT8U n, FP32 gain, FP32 offset)
{
    INT8U err;
    AIO   *paio;


    if (n < AIO_MAX_AO) {
        paio = &AOTbl[n];                /* Point to Analog Output structure         */
        OSSemPend(AIOSem, 0, &err);      /* Obtain exclusive access to AO channel     */
        paio->AIOCalGain = gain;         /* Store new cal. gain and offset into struct */
        paio->AIOCalOffset = offset;
        paio->AIOGain = paio->AIOCalGain * paio->AIOConvGain;      /* Compute overall
                                                                      gain            */
        paio->AIOOffset = paio->AIOCalOffset + paio->AIOConvOffset; /* Compute overall
                                                                       offset          */
        OSSemPost(AIOSem);                                         /* Release AO channel */
        return (0);
    } else {
        return (1);
    }
}

/*$PAGE*/
```

Listing 11.1 (Continued) AIO.C

```
/*
*********************************************************************************
*           CONFIGURE THE CONVERSION PARAMETERS OF AN ANALOG OUTPUT CHANNEL
*
* Description : This function is used to configure an analog output channel.
* Arguments   : n      is the analog channel to configure (0..AIO_MAX_AO-1).
*               gain   is the conversion gain
*               offset is the conversion offset
*               pass   is the value for the pass counts
* Returns     : 0      if successfull.
*               1      if you specified an invalid analog output channel number.
*********************************************************************************
*/

INT8U AOCfgConv (INT8U n, FP32 gain, FP32 offset, INT16S lim, INT8U pass)
{
    INT8U err;
    AIO   *paio;


    if (n < AIO_MAX_AO) {
        paio              = &AOTbl[n];          /* Point to Analog Output structure        */
        OSSemPend(AIOSem, 0, &err);             /* Obtain exclusive access to AO channel   */
        paio->AIOConvGain  = gain;              /* Store new conv. gain and offset into struct */
        paio->AIOConvOffset = offset;
        paio->AIOGain      = paio->AIOCalGain * paio->AIOConvGain; /* Compute overall
                                                                     gain                 */
        paio->AIOOffset    = paio->AIOCalOffset + paio->AIOConvOffset;
                                                           /* Compute overall
                                                              offset                      */
        paio->AIOLim       = lim;
        paio->AIOPassCnts  = pass;
        OSSemPost(AIOSem);                                 /* Release AO channel          */
        return (0);
    } else {
        return (1);
    }
}

/*$PAGE*/
```

Listing 11.1 (Continued) AIO.C

```
/*
*********************************************************************************
*               CONFIGURE THE SCALING PARAMETERS OF AN ANALOG OUTPUT CHANNEL
*
* Description : This function is used to configure the scaling parameters
*               associated with an analog output channel.
* Arguments   : n    is the analog output channel to configure (0..AIO_MAX_AO-1).
*               arg  is a pointer to arguments needed by the scaling function
*               fnct is a pointer to a scaling function
* Returns     : 0    if successfull.
*               1    if you specified an invalid analog output channel number.
*********************************************************************************
*/

INT8U AOCfgScaling (INT8U n, void (*fnct)(AIO *paio), void *arg)
{
    AIO *paio;


    if (n < AIO_MAX_AO) {
        paio              = &AOTbl[n];              /* Faster to use a pointer to the structure */
        OS_ENTER_CRITICAL();
        paio->AIOScaleFnct    = (void (*)())fnct;
        paio->AIOScaleFnctArg = arg;
        OS_EXIT_CRITICAL();
        return (0);
    } else {
        return (1);
    }
}

/*$PAGE*/
```

Listing 11.1 (Continued) AIO.C

```
/*
*********************************************************************************
*                       ANALOG OUTPUTS INITIALIZATION
*
* Description : This function initializes the analog output channels.
* Arguments   : None
* Returns     : None.
*********************************************************************************
*/

static void AOInit (void)
{
    INT8U i;
    AIO   *paio;


    paio = &AOTbl[0];
    for (i = 0; i < AIO_MAX_AO; i++) {
        paio->AIOBypassEn    = FALSE;        /* Analog channel is not bypassed            */
        paio->AIORaw         = 0x0000;       /* Raw counts of ADC or DAC                  */
        paio->AIOEU          = (FP32)0.0;    /* Engineering units of AI channel           */
        paio->AIOGain        = (FP32)1.0;    /* Total gain                                */
        paio->AIOOffset      = (FP32)0.0;    /* Total offset                              */
        paio->AIOLim         =      0;       /* Maximum count of an analog output channel */
        paio->AIOPassCnts    =      1;       /* Pass counts                               */
        paio->AIOPassCtr     =      1;       /* Pass counter                              */
        paio->AIOCalGain     = (FP32)1.0;    /* Calibration gain                          */
        paio->AIOCalOffset   = (FP32)0.0;    /* Calibration offset                        */
        paio->AIOConvGain    = (FP32)1.0;    /* Conversion gain                           */
        paio->AIOConvOffset  = (FP32)0.0;    /* Conversion offset                         */
        paio->AIOScaleIn     = (FP32)0.0;    /* Input to scaling function                 */
        paio->AIOScaleOut    = (FP32)0.0;    /* Output of scaling function                */
        paio->AIOScaleFnct   = (void *)0;    /* No function to execute                    */
        paio->AIOScaleFnctArg = (void *)0;   /* No arguments to scale function            */
        paio++;
    }
}

/*$PAGE*/
```

Listing 11.1 (Continued) AIO.C

```
/*
*********************************************************************************
*                    SET THE VALUE OF AN ANALOG OUTPUT CHANNEL
*
* Description : This function is used to set the currect value of an
*               analog output channel (in engineering units).
* Arguments   : n   is the analog output channel (0..AIO_MAX_AO-1).
*               val is the desired analog output value in Engineering Units
* Returns     : 0   if successfull.
*               1   if you specified an invalid analog output channel number.
*********************************************************************************
*/

INT8U AOSet (INT8U n, FP32 val)
{
    if (n < AIO_MAX_AO) {
        OS_ENTER_CRITICAL();
        AOTbl[n].AIOEU = val;           /* Set the engineering units of the analog
                                            output channel                        */
        OS_EXIT_CRITICAL();
        return (0);
    } else {
        return (1);
    }
}

/*$PAGE*/
```

Listing 11.1 (Continued) AIO.C

```
/*
*********************************************************************************************
*                     SET THE STATE OF THE BYPASSED ANALOG OUTPUT CHANNEL
*
* Description : This function is used to set the engineering units of a
*               bypassed analog output channel.
* Arguments   : n   is the analog output channel (0..AIO_MAX_AO-1).
*               val is the value of the bypassed analog output channel:
* Returns     : 0   if successfull.
*               1   if you specified an invalid analog output channel number.
*               2   if AIOBypassEn is not set to TRUE
*********************************************************************************************
*/

INT8U AOSetBypass (INT8U n, FP32 val)
{
    AIO *paio;


    if (n < AIO_MAX_AO) {
        paio = &AOTbl[n];                    /* Faster to use a pointer to the structure  */
        if (paio->AIOBypassEn == TRUE) { /* See if the analog output channel is
                                            bypassed                              */
            OS_ENTER_CRITICAL();
            paio->AIOScaleIn = val;          /* Yes, then set the new value of the channel */
            OS_EXIT_CRITICAL();
            return (0);
        } else {
            return (2);
        }
    } else {
        return (1);
    }
}

/*$PAGE*/
```

Listing 11.1 (Continued) AIO.C

```
/*
********************************************************************************
*                        SET THE STATE OF THE BYPASS SWITCH
*
* Description : This function is used to set the state of the bypass switch.
*               The analog output channel is bypassed when the 'switch' is
*               open (i.e. AIOBypassEn is set to TRUE).
* Arguments   : n     is the analog output channel (0..AIO_MAX_AO-1).
*               state is the state of the bypass switch:
*                       FALSE disables the bypass (i.e. the bypass 'switch' is closed)
*                       TRUE  enables the bypass (i.e. the bypass 'switch' is open)
* Returns     : 0     if successfull.
*               1     if you specified an invalid analog output channel number.
********************************************************************************
*/

INT8U AOSetBypassEn (INT8U n, BOOLEAN state)
{
    INT8U err;


    if (n < AIO_MAX_AO) {
        AOTbl[n].AIOBypassEn = state;
        return (0);
    } else {
        return (1);
    }
}

/*$PAGE*/
```

Listing 11.1 (Continued) AIO.C

```c
/*
**********************************************************************************
*                         UPDATE ALL ANALOG OUTPUT CHANNELS
*
* Description : This function processes all of the analog output channels.
* Arguments   : None.
* Returns     : None.
**********************************************************************************
*/

static void AOUpdate (void)
{
    INT8U    i;
    AIO      *paio;
    INT16S   raw;


    paio = &AOTbl[0];                          /* Point at first analog output channel */
    for (i = 0; i < AIO_MAX_AO; i++) {         /* Process all analog output channels    */
        if (paio->AIOBypassEn == FALSE) {      /* See if analog output channel is
                                                  bypassed                              */
            paio->AIOScaleIn = paio->AIOEU;    /* No                                    */
        }
        paio->AIOPassCtr--;                    /* Decrement pass counter                */
        if (paio->AIOPassCtr == 0) {           /* When pass counter reaches 0, read and
                                                  scale AI                              */
            paio->AIOPassCtr = paio->AIOPassCnts;        /* Reload pass counter         */
            if ((void *)paio->AIOScaleFnct != (void *)0) { /* See if function
                                                             defined                    */
                (*paio->AIOScaleFnct)(paio);             /* Yes, execute function       */
            } else {
                paio->AIOScaleOut = paio->AIOScaleIn; /* No, bypass scaling function    */
            }
            raw = (INT16S)(paio->AIOScaleOut * paio->AIOGain + paio->AIOOffset);
            if (raw > paio->AIOLim) {          /* Never output > maximum DAC counts     */
                raw = paio->AIOLim;
            } else if (raw < 0) {              /* DAC counts must always be >= 0         */
                raw = 0;
            }
            paio->AIORaw = raw;
            AOWr(i, paio->AIORaw);             /* Write counts to DAC                   */
        }
        paio++;                                /* Point at next AO channel               */
    }
}

/*$PAGE*/
```

Listing 11.1 (Continued) AIO.C

```
#ifndef CFG_C
/*
********************************************************************************
*                         INITIALIZE PHYSICAL I/Os
*
* Description : This function is called by AIOInit() to initialize the
*               physical I/O used by the AIO driver.
* Arguments   : None.
* Returns     : None.
********************************************************************************
*/

void AIOInitIO (void)
{
    /* This is where you will need to put you initialization code for the
       ADCs and DACs */
    /* You should also consider initializing the contents of your DAC(s)
       to a known value. */
}


/*
********************************************************************************
*                         READ PHYSICAL INPUTS
*
* Description : This function is called to read a physical ADC channel.
*               The function is assumed to also control a multiplexer if
*               more than one analog input is connected to the ADC.
* Arguments   : ch   is the ADC logical channel number (0..AIO_MAX_AI-1).
* Returns     : The raw ADC counts from the physical device.
********************************************************************************
*/

INT16S AIRd (INT8U ch)
{
    /* This is where you will need to provide the code to read your ADC(s).    */
    /* AIRd() is passed a 'LOGICAL' channel number. You will have to convert this */
       logical channel                                                         */
    /* number into actual physical port locations (or addresses) where your MUX. */
       and ADCs are located.                                                    */
    /* AIRd() is responsible for:                                               */
    /*     1) Selecting the proper MUX. channel,                                */
    /*     2) Waiting for the MUX. to stabilize,                                */
    /*     3) Starting the ADC,                                                 */
    /*     4) Waiting for the ADC to complete its conversion,                   */
    /*     5) Reading the counts from the ADC and,                              */
    /*     6) Returning the counts to the calling function.                     */

    return (ch);
}

/*$PAGE*/
```

Listing 11.1 (Continued) AIO.C

```
/*
*********************************************************************************
*                           UPDATE PHYSICAL OUTPUTS
*
* Description : This function is called to write the 'raw' counts to the
*               proper analog output device (i.e. DAC). It is up to this
*               function to direct the DAC counts to the proper DAC if more
*               than one DAC is used.
* Arguments   : ch    is the DAC logical channel number (0..AIO_MAX_AO-1).
*               cnts  are the DAC counts to write to the DAC
* Returns     : None.
*********************************************************************************

*/

void AOWr (INT8U ch, INT16S cnts)
{
    ch = ch;
    cnts = cnts;

    /* This is where you will need to provide the code to update your DAC(s).  */
    /* AOWr() is passed a 'LOGICAL' channel number. You will have to convert
       this logical channel number into actual physical port locations (or     */
    /* addresses) where your DACs are located.                                 */
    /* AOWr() is responsible for writing the counts to the selected DAC based
       on a logical number.                                                    */
}
#endif
```

Listing 11.2 AIO.H

```
/*
********************************************************************************
*                           Analog I/O Module
*
*                (c) Copyright 1999, Jean J. Labrosse, Weston, FL
*                            All Rights Reserved
*
* Filename    : AIO.H
* Programmer : Jean J. Labrosse
********************************************************************************
*/

#ifdef AIO_GLOBALS
#define AIO_EXT
#else
#define AIO_EXT extern
#endif

/*
********************************************************************************
*                           CONFIGURATION CONSTANTS
********************************************************************************
*/

#ifndef CFG_H

#define AIO_TASK_PRIO      40
#define AIO_TASK_DLY       100
#define AIO_TASK_STK_SIZE  512

#define AIO_MAX_AI          8 /* Maximum number of Analog Input Channels
                                 (1..250)                                    */
#define AIO_MAX_AO          8 /* Maximum number of Analog Output Channels
                                 (1..250)                                    */

#endif
/*$PAGE*/
```

Listing 11.2 (Continued) AIO.H

```
/*
*****************************************************************************
*                               DATA TYPES
*****************************************************************************
*/

typedef struct aio {          /* ANALOG I/O CHANNEL DATA STRUCTURE              */
    BOOLEAN AIOBypassEn;      /* Bypass enable switch (Bypass when TRUE)        */
    INT16S  AIORaw;           /* Raw counts of ADC or DAC                       */
    FP32    AIOEU;            /* Engineering units of AI channel                */
    FP32    AIOGain;          /* Total gain (AIOCalGain * AIOConvGain)          */
    FP32    AIOOffset;        /* Total offset (AIOCalOffset + AIOConvOffset)    */
    INT16S  AIOLim;           /* Maximum count of an analog output channel      */
    INT8U   AIOPassCnts;      /* Pass counts                                    */
    INT8U   AIOPassCtr;       /* Pass counter (loaded from PassCnts)            */
    FP32    AIOCalGain;       /* Calibration gain                               */
    FP32    AIOCalOffset;     /* Calibration offset                             */
    FP32    AIOConvGain;      /* Conversion gain                                */
    FP32    AIOConvOffset;    /* Conversion offset                              */
    FP32    AIOScaleIn;       /* Input to scaling function                      */
    FP32    AIOScaleOut;      /* Output from scaling function                   */
    void  (*AIOScaleFnct)(struct aio *paio);  /* Function to execute for further
                                      processing                                */
    void    *AIOScaleFnctArg;              /* Pointer to argument to pass to
                                      'AIOScaleFnct'                            */
} AIO;

/*
*****************************************************************************
*                            GLOBAL VARIABLES
*****************************************************************************
*/

AIO_EXT AIO    AITbl[AIO_MAX_AI];
AIO_EXT AIO    AOTbl[AIO_MAX_AO];

/*$PAGE*/
```

Listing 11.2 (Continued) AIO.H

```
/*
*********************************************************************************
*                               FUNCTION PROTOTYPES
*********************************************************************************
*/

void    AIOInit(void);

INT8U   AICfgCal(INT8U n, FP32 gain, FP32 offset);
INT8U   AICfgConv(INT8U n, FP32 gain, FP32 offset, INT8U pass);
INT8U   AICfgScaling(INT8U n, void (*fnct)(AIO *paio), void *arg);
INT8U   AISetBypass(INT8U n, FP32 val);
INT8U   AISetBypassEn(INT8U n, BOOLEAN state);
INT8U   AIGet(INT8U n, FP32 *pval);

INT8U   AOCfgCal(INT8U n, FP32 gain, FP32 offset);
INT8U   AOCfgConv(INT8U n, FP32 gain, FP32 offset, INT16S lim, INT8U pass);
INT8U   AOCfgScaling(INT8U n, void (*fnct)(AIO *paio), void *arg);
INT8U   AOSet(INT8U n, FP32 val);
INT8U   AOSetBypass(INT8U n, FP32 val);
INT8U   AOSetBypassEn(INT8U n, BOOLEAN state);

void    AIOInitIO(void);                          /* Hardware dependant functions   */
INT16S  AIRd(INT8U ch);
void    AOWr(INT8U ch, INT16S cnts);
```

CHAPTER 12

Optimizing DSP Software

Robert Oshana

Digital Signal Processors (DSPs), once a niche product used only in expensive, high-end apps with extreme computing requirements, are now found practically everywhere, from cell phones to digital cameras and in all sorts of data acquisition equipment. A DSP is a general purpose microprocessor, but it has special instructions that handle certain common signal processing algorithms efficiently, and they generally use two different data busses that can exchange data simultaneously. DSPs are found solely in those applications that have severe performance requirements.

In this chapter Rob Oshana covers three intertwined subjects:

- *What a DSP can do*

- *How the compiler converts C code to DSP-specific sequences of instructions*

- *How the programmer can help the compiler do its thing more—much more—efficiently.*

Using a DSP is not like writing Visual C++ code. You need to be intimately familiar with many concepts that are totally alien to the IT or PC programmer. The way one manages DMA, interrupts, various flavors of pipelines, all drastically affect system performance.

Dispelling one myth, Rob shows that cache is not always a good answer to speed problems, as it invariably leads to determinacy issues that, while not even noticeable when building a spreadsheet or web browser, can destroy real-time response. He then takes on the old chestnut that a good assembly programmer can beat any compiler. With instruction queues, pipelined execution, and multiple execution units this just isn't so. A compiler can guarantee that the processor's pipeline is always full, which is perhaps possible for the human developer, but ineffably difficult and tedious.

Yet, he shows though example after example how important it is for the DSP programmer to know assembly language. As he says, compilers are pessimistic creatures. Without some form of a priori information more extensive than can be gleaned from the code itself, they must *make conservative decisions to insure the program compiles correctly. In the real-time world, correct is not enough. The compiler will need some expert guidance from time to time to help it understand the program's relationship to externalities, in order to produce code that both works and runs as fast as possible.*

Rob gives a great list of the optimizations made by compilers, what each does, and what to look out for when designing real-world, real-time systems. Though other tomes talk about these optimizations, few give guidance on their effects when using the multiple execution units that are increasingly common in DSPs.

Like any good work of pedagogy, Rob's chapter concludes with a step-by-step guide to optimizing your DSP code. All developers working with these sorts of processors need this information.

—Jack Ganssle

12.1   Introduction

Many of today's DSP applications are subject to real-time constraints. Many embedded DSP applications will eventually grow to a point where they are stressing the available CPU, memory, or power resources. Understanding the workings of the DSP architecture, compiler, and application algorithms can speed up applications, sometimes by an order of magnitude. This chapter will summarize some of the techniques that can improve the performance of your code in terms of cycle count, memory use, and power consumption.

12.2   What Is Optimization?

Optimization is a procedure that seeks to maximize or minimize one or more performance indices. These indices include:

- Throughput (execution speed)

- Memory usage

- I/O bandwidth

- Power dissipation

Since many DSP systems are real-time systems, at least one (and probably more) of these indices must be optimized. It is difficult (and usually impossible) to optimize all these performance indices at the same time. For example, to make the application faster, the developer may require more memory to achieve the goal. The designer must weigh each of these indices and make the best trade-off. The tricky part to optimizing DSP applications is understanding the trade-off between the various performance indices. For example, optimizing an application for speed often means a corresponding decrease in power consumption but an increase in memory usage. Optimizing for memory may also result in a decrease in power consumption due to fewer memory accesses but an offsetting decrease in code performance. The various trade-offs and system goals must be understood and considered before attempting any form of application optimization.

Determining which index or set of indices is important to optimize depends on the goals of the application developer. For example, optimizing for performance means that the developer can use a slow or less expensive DSP to do the same amount of work. In some embedded systems, cost savings like this can have a significant impact on the success of the product. The developer can alternatively choose to optimize the application to allow the addition of more functionality. This may be very important if the additional functionality improves the overall performance of the system, or if the developer can add more capability to the system such as an additional channel of a base station system. Optimizing for memory use can also lead to overall system cost reduction. Reducing the application size leads to a lower demand for memory, which reduces overall system cost. Finally, optimizing for power means that the application can run longer on the same amount of power. This is important for battery powered applications. This type of optimization also reduces the overall system cost with respect to power supply requirements and other cooling functionality required.

12.3 The Process

Generally, DSP optimization follows the 80/20 rule. This rule states that 20% of the software in a typical application uses 80% of the processing time. This is especially true for DSP applications that spend much of their time in tight inner loops of DSP algorithms. Thus, the real issue in optimization isn't how to optimize, but where to optimize. The first rule of optimization is "Don't!" Do not start the optimization process until you have a good understanding of where the execution cycles are being spent.

The best way to determine which parts of the code should be optimized is to profile the application. This will answer the question as to which modules take the longest to execute. These will become the best candidates for performance-based optimization. Similar questions can be asked about memory usage and power consumption.

DSP application optimization requires a disciplined approach to get the best results. To get the best results out of your DSP optimization effort, the following process should be used:

- **Do your homework**: Make certain you have a thorough understanding of the DSP architecture, the DSP compiler, and the application algorithms. Each target processor and compiler has different strengths and weaknesses and understanding them is critical to successful software optimization. Today's DSP optimizing compilers are advanced. Many allow the developer to use a higher order language such as C and very little, if any, assembly language. This allows for faster code development, easier debugging, and more reusable code. But the developer must understand the "hints" and guidelines to follow to enable the compiler to produce the most efficient code.

- **Know when to stop**: Performance analysis and optimization is a process of diminishing returns. Significant improvements can be found early in the process with relatively little effort. This is the "low hanging fruit." Examples of this include accessing data from fast on-chip memory using the DMA and pipelining inner loops. However, as the optimization process continues, the effort expended will increase dramatically and further improvements and results will fall dramatically.

- **Change one parameter at a time**: Go forward one step at a time. Avoid making several optimization changes at the same time. This will make it difficult to determine what change led to which improvement percentage. Retest after each significant change in the code. Keep optimization changes down to one change per test in order to know exactly how that change affected the whole program. Document these results and keep a history of these changes and the resulting improvements. This will prove useful if you have to go back and understand how you got to where you are.

- **Use the right tools**: Given the complexity of modern DSP CPUs and the increasing sophistication of optimizing compilers, there is often little correlation between what a programmer thinks is optimized code and what actually performs well. One of the most useful tools to the DSP programmer is the profiler. This is a tool that allows the developer to run an application and get a "profile" of where cycles are being used throughput the program. This allows the developer to identify and focus on the core bottlenecks in the program quickly. Without a profiler, gross performance issues as well as minor code modifications can go unnoticed for long periods of time and make the entire code optimization process less disciplined.

- **Have a set of regression tests and use them after each iteration**: Optimization can be difficult. More difficult optimizations can result in subtle changes to the program behavior that lead to wrong answers. More complex code optimizations in the compiler can, at times, produce incorrect code (a compiler, after all, is a software program with its own bugs!). Develop a test plan that compares the expected results to the actual results of the software program. Run the test regression often enough to catch problems early. The programmer must verify that program optimizations have not broken the application. It is extremely difficult to backtrack optimized changes out of a program when a program breaks.

A general code optimization process (see Figure 12-1) consists of a series of iterations. In each iteration, the programmer should examine the compiler-generated code and look for optimization opportunities. For example, the programmer may look for an abundance of NOPs or other inefficiencies in the code due to delays in accessing memory and/or another processor resource. These are the areas that become the focus of improvement. The programmer will apply techniques such as software pipelining, loop unrolling, DMA resource utilization, etc., to reduce the processor cycle count (we will talk more about

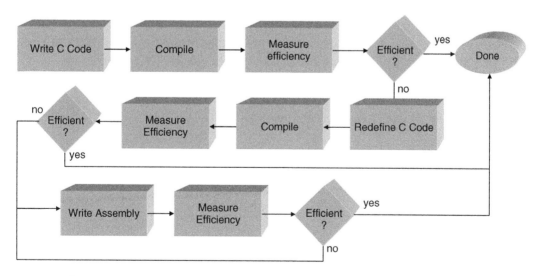

Figure 12-1: A General DSP Code Optimization Process (courtesy of Texas Instruments).

these specific techniques later). As a last resort the programmer can consider hand-tuning the algorithms using assembly language.

Many times, the C code can be modified slightly to achieve the desired efficiency, but finding the right "tweak" for the optimal (or close to optimal) solution can take time and several iterations. Keep in mind that the software engineer/programmer must take responsibility for at least a portion of this optimization. There have been substantial improvements in production DSP compilers with respect to advanced optimization techniques. These optimizing compilers have grown to be quite complex due to the advanced algorithms used to identify optimization opportunities and make these code transformations. With this increased complexity comes the opportunity for errors in the compiler. You still need to understand the algorithms and the tools well enough that you can supply the necessary improvements when the compiler can't. In this chapter we will discuss how to optimize DSP software in the the context of this process.

12.4 Make the Common Case Fast

The fundamental rule in computer design as well as programming real-time DSP-based systems is "make the common case fast, and favor the frequent case." This is really just Amdahl's Law that says the performance improvement to be gained using some faster mode of execution is limited by how often you use that faster mode of execution. So don't spend time trying to optimize a piece of code that will hardly ever run. You won't get much out of it, no matter how innovative you are. Instead, if you can eliminate just one cycle from a loop that executes thousands of times, you will see a bigger impact on the bottom line. I will now discuss three different approaches to making the common case fast (by common case, I am referring to the areas in the code that consume the most resources in terms of cycles, memory, or power):

- Understand the DSP architecture.

- Understand the DSP algorithms.

- Understand the DSP compiler.

12.5 Make the Common Case Fast: DSP Architectures

DSP architectures are designed to make the common case fast. Many DSP applications are composed from a standard set of DSP building blocks such as filters, Fourier transforms, and convolutions. Table 12-1 contains a number of these common DSP algorithms. Notice the common structure of each of the algorithms:

- They all accumulate a number of computations.

- They all sum over a number of elements.

- They all perform a series of multiplies and adds.

Table 12-1 DSP algorithms share common characteristics

Algorithm	Equation
Finite Impulse Response Filter	$y(n) = \sum\limits_{k=0}^{M} a_k x(n-k)$
Infinite Impulse Response Filter	$y(n) = \sum\limits_{k=0}^{M} a_k x(n-k) + \sum\limits_{k=1}^{N} b_k y(n-k)$
Convolution	$y(n) = \sum\limits_{k=0}^{N} x(k) h(n-k)$
Discrete Fourier Transform	$X(k) = \sum\limits_{n=0}^{N-1} x(n) \exp[-j(2\pi/N)nk]$
Discrete Cosine Transform	$F(u) = \sum\limits_{x=0}^{N-1} c(u).f(x).\cos\left[\frac{\pi}{2N} u(2x+1)\right]$

These algorithms all share some common characteristics; they perform multiplies and adds over and over again. This is generally referred to as the **sum of products (SOP)**.

Like we discussed earlier, a DSP is, in many ways, an application specific microprocessor. DSP designers have developed hardware architectures that allow the efficient execution of algorithms to take advantage of the algorithmic specialty in signal processing. For example, some of the specific architectural features of DSPs to accommodate the algorithmic structure of DSP algorithms include:

- Special instructions such as a single cycle multiple and accumulate (MAC). Many signal processing algorithms perform many of these operations in tight loops. Figure 12-2 shows the savings from computing a multiplication in hardware instead of microde in the DSP processor. A savings of four cycles is significant when multiplications are performed millions of times in signal processing applications.

Hardware		Software/microcode	
1011		1001	
x 1110		x 1010	
1011010		0000	Cycle 1
		1001.	Cycle 2
		0000..	Cycle 3
		1001...	Cycle 4
		1011010	Cycle 5

Figure 12-2: Special Multiplication Hardware Speeds Up DSP Processing (courtesy of Texas Instruments).

- Large accumulators to allow for accumulating a large number of elements.

- Special hardware to assist in loop checking so this does not have to be performed in software, which is much slower.

- Access to two or more data elements in the same cycle. Many signal processing algorithms multiple two arrays of data and coefficients. Being able to access two operands at the same time makes these operations very efficient. The DSP Harvard architecture shown in Figure 12-3 allows for access of two or more data elements in the same cycle.

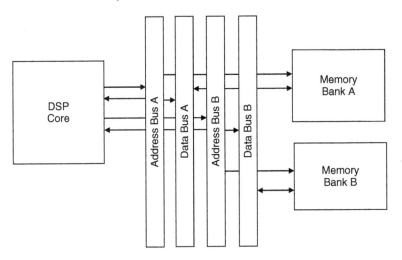

Figure 12-3: DSP Harvard Architecture: Multiple Address and Data Busses Accessing Multiple Banks of Memory Simultaneously

The DSP developer must choose the right DSP architecture to accommodate the signal processing algorithms required for the application as well as the other selection factors such as cost, tools support, etc.

12.6 Make the Common Case Fast: DSP Algorithms

DSP algorithms can be made to run faster using techniques of algorithmic transformation. For example, a common algorithm used in DSP applications is the Fourier transform. The Fourier transform is a mathematical method of breaking a signal in the time domain into all of its individual frequency components. The process of examining a time signal broken down into its individual frequency components is also called **spectral analysis** or **harmonic analysis**.

> Brigham, E. Oren, *The Fast Fourier Transform and Its Applications*. Englewood Cliffs, NJ: Prentice-Hall, Inc., 1988, p. 448.

There are different ways to characterize a Fourier transform:

- The Fourier transform (FT) is a mathematical formula using integrals:

$$F(u) = \int_{-\infty}^{\infty} f(x)e^{-x\pi ixu}d\omega$$

- The discrete Fourier transform (DFT) is a discrete numerical equivalent using sums instead of integrals, which maps well to a digital processor like a DSP:

$$F(u) = \frac{1}{N}\sum_{x=0}^{N-1} f(x)e^{-x\pi i\omega ufN}{}_j$$

- The fast Fourier transform (FFT) is just a computationally fast way to calculate the DFT, which reduces many of the redundant computations of the DFT.

How these are implemented on a DSP has a significant impact on overall performance of the algorithm. The FFT, for example, is a fast version of the DFT. The FFT makes use of periodicities in the sines that are multiplied to perform the transform. This significantly reduces the amount of calculations required. A DFT implementation requires N^2 operations to calculate an N point transform. For the same N point data set, using an FFT algorithm requires $N*\log 2(N)$ operations. The FFT is therefore faster than the DFT by a factor of N/log2(n). The speedup for an FFT is more significant as N increases (Figure 12-4).

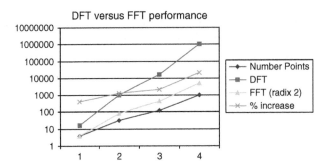

Figure 12-4: DFT vs. FFT for Various Sizes of Transforms (logarithmic scale).

Recognizing the significant impact that efficiently implemented algorithms have on overall system performance, DSP vendors and other providers have developed libraries of efficient DSP algorithms optimized for specific DSP architectures. Depending on the type of algorithm, these can be downloaded from web sites (be careful of obtaining free software like this—the code may be buggy as there is no guarantee of quality) or bought from DSP solution providers.

12.7   Make the Common Case Fast: DSP Compilers

Just a few years ago, it was an unwritten rule that writing programs in assembly would usually result in better performance than writing in higher level languages like C or C++. The early "optimizing" compilers produced code that was not as good as what one could get by programming in assembly language, where an experienced programmer generally achieves better performance. Compilers have gotten much better and today there are very specific high performance optimizations performed that compete well with even the best assembly language programmers.

Optimizing compilers perform sophisticated program analysis including intraprocedural and interprocedural analysis. These compilers also perform data and control flow analysis as well as dependence analysis and often employ provably correct methods for modifying or transforming code. Much of this analysis is to prove that the transformation is correct in the general sense. Many optimization strategies used in DSP compilers are also strongly heuristic.

Heuristics involves problem-solving by experimental and especially trial-and-error methods or relating to exploratory problem-solving techniques that utilize self-educating techniques (as the evaluation of feedback) to improve performance (*Webster's English Language Dictionary*).

One effective code optimization strategy is to write DSP application code that can be **pipelined** efficiently by the compiler. **Software** pipelining is an optimization strategy to schedule loops and functional units efficiently. In modern DSPs there are multiple functional units that are orthogonal and can be used at the same time (Figure 12-5). The compiler is given the burden of figuring out how to schedule instructions so that these functional units can be used in parallel whenever possible. Sometimes this is a matter of a subtle change in the way the C code is structured that makes all the difference. In software pipelining, multiple iterations of a loop are scheduled to execute in parallel. The loop is reorganized in a way that each iteration in the pipelined code is made from instruction sequences selected from different iterations in the original loop. In the example in Figure 12-6, a five-stage loop with three iterations is shown. There is an initial period (cycles n and n + 1), called the **prolog** when the pipes are being "primed" or initially loaded with operations. Cycles n + 2 to n + 4 are the actual pipelined section of the code. It is in this section that the processor is performing three different operations

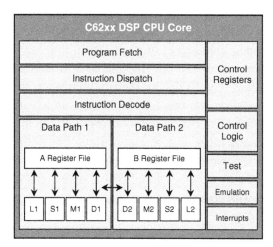

Figure 12-5: DSP architectures may have orthogonal execution units and data paths used to execute DSP algorithms more efficiently. In this figure, units L1, S1, M1, D1, and L2, S2, M2, and D2 are all orthogonal execution units that can have instructions scheduled for execution by the compiler in the same cycle if the conditions are right.

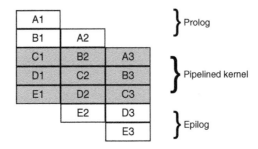

Figure 12-6: A Five-Stage Instruction Pipeline That Is Scheduled to be Software Pipelined by the Compiler.

(C, B, and A) for three different loops (1, 2, and 3). There is an epilog section where the last remaining instructions are performed before exiting the loop. This is an example of a fully utilized set of pipelines that produces the fastest, most efficient code.

Figure 12-7 shows a sample piece of C code and the corresponding assembly language output. In this case the compiler was asked to attempt to pipeline the code. This is evident by the piped loop prolog and piped loop kernel sections in the assembly language output. Keep in mind that the prolog and epilog sections of the code prime the pipe and flush the pipe, respectively, as shown in Figure 12-6. In this case, the pipelined code is not as good as it could be. You can spot inefficient code by looking for how many NOPs are in the piped loop kernel of the code. In this case the piped loop kernel has a total of five NOP cycles, two in line 16, and three in line 20. This loop takes a total of 10 cycles to execute. The NOPs are the first indication that a more efficient loop may be possible. But how short can this loop be? One way to estimate the minimum loop size is to determine what execution unit is being used the most. In this example, the D unit is used more than any other unit, a total of three times (lines 14, 15, and 21). There are two sides to a superscalar device, enabling each unit to be used twice (D1 and D2) per clock for a minimum two clock loop; two D operations in one clock and one D unit in the second clock. The compiler was smart enough to use the D units on both sides of the pipe (lines 14 and 15), enabling it to parallelize the instructions and only use one clock. It should be possible to perform other instructions while waiting for the loads to complete, instead of delaying with NOPs.

In the simple "for" loop, the two input arrays (array1 and array2) may or may not be dependent or overlap in memory. The same with the output array. In a language such as C/C++ this is something that is allowed and therefore the compiler must be able to

```
1      void  example1(float *out, float *input1, loat *input2)
2      {
3        int i;
4
5        for(i = 0; i < 100; i++)
6          {
7            out[i] = input1[i] * input2[i];
8          }
9        }
```

```
1      _example1:
2      ;** ---------------------------------------------------*
3              MVK           .S2           0x64,B0
4
5              MVC           .S2           CSR, B6
6      ||      MV            .L1X          B4,A3
7      ||      MV            .L2X          A6,B5
8
8              AND           .L1X          -2,B6,A0
9              MVC            .S2X         A0,CSR
10     ;** ---------------------------------------------------*
11     L11:       ; PIPED LOOP PROLOG
12     ;** ---------------------------------------------------*
13     L12:       ; PIPED LOOP  KERNEL
14
14              LDW           .D2           *B5++,B4      ;
15     ||      LDW           .D1           *A3++,A0      ;
16
16              NOP                         2
17     [B0]     SUB           .L2           B0,1,B0       ;
18     [B0]     B             .S2           L12           ;
19              MPYSP         .M1X          B4,A0,A0      ;
20              NOP                         3
21              STW           .D1           A0,*A4++      ;
22     ;** ---------------------------------------------------*
23              MVC           .S2           B6,CSR
24              B             .S2           B3
25              NOP                         5
26              ; BRANCH OCCURS
```

**Figure 12-7: C Example and the Corresponding Pipelined Assembly
Language Output.**

handle this correctly. Compilers are generally pessimistic creatures. They will not attempt an optimization if there is a case where the resultant code will not execute properly. In these situations, the compiler takes a conservative approach and assumes the inputs can be dependent on the previous output each time through the loop. If it is known that the inputs are not dependent on the output, we can hint to the compiler by declaring input1 and input2 as "restrict," indicating that these fields will not change. In this example, "restrict" is a keyword in C that can be used for this purpose. This is also a trigger for enabling software pipelining which can improve throughput. This C code is shown in Figure 12-8 with the corresponding assembly language.

There are a few things to notice in looking at this assembly language. First, the piped loop kernel has become smaller. In fact, the loop is now only two cycles long. Lines 44–47 are all executed in one cycle (the parallel instructions are indicated by the || symbol) and lines 48–50 are executed in the second cycle of the loop. The compiler, with the additional dependency information we supplied it with the "restrict" declaration, has been able to take advantage of the parallelism in the execution units to schedule the inner part of the loop very efficiently. The prolog and epilog portions of the code are much larger now. Tighter piped kernels will require more priming operations to coordinate all of the execution based on the various instruction and branching delays. But once primed, the kernel loop executes extremely fast, performing operations on various iterations of the loop. The goal of software pipelining is, like we mentioned earlier, to make the common case fast. The kernel is the common case in this example, and we have made it very fast. Pipelined code may not be worth doing for loops with a small loop count. But for loops with a large loop count, executing thousands of times, software pipelining produces significant savings in performance while also increasing the size of the code.

In the two cycles the piped kernel takes to execute, there are a lot of things going on. The right-hand column in the assembly listing indicates what iteration is being performed by each instruction (Each "@" symbol is an iteration count. So, in this kernel, line 44 is performing a branch for iteration $n + 2$, lines 45 and 46 are performing loads for iteration $n + 4$, line 48 is storing a result for iteration n, line 49 is performing a multiply for iteration $n + 2$, and line 50 is performing a subtraction for iteration $n + 3$, all in two cycles!). The epilog is completing the operations once the piped kernel stops executing. The compiler was able to make the loop two cycles long, which is what we predicted by looking at the inefficient version of the code.

The code size for a pipelined function becomes larger, as is obvious by looking at the code produced. This is one of the trade-offs for speed that the programmer must make.

```
1      void example2(float *out, restrict float *input1,
         restrict float *input2)
2      {
3        int i;
4
5        for(i = 0; i < 100; i++)
6          {
7            out[i] = input1[i] * input2[i];
8          }
9      }


1      _example2:
2      ;** -------------------------------------------------*
3                      MVK         .S2         0x64,B0
4
5                      MVC         .S2         CSR,B6
6      ||             MV          .L1X        B4,A3
7      ||             MV          .L2X        A6,B5
8
9                      AND         .L1X        -2,B6,A0
10
11                     MVC         .S2X        A0,CSR
12     ||             SUB         .L2         B0,4,B0
13
14     ;** --------------------------------------------------*
15     L8:             ; PIPED LOOP PROLOG
16
17                     LDW         .D2         *B5++,B4      ;
18     ||             LDW         .D1         *A3++,A0      ;
19
20                     NOP                     1
21
22                     LDW         .D2         *B5++,B4      ;@
23     ||             LDW         .D1         *A3++,A0      ;@
24
25         [ B0]      SUB         .L2         B0,1,B0       ;
26
27         [ B0]      B           .S2         L9            ;
28     ||             LDW         .D2         *B5++,B4      ;@@
29     ||             LDW         .D1         *A3++,A0      ;@@
30
31                     MPYSP       .M1X        B4,A0,A5      ;
```

Figure 12-8: Corresponding Pipelined Assembly Language Output.

```
32   ||    [ B0]   SUB       .L2      B0,1,B0         ;@
33
34         [ B0]   B         .S2      L9              ;@
35   ||            LDW       .D2      *B5++,B4        ;@@@
36   ||            LDW       .D1      *A3++,A0        ;@@@
37
38               MPYSP       .M1X     B4,A0,A5        ;@
39   ||    [ B0]   SUB       .L2      B0,1,B0         ;@@
40
41   ;** ------------------------------------------------*
42   L9:           ; PIPED LOOP KERNEL
43
44         [ B0]   B         .S2      L9              ;@@
45   ||            LDW       .D2      *B5++,B4        ;@@@@
46   ||            LDW       .D1      *A3++,A0        ;@@@@
47
48               STW        .D1      A5,*A4++        ;
49   ||          MPYSP       .M1X     B4,A0,A5        ;@@
50   ||    [ B0]   SUB       .L2      B0,1,B0         ;@@@
51
52   ;** ------------------------------------------------*
53   L10:          ; PIPED LOOP EPILOG
54               NOP                  1
55
56               STW        .D1      A5,*A4++        ;@
57   ||          MPYSP       .M1X     B4,A0,A5        ;@@@
58
59               NOP                  1
60
61               STW        .D1      A5,*A4++        ;@@
62   ||          MPYSP       .M1X     B4,A0,A5        ;@@@@
63
64               NOP                  1
65               STW        .D1      A5,*A4++        ;@@@
66               NOP                  1
67               STW        .D1      A5,*A4++        ;@@@@
68   ;** ------------------------------------------------*
69               MVC        .S2      B6,CSR
70               B          .S2      B3
71               NOP                  5
72               ; BRANCH OCCURS
```

Figure 12-8: Continued

Software pipelining does not happen without careful analysis and structuring of the code. Small loops that do not have many iterations may not be pipelined because the benefits are not realized. Loops that are large in the sense that there are many instructions per iteration that must be performed may not be pipelined because there are not enough processors resources (primarily registers) to hold the key data during the pipeline operation. If the compiler has to "spill" data to the stack, precious time will be wasted having to fetch this information from the stack during the execution of the loop.

12.8   An In-Depth Discussion of DSP Optimization

While DSP processors offer tremendous potential throughput, your application won't achieve that potential unless you understand certain important implementation techniques. We will now discuss key techniques and strategies that greatly reduce the overall number of DSP CPU cycles required by your application. For the most part, the main object of these techniques is to fully exploit the potential parallelism in the processor and in the memory subsystem. The specific techniques covered include:

- Direct memory access;

- Loop unrolling; and

- More on software pipelining.

12.9   Direct Memory Access

Modern DSPs are extremely fast; so fast that the processor can often compute results faster than the memory system can supply new operands—a situation known as "data starvation." In other words, the bottleneck for these systems becomes keeping the unit fed with data fast enough to prevent the DSP from sitting idle waiting for data. Direct memory access is one technique for addressing this problem.

Direct memory access (DMA) is a mechanism for accessing memory without the intervention of the CPU. A peripheral device (the DMA controller) is used to write data directly to and from memory, taking the burden off the CPU. The DMA controller is just another type of CPU whose only function is moving data around very quickly. In a DMA capable machine, the CPU can issue a few instructions to the DMA controller, describing what data is to be moved (using a data structure called a **transfer control block** (TCB)), and then go back to what it was doing, creating another opportunity for

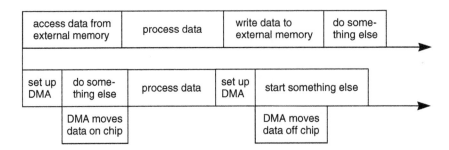

Figure 12-9: Using DMA instead of the CPU can offer big performance improvements because the DMA handles the movement of the data while the CPU is busy performing meaningful operations on the data.

parallelism. The DMA controller moves the data in parallel with the CPU operation (Figure 12-9), and notifies the CPU when the transfer is complete.

DMA is most useful for copying larger blocks of data. Smaller blocks of data do not have the payoff because the setup and overhead time for the DMA makes it worthwhile just to use the CPU. But when used smartly, the DMA can result in huge time savings. For example, using the DMA to stage data on- and off-chip allows the CPU to access the staged data in a single cycle instead of waiting multiple cycles while data is fetched from slower external memory.

12.9.1 Using DMA

Because of the large penalty associated with accessing external memory, and the cost of getting the CPU involved, the DMA should be used wherever possible. The code for this is not too overwhelming. The DMA requires a data structure to describe the data it is going to access (where it is, where its going, how much, etc.). A good portion of this structure can be built ahead of time. Then it is simply a matter of writing to a memory-mapped DMA enable register to start the operation (Figure 12-10). It is best to start the DMA operation well ahead of when the data is actually needed. This gives the CPU something to do in the meantime and does not force the application to wait for the data to be moved. Then, when the data is actually needed, it is already there. The application should check to verify the operation was successful and this requires checking a register. If the operation was done ahead of time, this should be a one time poll of the register, and not a spin on the register, chewing up valuable processing time.

```
----------------------------------------------------------------
/* Addresses of some of the important DMA registers */
#define DMA_CONTROL_REG    (*(volatile unsigned*)0x40000404)
#define DMA_STATUS_REG     (*(volatile unsigned*)0x40000408)
#define DMA_CHAIN_REG        (*(volatile unsigned*)0x40000414)

/* macro to wait for the DMA to complete and signal the
   status register */
#define DMA_WAIT                  while(DMA_STATUS_REG&1) {}

/* pre-built tcb structure */
typedef struct {

    tcb setup fields

} DMA_TCB;
----------------------------------------------------------------
extern DMA_TCB tcb;

/* set up the remaining fields of the tcb structure -
   where you want the data to go, and how much you want to send */
tcb.destination_address = dest_address;
tcb.word_count = word_count;

/* writing to the chain register kicks off the DMA operation */
DMA_CHAIN_REG = (unsigned)&tcb;

Allow the CPU to do other meaningful work....

/* wait for the DMA operation to complete */
DMA_WAIT;
```

Figure 12-10: Code to set up and enable a DMA operation is pretty simple. The main operations include setting up a data structure (called a TCB in this example) and performing a few memory mapped operations to initialize and check the results of the operation.

12.9.2 Staging Data

The CPU can access on-chip memory much faster than off-chip or external memory. Having as much data as possible on chip is the best way to improve performance. Unfortunately, because of cost and space considerations most DSPs do not have a lot of on-chip memory. This requires the programmer to coordinate the algorithms in such a way to efficiently use the available on-chip memory. With limited on-chip memory, data must be staged on- and off-chip using the DMA. All of the data transfers can be happening in the background, while the CPU is actually crunching the data. Once the data is in internal memory, the CPU can access the data in on-chip memory very quickly (Figure 12-12).

Smart layout and utilization of on-chip memory, and judicious use of the DMA can eliminate most of the penalty associated with accessing off-chip memory. In general, the rule is to stage the data in and out of on-chip memory using the DMA and generate the results on chip. Figure 12-11 shows a template describing how to use the DMA to stage blocks of data on and off chip. This technique uses a double-buffering mechanism to stage the data. This way the CPU can be processing one buffer while the DMA is staging the other buffer. Speed improvements over 90% are possible using this technique.

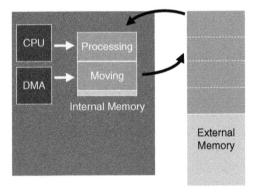

Figure 12-11: Template for Using the DMA to Stage Data On and Off Chip

Writing DSP code to use the DMA does have some cost penalties. Code size will increase, depending on how much of the application uses the DMA. Using the DMA also adds increased complexity and synchronization to the application. Code portability is reduced when you add processor specific DMA operations. Using the DMA should only be done in areas requiring high throughput.

```
INITIALIZE TCBS

DMA SOURCE DATA 0 INTO SOURCE BUFFER 0
WAIT FOR DMA TO COMPLETE

DMA SOURCE DATA 1 INTO SOURCE BUFFER 1
PERFORM CALCULATION AND STORE IN RESULT BUFFER

FOR LOOP_COUNT =1 TO N-1
   WAIT FOR DMA TO COMPLETE
   DMA SOURCE DATA I+1 INTO SOURCE BUFFER [(I+1)%2]
   DMA RESULT BUFFER[(I-1)%2] TO DESTINATION DATA
   PERFORM CALCULATION AND STORE IN RESULT BUFFER
END FOR

WAIT FOR DMA TO COMPLETE
DMA RESULT BUFFER[(I-1)%2] TO DESTINATION DATA
PERFORM CALCULATION AND STORE IN RESULT BUFFER

WAIT FOR DMA TO COMPLETE
DMA LAST RESULT BUFFER TO DESTINATION DATA
```

Figure 12-12: With limited on-chip memory, data can be staged in and out of on-chip memory using the DMA and leaving the CPU to perform other processing.

An Example

As an example of this technique, consider the code in Figure 12-13. This code snippet sums a data field and computes a simple percentage before returning. The code in Figure 12-13 consists of 5 executable lines of code. In this example, the "processed_data" field is assumed to be in external memory of the DSP. Each access of a processed_data element in the loop will cause an external memory access to fetch the data value.

The code shown in Figure 12-14 is the same function shown in Figure 12-13 but implemented to use the DMA to transfer blocks of data from external memory to internal or on-chip memory. This code consists of 36 executable lines of code, but runs much faster than the code in Figure 12-13. The overhead associated with setting up the

```
        int i;
        float sum ;

        /*
        **sum data field
        */
        sum = 0.0f;
        for(i=0; i<num_data_points; i++;)
        {
                sum += processed_data[i];
        }

        /*
        **Compute percentage and return
        */
        return(MTH_divide(sum,num_data_points));
        } /* end */
```

Figure 12-13: A Simple Function Consisting of Five Executable Lines of Code.

DMA, building the transfer packets, initiating the DMA transfers, and checking for completion are relatively small compared to the fast memory accesses of the data from on-chip memory. This code snippet was instrumented following the guidelines in the template of Figure 12-11. This code also performs a loop unrolling operation when summing the data at the end of the computation (we'll talk more about loop unrolling later). This also adds to the speedup of this code. This code snippet uses semaphores to protect the on-chip memory and DMA resources. Semaphores and other operating system functions are discussed in a separate chapter on real-time operating systems.

The code in Figure 12-14 runs much faster than the code in Figure 12-13. The penalty is an increased number of lines of code, which takes up memory space. This may or may not be a problem, depending on the memory constraints of the system. Another drawback to the code in Figure 12-14 is that it is a bit less understandable and portable than the code in Figure 12-13. Implementing the code to use the DMA requires the programmer to make the code less readable, which could possibly lead to maintainability problems. The code is now also tuned for a specific DSP. Porting the code to another DSP family may require the programmer to re-write this code to use the different resources on the new DSP.

```
#define NUM_BLOCKS 2
#define BLOCK_SIZE(DATA_SIZE/NUM_BLOCKS)

int i, block;
float sum = 0;
float sum0, sum1, sum2, sum3, sum4, sum5,
      sum6, sum7;

/* array contains pointers to internal memory buffers*/
float *processed_data[2];
unsigned index = 0;

/* wait for on chip memory semaphore */
SEM_ pend(g_aos_onchip_avail_sem, SYS_FOREVER);

MDMA_build_1d (tcb,                /* tcb pointer */
    0,                             /* prev tcb in chain */
(void*)processed_data[0]           /* destination address */
(void*)p_processed_data,           /* source addr */
BLOCK_SIZE);                       /* num words to tran */

MDMA_update (tcb,                  /* tcb pointer */
  MDMA_ADD,BLOCK_SIZE              /* src update mode */
  MDMA_TOGGLE,BLOCK_SIZE           /* dst update mode*/

MDMA_initiate_chain(1_tcb);

for(block=0, block<NUM_BLOCKS; block++)
{
  /* point to current buffers */
  internal_processed_data = processed_data[index];

  /* swap buffers */
  index ^= 1;
```

Figure 12-14: The same function enhanced to use the DMA. This function is 36 executable lines of code but runs much faster than the code in Figure 12-13.

```
   if(block < (NUM_BLOCKS - 1))
   {
      MDMA_initiate_chain(1-tcb);
   }
   else
   {
      MDMA_wait();
   }
/*
** sum data fields to compute percentages
*/
   sum0 = 0.0;
   sum1 = 0.0;
   sum2 = 0.0;
   sum3 = 0.0;
   sum4 = 0.0;
   sum5 = 0.0;
   sum6 = 0.0;
   sum7 = 0.0;

for (i=0; i<BLOCK_SIZE; i+=8)
{
   sum0 += internal_processed_data[i  ];
   sum1 += internal_processed_data[i + 1];
   sum2 += internal_processed_data[i + 2];
   sum3 += internal_processed_data[i + 3];
   sum4 += internal_processed_data[i + 4];
   sum5 += internal_processed_data[i + 5];
   sum6 += internal_processed_data[i + 6];
   sum7 += internal_processed_data[i + 7];
}

  sum += sum0 + sum1 + sum2 + sum3 + sum4 +
         sum5 + sum6 + sum7;


} /* block loop */

/* release on chip memory semaphore */
SEM_post(g_aos_onchip _avail_ sem);
```

Figure 12-14: Continued

12.9.3 Pending vs. Polling

The DMA can be considered a resource, just like memory and the CPU. When a DMA operation is in progress, the application can either wait for the DMA transfer to complete or continue processing another part of the application until the data transfer is complete. There are advantages and disadvantages to each approach. If the application waits for the DMA transfer to complete, it must poll the DMA hardware status register until the completion bit is set. This requires the CPU to check the DMA status register in a looping operation that wastes valuable CPU cycles. If the transfer is short enough, this may only require a few cycles to do and may be appropriate. For longer data transfers, the application engineer may want to use a synchronization mechanism like a semaphore to signal when the transfer is complete. In this case, the application will **pend** on a semaphore through the operating system while the transfer is taking place. The application will be swapped with another application that is ready to run. This swapping of tasks incurs overhead as well and should not be performed unless the overhead associated with swapping tasks is less than the overhead associated with simply polling on the DMA completion. The wait time depends on the amount of data being transferred.

Figure 12-15 shows some code that checks for the transfer length and performs either a DMA polling operation (if there are only a few words to transfer), or a semaphore pend operation (for larger data transfer sizes). The "break even" length for data size is dependent on the processor and the interface structure and should be prototyped to determine the optimal size.

The code for a pend operation is shown in Figure 12-16. In this case, the application will perform a SEM_pend operation to wait for the DMA transfer to complete. This allows the application to perform other meaningful work by temporarily suspending the currently executing task and switching to another task to perform other processing. When the operating system suspends one task and begins executing another task, a

```
if (transfer_length < LARGE_TRANSFER)
        IO_DRIVER();
else
        IO_LARGE_DRIVER();
endif
```

Figure 12-15: A code snippet that checks for the transfer length and calls a driver function that will either poll the DMA completion bit in the DSP status register or pend on an operating system semaphore.

```
/* wait for port to become available */
while(g_io_channel_status[ dir] & ACTIVE_MASK)
{
            /* poll */
}

/* submittcb */
*(g_io_chain_queue_a[ dir]) = (unsigned int)tcb;

/* wait for transfer to complete */
sem_status = SEM_pend(handle, SYS_FOREVER);
```

Figure 12-16: A Code Snippet that Pends on a Semaphore for DMA Completion.

certain amount of overhead is incurred. The amount of this overhead is dependent on the DSP and operating system.

Figure 12-17 shows the code for the polling operation. In this example, the application will continue polling the DMA completion status register for the operation to complete.

```
/* wait for port to become available */
while (g_io_channel_status[ dir] & ACTIVE_MASK)
{
            /* poll */
}

/* submittcb */
*(g_io_chain_queue_a[ dir]) = (unsigned int)tcb;

/* wait for transfer to complete by polling the
   DMA status register */

status = *((vol_uint)*)g_io_channel_status[ dir];
while ((status & DMA_ACTIVE_MASK) ==
            DMA_CHANNEL_ACTIVE_MASK)
{
            status = *((vol_uint*)g_io_channel_status[ dir];
}
.
.
.
```

Figure 12-17: A Code Snippet that Polls for DMA Completion.

This requires the use of the CPU to perform the polling operation. Doing this prevents the CPU from doing other meaningful work. If the transfer is short enough and the CPU only has to poll the status register for a short period of time, this approach may be more efficient. The decision is based on how much data is being transferred and how many cycles the CPU must spend polling. If the poll takes less time than the overhead to go through the operating system to swap out one task and begin executing another, this approach may be more efficient. In that context, the code snippet below checks for the transfer length and, if the length is less than the breakeven transfer length, a function will be called to poll the DMA transfer completion status. If the length is greater than the predetermined cutoff transfer length, a function will be called to set-up the DMA to interrupt on completion of the transfer. This ensures the most efficient processing of the completion of each DMA operation.

12.10   Managing Internal Memory

One of the most important resources for a DSP is its on-chip or internal memory. This is the area where data for most computations should reside because access to this memory is so much faster than off-chip or external memory. Since many DSPs do not have a data cache because of determinism unpredictability, software designers should think of a DSP internal memory as a sort of programmer managed cache. Instead of the hardware on the processor caching data for performance improvements with no control by the programmer, the DSP internal data memory is under full control of the DSP programmer. Using the DMA, data can be cycled in and out of the internal memory in the background, with little or no intervention by the DSP CPU. If managed correctly and efficiently, this internal memory can be a very valuable resource.

It is important to map out the use of internal memory and manage where data is going in the internal memory at all times. Given the limited amount of internal memory for many applications, not all the program's data can reside in internal memory for the duration of the application timeline. Over time, data will be moved to internal memory, processed, perhaps used again, and moved to external memory when it is no longer needed. Figure 12-18 shows an example of how a memory map of internal DSP memory might look during the timeline of the application. During the execution of the application, different data structures will be moved to on-chip memory, processed to form additional structures on chip, and eventually be moved off chip to external memory to be saved, or overwritten in internal memory when the data is no longer needed.

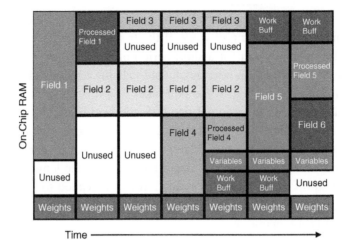

Figure 12-18: Internal Memory of a DSP Must Be Managed by the Programmer.

12.11 Loop Unrolling

The standard rule when programming superscalar and VLIW devices is "Keep the pipelines full!" A full pipe means efficient code. In order to determine how full the pipelines are, you need to spend some time inspecting the assembly language code generated by the compiler.

To demonstrate the advantages of parallelism in VLIW-based machines, let's start with a simple looping program shown in Figure 12-19. If we were to write a serial assembly language implementation of this, the code would be similar to that in Figure 12-20. This

```
1       void example1(float *out, float *input1, float *input2)
2       {
3         int i;
4
5         for(i = 0; i < 100; i++)
6           {
7             out[i] = input1[i] * input2[i];
8           }
9       }
```

Figure 12-19: Simple for Loop in C.

```
1    ;
2    ; serial implementation of loop (26 cycles per iteration)
3    ;
4    L1:    LDW      *B++,B5      ; load B[i] into B5
5           NOP      4            ; wait for load to complete
6
7           LDW      *A++,A4      ; load A[i] into A4
8           NOP      4            ; wait for load to complete
9
10          MPYSP    B5,A4,A4     ; A4 = A4 * B5
11          NOP      3            ; wait for mult to complete
12
13          STW      A4,*C++      ; store A4 in C[i]
14          NOP      4            ; wait for store to complete
15
16          SUB      i,1,i        ; decrement i
17   [i]    B        L1           ; if i != 0, goto L1
18          NOP      5            ; delay for branch
```

Figure 12-20: Serial Assembly Language Implementation of C Loop.

loop uses one of the two available sides of the superscalar machine. By counting up the instructions and the NOPs, it takes 26 cycles to execute each iteration of the loop. We should be able to do much better.

12.11.1 Filling the Execution Units

There are two things to notice in this example. Many of the execution units are not being used and are sitting idle. This is a waste of processor hardware. Second, there are many delay slots in this piece of assembly (20 to be exact) where the CPU is stalled waiting for data to be loaded or stored. When the CPU is stalled, nothing is happening. This is the worst thing you can do to a processor when trying to crunch large amounts of data.

There are ways to keep the CPU busy while it is waiting for data to arrive. We can be doing other operations that are not dependent on the data we are waiting for. We can also use both sides of the VLIW architecture to help us load and store other data values. The code in Figure 12-21 shows an implementation designed for a CPU with multiple execution units. While the assembly looks very much like conventional serial assembly, the run-time execution is very unconventional. Instead of each line representing a single instruction that is completed before the next instruction is begun, each line in Figure 12-21 represents an individual operation, an operation that might be scheduled to execute in parallel with some other operation. The assembly format has been extended to allow the programmer to specify which execution unit should perform a particular operation and which operations may be scheduled concurrently. The DSP compiler automatically determines which execution unit to use for an operation and indicates this by naming the target unit in the extra column that precedes the operand fields (the column containing D1, D2, etc.) in the assembly listing. To indicate that two or more operations may proceed in parallel, the lines describing the individual operations are "joined" with a parallel bar (as in lines 4 and 5 of Figure 12-21). The parallelism rules are also determined by the compiler. Keep in mind that if the programmer decides to program the application using assembly language, the responsibility for scheduling the instructions on each of the available execution units as well as determining the parallelism rules falls on the programmer. This is a difficult task and should only be done when the compiler-generated assembly does not have the required performance.

The code in Figure 12-21 is an improvement over the serial version. We have reduced the number of NOPs from 20 to 5. We are also performing some steps in parallel. Lines 4 and 5 are executing two loads at the same time into each of the two load units (D1 and

```
1        ; using delay slots and duplicate execution units
           of the device
2        ; 10 cycles per iteration
3
4    L1:          LDW    .D2    *B++,B5      ; load B[i] into B5
5    ||           LDW    .D1    *A++,A4      ; load A[i] into A4
6
7                 NOP           2            ; wait load to complete
8                 SUB    .L2    i,1,i        ; decrement i
9         [i]     B      .S1    L1           ; if i != 0, goto L1
10
11                MPYSP  .M1X   B5,A4,A4     ; A4 = A4 * B5
12                NOP    3                   ; wait mpy to complete
13
14                STW    .D1    A4,*C++      ; store A4 into C[i]
```

Figure 12-21: A More Parallel Implementation of the C Loop.

D2) of the device. This code is also performing the branch operation earlier in the loop and then taking advantage of the delays associated with that operation to complete operations on the current cycle.

12.11.2   Reducing Loop Overhead

Loop unrolling is a technique used to increase the number of instructions executed between executions of the loop branch logic. This reduces the number of times the loop branch logic is executed. Since the loop branch logic is overhead, reducing the number of times this has to execute reduces the overhead and makes the loop body, the important part of the structure, run faster. A loop can be unrolled by replicating the loop body a number of times and then changing the termination logic to comprehend the multiple iterations of the loop body (Figure 12-22). The loops in Figures 12-22a and 12-22b each take four cycles to execute, but the loop in Figure 12-22b is doing four times as much work! This is illustrated in Figure 12-23. The assembly language kernel of this loop is shown in Figure 12-23a. The mapping of variables from the loop to the processor is shown in Figure 12-23b. The compiler is able to structure this loop such that all required resources are stored in the register file, and the work is spread across

(a)

```
for (i = 0; i < 128; i ++)
{
        sum1 + = const[i] * input[128 - i];
}
```

(b)

```
for (i = 0; i < 32; i ++)
{
        sum1 += const[i] * input[128 - i];
        sum2 += const[2*i] * input[128 - (2*i)];
        sum3 += const[3*i] * input[128 - (3*i)];
        sum4 += const[4*i] * input[128 - (4*i)];
}
```

Figure 12-22: Loop Unrolling: a) a simple loop; b) the same loop unrolled 4 times.

```
     NOP            1
     SUB     .L2    B0,1,B0

     B       .S2    L3
||   LDW     .D2    *B5--,B4
||   LDW     .D1    *A3++,A0

     ADDSP   .L1    A5,A4,A4
||   MPYSP   .M1X   B4,A0,A5
```

a. The assembly language kernel

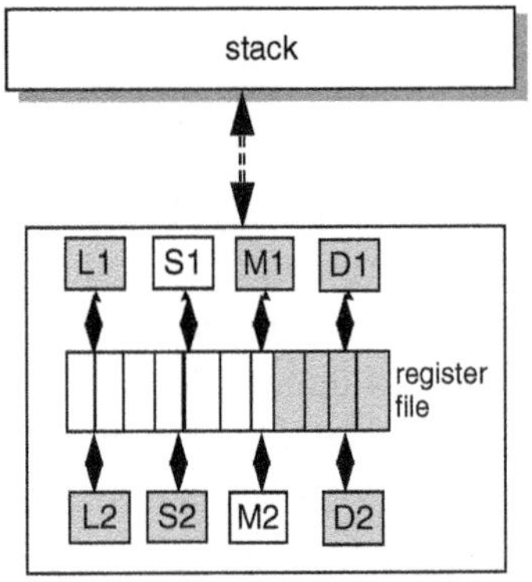

b. The resources used (shaded)
on the processor

1 POINT								
CYCLE	D1	S1	L1	M1	D2	S2	L2	M2
1								
2								SUB	
3	LOAD				LOAD	BRANCH		
4			ADD	MPY				

c. Resources use by cycle

Figure 12-23: Implementation of a Simple Loop.

several of the execution units. The work done by cycle for each of these units is shown in Figure 12-23c.

Now look at the implementation of the loop unrolled four times in Figure 12-24. Again, only the assembly language for the loop kernel is shown. Notice that more of the register file is being used to store the variables needed in the larger loop kernel. An additional execution unit is also being used, as well as a several bytes from the stack in external memory (Figure 12-24a). Also, the execution unit utilization shown in Figure 12-24c indicates the execution units are being used more efficiently while still maintaining a four cycle latency to complete the loop. This is an example of using all the available resources of the device to gain significant speed improvements. Although the code size looks bigger, it actually runs faster than the loop in Figure 12-22a.

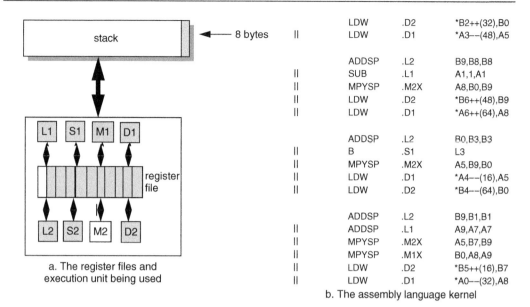

	LDW	.D2	*B2++(32),B0
‖	LDW	.D1	*A3--(48),A5
	ADDSP	.L2	B9,B8,B8
‖	SUB	.L1	A1,1,A1
‖	MPYSP	.M2X	A8,B0,B9
‖	LDW	.D2	*B6++(48),B9
‖	LDW	.D1	*A6++(64),A8
	ADDSP	.L2	B0,B3,B3
‖	B	.S1	L3
‖	MPYSP	.M2X	A5,B9,B0
‖	LDW	.D1	*A4--(16),A5
‖	LDW	.D2	*B4--(64),B0
	ADDSP	.L2	B9,B1,B1
‖	ADDSP	.L1	A9,A7,A7
‖	MPYSP	.M2X	A5,B7,B9
‖	MPYSP	.M1X	B0,A8,A9
‖	LDW	.D2	*B5++(16),B7
‖	LDW	.D1	*A0--(32),A8

a. The register files and execution unit being used

b. The assembly language kernel

4 POINT								
CYCLE	D1	S1	L1	M1	D2	S2	L2	M2
1	LOAD				LOAD			
2	LOAD		SUB		LOAD		ADD	MPY
3	LOAD	BRANCH			LOAD		ADD	MPY
4	LOAD		ADD	MPY	LOAD		ADD	MPY

c. Utilization of the execution units

Figure 12-24: Implementation of a Loop Unrolled Four Times.

12.11.3 Fitting the Loop to Register Space

Unrolling too much can cause performance problems. In Figure 12-25, the loop is unrolled eight times. At this point, the compiler cannot find enough registers to map all the required variables. When this happens, variables start getting stored on the stack, which is usually in external memory somewhere. This is expensive because instead of a single cycle read, it can now take many cycles to read each of the variables each time it is needed. This causes things to break down, as shown in Figure 12-25. The obvious problems are the number of bytes that are now being stored in external memory (88 vs. 8 before) and the lack of parallelism in the assembly language loop kernel. The actual kernel assembly language was very long and inefficient. A small part of it is shown in Figure 12-25b. Notice the lack of "‖" instructions and the new "NOP" instructions. These are, effectively, stalls to the CPU when nothing else can happen. The CPU is waiting for data to arrive from external memory.

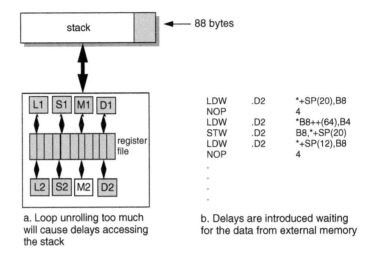

a. Loop unrolling too much
will cause delays accessing
the stack

b. Delays are introduced waiting
for the data from external memory

Figure 12-25: Loop unrolled eight times. Too much of a good thing!

12.11.4 Trade-offs

The drawback to loop unrolling is that it uses more registers in the register file as well
as execution units. Different registers need to be used for each iteration. Once the
available registers are used, the processor starts going to the stack to store required data.
Going to the off-chip stack is expensive and may wipe out the gains achieved by
unrolling the loop in the first place. Loop unrolling should only be used when the
operations in a single iteration of the loop do not use all of the available resources of the
processor architecture. Check the assembly language output if you are not sure of this.
Another drawback is the code size increase. As you can see in Figure 12-24, the unrolled
loop, albeit faster, requires more instructions and, therefore, more memory.

12.12 Software Pipelining

One of the best performance strategies for the DSP programmer is writing code that can
be pipelined efficiently by the compiler. Software pipelining is an optimization strategy
to schedule loops and functional units efficiently. In software pipelining, operations from
different iterations of a software loop are performed in parallel. In each iteration,
intermediate results generated by the previous iteration are used. Each iteration will also
perform operations whose intermediate results will be used in the next iteration. This
technique produces highly optimized code through maximum use of the processor

functional units. The advantage to this approach is that most of the scheduling associated with software pipelining is performed by the compiler and not by the programmer (unless the programmer is writing code at the assembly language level). There are certain conditions which must be satisfied for this to work properly and we will talk about that shortly.

DSPs may have multiple functional units available for use while a piece of code is executing. In the case of the TMS320C6X family of VLIW DSPs, there are eight functional units that can be used at the same time, if the compiler can determine how to utilize all of them efficiently. Sometimes, subtle changes in the way the C code is structured can make all the difference. In software pipelining, multiple iterations of a loop are scheduled to execute in parallel. The loop is reorganized so that each iteration in the pipelined code is made from instruction sequences selected from different iterations in the original loop. In the example in Figure 12-26, a five-stage loop with three iterations is shown. As we discussed earlier, the initial period (cycles n and n + 1), called the **prolog**, is when the pipes are being "primed" or initially loaded with operations. Cycles n + 2 to n + 4 are the actual pipelined section of the code. It is in this section that the processor is performing three different operations (C, B, and A) for three different loops (1, 2, and 3). The epilog section is where the last remaining instructions are performed before exiting the loop. This is an example of a fully utilized set of pipelines that produces the fastest, most efficient code.

We saw earlier how loop unrolling offers speed improvements over simple loops. Software pipelining can be faster than loop unrolling for certain sections of code because, with loop unrolling, the prolog and epilog are only performed once (Figure 12-27).

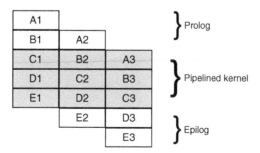

Figure 12-26: A Five-Stage Pipe that Is Software Pipelined.

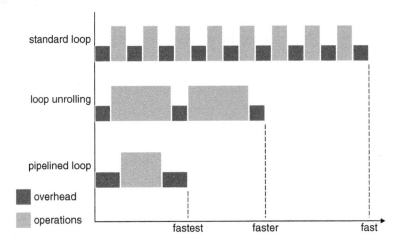

Figure 12-27: Standard Loop Overhead vs. Loop Unrolling and Software Pipelining. The standard loop uses a loop check each iteration through the loop. This is considered overhead. Loop unrolling still checks the loop count but less often. Software pipelining substitutes the loop check with prolog and epilog operations that prime and empty a pipelined loop operation which performs many iterations of the loop in parallel.

12.12.1 An Example

To demonstrate this technique, lets look at an example. In Figure 12-28, a simple loop is implemented in C. This loop simply multiplies elements from two arrays and stores the result in another array.

```
1     void example1(float *out, float *input1, float *input2)
2     {
3       int i;
4
5       for(i = 0; i < 100; i++)
6       {
7         out[i] = input1[i] * input2[i];
8       }
9     }
```

Figure 12-28: A Simple Loop in C.

```
 1      ;
 2      ; serial implementation of loop (26 cycles per iteration)
 3      ;
 4      L1:   LDW    *B++,B5     ; load B[i] into B5
 5            NOP    4           ; wait for load to complete
 6
 7            LDW    *A++,A4     ; load A[i] into A4
 8            NOP    4           ; wait for load to complete
 9
10            MPYSP  B5,A4,A4    ; A4 = A4 * B5
11            NOP    3           ; wait for mult to complete
12
13            STW    A4,*C++     ; store A4 in C[i]
14            NOP    4           ; wait got store to complete
15
16            SUB    i,1,i       ; decrement i
17      [i]   B      L 1         ; if i != 0, goto L1
18            NOP    5           ; delay for branch
```

Figure 12-29: Assembly Language Output of the Simple Loop.

A Serial Implementation

The assembly language output for this simple loop is shown in Figure 12-29. This is a serial assembly language implementation of the C loop; serial in the sense that there is no real parallelism (use of the other processor resources) taking place in the code. This is easy to see by the abundance of NOP operations in the code. These NOPs are instructions that do nothing but wait and burn CPU cycles. These are required and inserted in the code when the CPU is waiting for memory fetches or writes to complete. Since a load operation takes five CPU cycles to complete, the load word operation in line 4 (LDW) is followed by an NOP with a length of 4 indicating that the CPU must now wait four additional cycles for the data to arrive in the appropriate register before it can be used.

These NOPs are extremely inefficient and the programmer should endeavor to remove as many of these delay slots as possible to improve performance of the system. One way of doing this is to take advantage of the other DSP functional resources.

A Minimally Parallel Implementation

A more parallel implementation of the same C loop is shown in Figure 12-30. In this implementation, the compiler is able to use more of the functional units on the DSP. Line 4 shows a load operation, similar to the previous example. However, this load is explicitly loaded into the D2 functional unit on the DSP. This instruction is followed by another load into the D1 unit of the DSP. What is going on here? While the first load is taking place, moving data into the D2 unit, another load is initiated that loads data into the D1 register. These operations can be performed during the same clock cycle because the destination of the data is different. These values can be preloaded for use later and we do not have to waste as many clock cycles waiting for data to move. As you can see in this code listing, there are now only two NOP cycles required instead of four. This is a step in the right direction. The "||" symbol means that the two loads are performed during the same clock cycle.

```
1      ; using delay slots and duplicate execution units of
         the device
2      ; 10 cycles per iteration
3
4   L1:        LDW    .D2    *B++,B5      ;load B[i] into B5
5   ||         LDW    .D1    *A++,A4      ;load A[i] into A4
6
7              NOP           2            ; wait load to complete
8              SUB    .L2    i,1,i        ; decrement i
9         [i]  B      .S1    L1           ; if i != 0, goto L1
10
11             MPYSP  .M1X   B5,A4,A4     ; A4 = A4 * B5
12             NOP           3            ; wait mpy to complete
13
14             STW    .D1    A4,*C++      ;store A4 into C[i]
```

Figure 12-30: Assembly Language Output of the Simple Loop Exploiting the Parallel Orthogonal Execution Units of the DSP.

Compiler-Generated Pipeline

Figure 12-31 shows the same sample piece of C code and the corresponding assembly language output. In this case, the compiler was asked (via a compile switch) to attempt to pipeline the code. This is evident by the piped loop prolog and piped loop kernel sections in the assembly language output. In this case, the pipelined code is not as good as it could be. Inefficient code can be located by looking for how many NOPs there are in the piped loop kernel of the code. In this case the piped loop kernel has a total of five NOP cycles, two in line 16, and three in line 20. This loop takes a total of ten cycles to execute. The NOPs are the first indication that a more efficient loop may be possible. But how short can this loop be? One way to estimate the minimum loop size is to determine what execution unit is being used the most. In this example, the D unit is used more than any other unit, a total of three times (lines 14, 15, and 21). There are two sides to this VLIW device, enabling each unit to be used twice (D1 and D2) per clock for a minimum two clock loop; two D operations in one clock and one D unit in the second clock. The compiler was smart enough to use the D units on both sides of the pipe (lines 14 and 15), enabling it to parallelize the instructions and only use one clock.

```
1       void example1(float *out, float *input1, float *input2)
2       {
3         int i;
4
5         for(i = 0; i < 100; i++)
6           {
7               out[i] = input1[i] * input2[i];
8           }
9       }


1       _example1:
2       ;** ---------------------------------------------------*
3                       MVK       .S2          0x64,B0
4
5                       MVC       .S2          CSR,B6
6       ||              MV        .L1X         B4,A3
7       ||              MV        .L2X         A6,B5
8                       AND       .L1X         -2,B6,A0
9                       MVC       .S2X         A0,CSR
10      ;** ---------------------------------------------------*
11      L11:    ; PIPED LOOP PROLOG
12      ;** ---------------------------------------------------*
13      L12:    ; PIPED LOOP KERNEL
```

Figure 12-31: C Example and the Corresponding Pipelined Assembly Language Output.

```
14                    LDW       .D2        *B5++,B4  ;
15      ||            LDW       .D1        *A3++,A0  ;

16                    NOP                  2
17      [ B0]         SUB       .L2        B0,1,B0   ;
18      [ B0]         B         .S2        L12       ;
19                    MPYSP     .M1X       B4,A0,A0  ;
20                    NOP                  3
21                    STW       .D1        A0,*A4++  ;
22      ;** --------------------------------------------------*
23                    MVC       .S2        B6,CSR
24                    B         .S2        B3
25                    NOP                  5
26                    ; BRANCH OCCURS
```

Figure 12-31: Continued

It should be possible to perform other instructions while waiting for the loads to complete, instead of delaying with NOPs.

In summary, when processing arrays of data (which is common in many DSP applications) the programmer must inform the compiler when arrays are not dependent on each other. The compiler must assume that the data arrays can be anywhere in memory, even overlapping each other. Unless informed of array independence, the compiler will assume the next load operation requires the previous store operation to complete (as to not load stale data). Independent data structures allow the compiler to structure the code to load from the input array before storing the last output. Basically, if two arrays are not pointing to the same place in memory, using the "restrict" keyword to indicate this independence will improve performance. Another term for this technique is **memory disambiguation**.

12.12.2 Enabling Software Pipelining

The compiler must decide what variables to put on the stack (which take longer to access) and which variables to put in the fast on-chip registers. This is part of the register allocator of a compiler. If a loop contains too many operations to make efficient use of the processor registers, the compiler may decide to not pipeline the loop. In cases like that, it may make sense to break up the loop into smaller loops that will enable the compiler to pipeline each of the smaller loops (Figure 12-32).

Instead of:

```
for (expression)
{
        Do A
        Do B
        Do C
        Do D
}
```

Try:

```
for (expression)
        Do A

for (expression)
        Do B

for (expression)
        Do C

for (expression)
        Do D
```

Figure 12-32: Breaking up larger loops into smaller loops may enable each loop to be pipelined more efficiently.

The compiler will not attempt to software pipeline a loop when there are not enough resources (execution units, registers, etc) to allow it, or if the compiler determines that it is not worth the effort to pipeline a loop because the benefit does not outweigh the gain (for example, the amount of cycles required to produce the prolog and epilog far outweighs the amount of cycles saved in the loop kernel). But the programmer can intervene in some cases to improve the situation. With careful analysis and structuring of the code, the programmer can make the necessary modification at the high language level to allow certain loops to pipeline. For example, some loops have so many processing requirements inside the loop that the compiler cannot find enough registers and execution units to map all the required data and instructions. When this happens, the compiler will not attempt to pipeline the loop. Also, function calls within a loop will not be pipelined because the compiler has a hard time resolving the function call. Instead, if you want a pipelined loop, replace the function call with an inline expansion of the function.

12.12.3   Interrupts and Pipelined Code

Because an interrupt in the middle of a fully primed pipe destroys the synergy in instruction execution, the compiler may protect a software pipelining operation by disabling interrupts before entering the pipelined section and enabling interrupts on the way out (Figure 12-33). Lines 11 and 69 of Figure 12-8 show interrupts being disabled prior to the prolog and enabled again just after completing the epilog. This means that the price of the efficiency in software pipelining is paid for in a nonpreemptible section of code. The programmer must be able to determine the impact of sections of nonpreemptible code on real time performance. This is not a problem for single task applications. But it may have an impact on systems built using a tasking architecture. Each of the software pipelined sections must be considered a blocking term in the overall tasking equation.

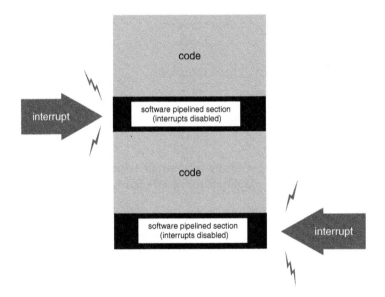

Figure 12-33: Interrupts may be disabled during a software pipelined section of code.

12.12.4   Compiler Architecture and Flow

The general architecture of a modern compiler is shown in Figure 12-34. The front end of the compiler reads in the DSP source code, determines whether the input is legal, detects and reports errors, obtains the meaning of the input, and creates an intermediate representation of the source code. The intermediate stage of the compiler is called the **optimizer**. The optimizer performs a set of optimization techniques on the code including:

- Control flow optimizations.

- Local optimizations.

- Global optimizations.

The back end of the compiler generates the target code from the intermediate code, performs the optimizations on the code for the specific target machine, and performs instruction selection, instruction scheduling and register allocation to minimize memory bandwidth, and finally outputs object code to be run on the target.

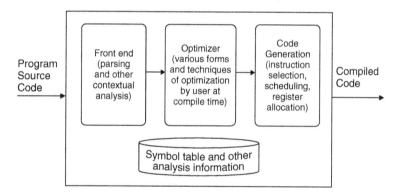

Figure 12-34: General Architecture of a Compiler.

12.12.5 Compiler Optimizations

Compilers perform what are called **machine independent** and **machine dependent** optimizations. Machine independent optimizations are those that are not dependent on the architecture of the device. Examples of this are:

- **Branch optimization**: This is a technique that rearranges the program code to minimize branching logic and to combine physically separate blocks of code.

- **Loop invariant code motion**: If variables used in a computation within a loop are not altered within the loop, the calculation can be performed outside of the loop and the results used within the loop (Figure 12-35).

```
do i = 1,100              j = 100
    j = 10                do i = 1,100
    x(i) = x(i) + j               x(i) = x(i) + j
enddo                     enddo
......                    ......
......                    ......
```

Figure 12-35: Example of Code Motion Optimization.

- **Loop unrolling**: In this technique, the compiler replicates the loop's body and adjusts the logic that controls the number of iterations performed. Now the code effectively performs the same useful work with less comparisons and branches. The compiler may or may not know the loop count. This approach reduces the total number of operations but also increases code size. This may or may not be an issue. If the resultant code size is too large for the device cache (if one is being used), then the resulting cache miss penalty can overcome any benefit in loop overhead. An example of loop unrolling was discussed earlier in this chapter.

- **Common subexpression elimination**: In common expressions, the same value is recalculated in a subsequent expression. The duplicate expression can be eliminated by using the previous value. The goal is to eliminate redundant or multiple computations. The compiler will compute the value once and store it in a temporary variable for subsequent reuse.

- **Constant propagation**: In this technique, constants used in an expression are combined, and new constants are generated. Some implicit conversions between integers and floating-point types may also be done. The goal is to save memory by the removal of these equivalent variables.

- **Dead code elimination**: This approach attempts to eliminate code that cannot be reached or where the results are not subsequently used.

- **Dead store elimination**: This optimization technique will try to eliminate stores to memory when the value stored will never be referenced in the future. An example of this approach is code that performs two stores to the same location without having an intervening load. The compiler will remove the first store because it is unnecessary.

- **Global register allocation**: This optimization technique allocates variables and expressions to available hardware registers using a "graph coloring" algorithm.

The problem of assigning data values to registers is a key challenge for compiler engineers. This problem can be converted into one of graph coloring. In this approach, attempting to color a graph with N colors is equivalent to attempting to allocate data into N registers. Graph coloring is then the partitioning of the vertices of a graph into a minimum number of independent sets.

- **Inlining**: Inlining replaces function calls with actual program code (Figure 12-36). This can speed up the execution of the software by not having to perform function calls with the associated overhead. The disadvantage to inlining is that the program size will increase.

```
do i = 1,n                            do i = 1,n
      j = k(i)                              j = k(i)
      call subroutine(a(i),j)               temp1 = a(i) * y
      call subroutine(b(i),j)               temp2 = a(i) / y
      call subroutine(c(i),j)               temp3 = temp1 + temp2
.........                                    temp1 = b(i) * y
subroutine INL(x,y)                          temp2 = b(i) / y
      temp1 = x * y                          temp3 = temp1 + temp2
      temp2 = x / y                          temp1 = c(i) * y
      temp3 = temp1 + temp2                  temp2 = c(i) / y
end                                          temp3 = temp1 + temp2
```

Figure 12-36: Inlining replaces function calls with actual code, which increases performance but may increase program size.

- **Strength reduction**: The basic approach with this form of optimization is to use cheaper operations instead of more expensive ones. A simple example of this is to use a compound assignment operator instead of an expanded one, since fewer instructions are needed:

  Instead of:

  $$\text{for } (i = 0; \ i < \text{array\_length}; \ i++)$$
  $$a[i] = a[i] + \text{constant};$$

  Use:

  $$\text{for } (i = 0; \ i < \text{array\_length}; \ i++)$$
  $$a[i] += \text{constant};$$

Another example of a strength reduction optimization is using shifts instead of multiplication by powers of two.

- **Alias disambiguation**: Aliasing occurs if two or more symbols, pointer references, or structure references refer to the same memory location. This situation can prevent the compiler from retaining values in registers because it cannot be certain that the register and memory continue to hold the same values over time. Alias disambiguation is a compiler technique that determines when two pointer expressions cannot point to the same location. This allows the compiler to freely optimize these expressions.

- **Inline expansion of runtime-support library functions**: This optimization technique replaces calls to small functions with inline code. This saves the overhead associated with a function call and provides increased opportunities to apply other optimizations.

The programmer has control over the various optimization approaches in a compiler, from aggressive to none at all. Some specific controls are discussed in the next section.

Machine dependent optimizations are those that require some knowledge of the target machine in order to perform the optimization. Examples of this type of optimization include:

- **Implementing special features**: This includes instruction selection techniques that produce efficient code, selecting an efficient combination of machine dependent instructions that implement the intermediate representation in the compiler.

- **Latency**: This involves selecting the right instruction schedules to implement the selected instructions for the target machine. There are a large number of different schedules that can be chosen and the compiler must select one that gives an efficient overall schedule for the code.

- **Resources**: This involves register allocation techniques which include analysis of program variables and selecting the right combination of registers to hold the variables for the optimum amount of time such that the optimal memory bandwidth goals can be met. This technique mainly determines which variables should be in which registers at each point in the program.

Instruction Selection

Instruction selection is important in generating code for a target machine for a number of reasons. As an example, there may be some instructions on the processor that the C compiler cannot implement efficiently. Saturation is a good example. Many DSP applications perform saturation checks on video and image processing applications. To manually write code to saturate requires a lot of code (check sign bits, determine proper limit, etc). Some DSPs, for example, can do a similar operation in one cycle or as part of another instruction (i.e., replace a multiply instruction, MPY with a saturate multiply, SMUL = 1). But a compiler is often unable to use these algorithm-specific instuctions which the DSP provides. So the programmer often has to force their use. To get the C compiler to use specific assembly language instructions like this, one approach is to use what are called **intrinsics**. Intrinsics are implemented with assembly language instructions on the target processor. Some examples of DSP intrinsics include:

- short _abs(short src); absolute value

- long _labs(long src); long absolute value

- short _norm(short src); normalization

- long _rnd(long src); rounding

- short _sadd(short src1, short src2); saturated add

One benefit to using intrinsics like this is that they are automatically inlined. Since we want to run a processor instruction directly, we would not want to waste the overhead of doing a call. Since intrinsics also require things like the saturation flag to be set, they may be longer than one processor instruction. Intrinsics are better than using assembly language function calls since the compiler is ignorant of the contents of these assembly language functions and may not be able to make some needed optimizations.

Figure 12-37 is an example of using C code to produce a saturated add function. The resulting assembly language is also shown for a C5x DSP. Notice the amount of C code and assembly code required to implement this basic function.

Now look at the same function in Figure 12-38 implemented with intrinsics. Notice the significant code size reduction using algorithm specific special instructions. The DSP designer should carefully analyze the algorithms required for the application and determine whether a specific DSP or family of DSPs supports the class of algorithm

```
C Code:                              Compiler Output:

int sadd(int a, int b)               _sadd:
{                                      MOV T1, AR1
  int result;                          XOR T0, T1
  result = a + b;                      BTST @#15, T1, TC1
  if (((a^b) & 0x8000) == 0)           ADD T0, AR1
  {                                     BCC L2,TC1
    if ((result ^ a) & 0x8000)         MOV T0, AR2
        result = ( a < 0)              XOR AR1, AR2
  ? 0x8000 : 0x7FFF;                    BTST @#15, AR2, TC1
  }                                     BCC L2,!TC1
  return result;                        BCC L1,T0 < #0
}                                       MOV #32767, T0
                                        B  L3
                                     L1:   MOV #-32768, AR1
                                     L2:   MOV AR1, T0
                                     L3:   return
```

Figure 12-37: Code for a Saturated Add Function.

```
C Code                   Compiler Output:

int sadd(int a, int b)   _sadd:
{                          BSET ST3_SATA
  return                   ADD T1, T0
  _sadd(a,b);              BCLR ST3_SATA
}                          return
```

Figure 12-38: Saturated Add Using DSP Intrinsics.

with these special instructions. The use of these special instructions in key areas of the application can have a significant impact on the overall performance of the system.

Latency and Instruction Scheduling

The order in which operations are executed on a DSP has a significant impact on length of time to execute a specific function. Different operations take different lengths of time due to differences in memory access times, and differences in the functional unit of the processor (different functional units in a DSP, for example, may require different lengths of time to complete a specific operation). If the conditions are not right, the processor may delay or stall. The compiler may be able to predict these unfavorable conditions and reorder some of the instructions to get a better schedule. In the worst case, the compiler may have to force the processor to wait for a period of time by inserting delays (sometimes called **NOP** for "no operation") into the instruction stream to force the processor to wait for a cycle or more for something to complete such as a memory transfer.

Optimizing compilers have instruction schedulers to perform one major function: to reorder operations in the compiled code in an attempt to decrease its running time. DSP compilers have sophisticated schedulers that search for the optimal schedule (within reason; the compiler has to eventually terminate and produce something for the programmer!). The main goals of the scheduler are to preserve the meaning of the code (it can't "break" anything), minimize the overall execution time (by avoiding extra register spills to main memory, for example), and operate as efficiently as possible from a usability standpoint.

From a DSP standpoint, loops are critical in many embedded DSP applications. Much of the signal processing performed by DSPs is in loops. Optimizing compilers for DSP often contain specialized loop schedulers to optimize this code. One of the most common examples of this is the function of software pipelining.

An example of software pipelining was given earlier in this chapter. Software pipelining is the execution of operations from different iterations of a software loop in parallel. In each iteration, intermediate results generated by the previous iteration are used and operations are also performed whose intermediate results will be used in the next iteration. This produces highly optimized code and makes maximum use of the processor functional units. Software pipelining is implemented by the compiler if the code structure is suited to making these transformations. In other words, the programmer must produce the right code structure to the compiler such that it can recognize the conditions are right to pipeline the loop. For example, when multiplying two arrays inside a loop, the programmer must inform the compiler when the two arrays do not point to the same space in memory. Compilers must assume arrays can be anywhere in memory, even overlapping one another. Unless informed of array independence, they will assume the next load requires the previous load to complete. By informing the compiler of this independent structure (something as simple as using a keyword in the C code) allows the compiler to load from the input array before storing last output, as shown in the code snippet below where the "restrict" keyword is used to show this independence:

```
void example (float *out, restrict float *input1,
restrict float *input2)
    {
        int i;
        for (i=0; i<100; i++)
        {
            out[ i ] = input1[ i ] * input2[ i ];
        }
    }
```

The primary goal of instruction scheduling is to improve running time of generated code. But be careful how this is measured. For example, measuring the quality of the produced code using a simple measure such as "Instructions per second" is misleading. Although this is a common metric in many advertisements, it may not be indicative of the quality of code produced for the specific application running on the DSP. That is why developers should spend time measuring the time to complete a fixed representative task for the application in question. Using industry benchmarks to measure overall system performance is not a good idea because the information is too specific to be used in a broad sense. In reality there is no single metric that can accurately measure quality of code produced by the compiler.

Register Allocation

On-chip DSP registers are the fastest locations in the memory hierarchy (see Figure 12-39). The primary responsibility of the register allocator is to make efficient use of the target registers. The register allocator works with the scheduled instructions generated by the instruction scheduler and finds an optimal arrangement for data and variables in the processors registers to avoid "spilling" data into main memory where it is more expensive to access (performance). By minimizing register spills the compiler will generate higher performing code by eliminating expensive reads and writes to main memory.

Sometimes the code structure forces the register allocator to use external memory. For example, the C code in Figure 12-40 (which does not do anything useful) shows what can happen when too many variables are required to perform a specific calculation. This function requires a number of different variables, x0...x9. When the compiler attempts to map these into registers, not all variables are accommodated and the compiler must spill some of the variables to the stack. This is shown in Figure 12-41.

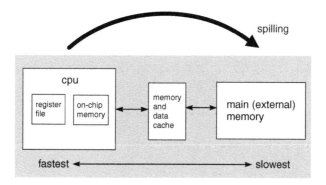

Figure 12-39: Processor Memory Hierarchy. On-chip registers and memory are the fastest way to access data, typically one cycle per access. Cache systems are used to increase the performance when requiring data from off chip. Main external memory is the slowest.

```
int foo(int a, int b, int c, int d)
{
      int x0, x1, x2, x3, x4, x5, x6, x7, x8, x9;

      x0 = (a&0xa);
      x1 = (b&0xa) + x0;

      x2 = (c&0xb) + x1;
      x3 = (d&0xb) + x2;

      x4 = (a&0xc) + x3;
      x5 = (b&0xc) + x4;

      x6 = (c&0xd) + x5;
      x7 = (d&0xd) + x6;

      x8 = (a&0xe);
      x9 = (b&0xe);

      return (x0&x1&x2&x3)|(x4&x5)|(x6&x7&x8+x9);
}
```

Figure 12-40: C Code Snippet with a Number of Variables.

```
;**************************************************************
;* FUNCTION NAME: foo                                        *
;*                                                           *
;* Regs Modified   : A1,A2,A3,A4,V1,V2,V3,V4,V9,SP,LR,SR *
;* Regs Used       : A1,A2,A3,A4,V1,V2,V3,V4,V9,SP,LR,SR *
;* Local Frame Size : 0 Args + 4 Auto + 20 Save = 24 byte *
;**************************************************************
;**************************************************************
        .compiler_opts --abi=ti_arm9_abi --code_state=16
        .state16
;       opt470 rob.if rob.opt

        .sect ".text"
        .clink
        .thumbfunc _foo
        .state16
        .global _foo
;**************************************************************
;* FUNCTION NAME: foo                                        *
;*                                                           *
;* Regs Modified   : A1,A2,A3,A4,V1,V2,V3,V4,V9,SP,LR,SR *
;* Regs Used       : A1,A2,A3,A4,V1,V2,V3,V4,V9,SP,LR,SR *
;* Local Frame Size : 0 Args + 4 Auto + 20 Save = 24 byte *
;**************************************************************
_foo:
;* --------------------------------------------*
    PUSH {A4, V1, V2, V3, V4, LR}
;** 21 -------------------- C$1 = a&(C$12 = 10);
     MOV V1,#10                ;|21|
     MOV V2,A1
     AND V2,V1
     MOV V9,V2
;** 21 -------------------- C$2 = (b&C$12)+C$1;
     AND V1,A2
     ADD V3,V2,V1              ;|21|
;** 21 -------------------- C$3 = (c&(C$11 = 11))+C$2;
     MOV V1,#11                ;|21|
     MOV V2,A3
     AND V2,V1
     ADD V2,V3,V2              ;|21|
     STR V2,[SP,#0] **** this is an example of register spilling.
     SP indicates the stack pointer
```

Figure 12-41: Register Spilling Caused by Lack of Register Resources.

```
;** 21 --------------------- C$4 = (d&C$11)+C$3;
     AND  V1,A4
     ADD  V1,V2,V1              ;|21|
     MOV  LR,V1
;** 21 --------------------- C$5 = (a&(C$10 = 12))+C$4;
     MOV  V1,#12               ;|21|
     MOV  V4,A1
     AND  V4,V1
     MOV  V2,LR
     ADD  V2,V2,V4             ;|21|
;** 21 --------------------- C$6 = (b&C$10)+C$5;
     AND  V1,A2
     ADD  V1,V2,V1             ;|21|
;** 21 --------------------- C$8 = (c&(C$9 = 13))+C$6;
     MOV  V4,#13               ;|21|
     AND  A3,V4
     ADD  A3,V1,A3             ;|21|
;** 21 --------------------- C$7 = 14;
;** 21 --------------------- return C$1&C$2&C$3&C$4|C$5&C$6|
(a&C$7)+(b&C$7)&(d&C$9)+C$8&C$8;
     MOV  V4,V9
     AND  V4,V3
     LDR  V3,[SP,#0] **** this is an example of
     register spilling. SP indicates the stack pointer
     AND  V4,V3                ;|21|
     MOV  V3,LR
     AND  V4,V3                ;|21|
     AND  V2,V1
     ORR  V2,V4                ;|21|
     MOV  V1,#14               ;|21|
     AND  A1,V1
     AND  A2,V1
     ADD  A2,A2,A1             ;|21|
     MOV  A1,#13               ;|21|
     AND  A4,A1
     ADD  A1,A3,A4             ;|21|
     AND  A1,A2                ;|21|
     AND  A1,A3                ;|21|
     ORR  A1,V2
     POP  {A4,V1,V2,V3,V4}
     POP  {A3}
     BX   A3
```

Figure 12-41: Continued

One way to reduce or eliminate register spilling is to break larger loops into smaller loops, if the correctness of the code can be maintained. This will enable the register allocator to treat each loop independently, thereby increasing the possibility of finding a suitable register allocation for each of the sub-functions. Figure 12-42 is a simple example of breaking larger loops into smaller loops to enable this improvement.

```
Instead of:              Try:

for (expression)         for (expression) {
{                                Do A    }
        Do A             for (expression) {
        Do B                     Do B    }
        Do C             for (expression) {
        Do D                     Do C    }
}                        for (expression) {
                                 Do D    }
```

Figure 12-42: Some loops are too large for the compiler to pipeline. Reducing the computational load within a loop may allow the compiler to pipeline the smaller loops!

Eliminating embedded loops as shown below can also free up registers and allow for more efficient register allocation in DSP code.

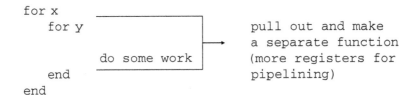

```
for x
    for y                          pull out and make
                                   a separate function
        do some work               (more registers for
    end                            pipelining)
end
```

12.12.6 Compile Time Options

Many DSP optimizing compilers offer several options for code size vs. performance. Each option allows the programmer to achieve a different level of performance vs. code size. These options allow more and more aggressive code size reduction from the compiler. DSP compilers support different levels of code performance. Each option allows the compiler to perform different DSP optimization techniques, for example:

1. **First level of optimization: Register level optimizations**. This level of optimization may include techniques such as:

    *   Simplification of control flow.

    *   Allocation of variables to registers.

    *   Elimination of unused code.

    *   Simplification of expressions and statements.

    *   Expand calls to inline functions.

2. **Second level of optimization: Local optimization**. This level of optimization may include techniques such as:

    *   Local copy/constant propagation.

    *   Removal of unused assignments.

    *   Elimination of local common expressions.

3. **Third level of optimization: Global optimization**. This level of optimization may include techniques such as:

    *   Loop optimizations.

    *   Elimination of global common sub-expressions.

    *   Elimination of global unused assignments.

    *   Loop unrolling.

4. **Highest level of optimization: File optimizations**. This level of optimization may include techniques such as:

    *   Removal of functions that are never called.

    *   Simplification of functions with return values that are never used.

    *   Inlines calls to small functions (regardless of declaration).

    *   Reordering of functions so that attributes of called function are known when caller is optimized.

    *   Identification of file-level variable characteristics.

The different levels of optimization and the specific techniques used at each level obviously vary by vendor and compiler. The DSP programmer should study the compiler manual to understand what each level does and experiment to see how the code is modified at each level.

Understanding What the Compiler Is Thinking

There will always be situations where the DSP programmer will need to get pieces of information from the compiler to help understand why an optimization was or was not made. The compiler will normally generate information on each function in the generated assembly code. Information regarding register usage, stack usage, frame size, and how memory is used is usually listed. The programmer can usually get information concerning optimizer decisions (such as to inline or not) by examining information written out by the compiler usually stored in some form of information file (you may have to explicitly ask the compiler to produce this information for you using a compiler command which should be documented in the users manual). This output file contains information as to how optimizations were done or not done (check your compiler—in some situations, asking for this information can sometimes reduce the amount of optimization actually performed).

12.13   Programmer Helping Out the Compiler

Part of the job of an optimizing compiler is figuring out what the programmer is trying to do and then helping the programmer achieve that goal as efficiently as possible. That's why well structured code is better for a compiler—it's easier to determine what the programmer is trying to do. This process can be aided by certain "hints" the programmer can provide to the compiler. The proper "hints" allow the compiler to be more aggressive in the optimizations it makes to the source code. Using the standard compiler options can only get you so far towards achieving optimal performance. To get even more optimization, the DSP programmer needs to provide helpful information to the compiler and optimizer, using mechanisms called *pragmas, intrinsics*, and *keywords*. We already discussed intrinsics as a method of informing the compiler about special instructions to use in the optimization of certain algorithms. These are special function names that map directly to assembly instructions. Pragmas provide extra information about functions and variables to the preprocessor part of the compiler. Helpful keywords are usually type modifiers that give the compiler information about how a variable is used. *Inline* is a special keyword that causes a function to be expanded in place instead of being called.

Pragmas

Pragmas are special instructions to the compiler that tell it how to treat functions and variables. These pragmas must be listed before the function is declared or referenced. Examples of some common pragmas the the TI DSP include:

Pragma	Description
CODE_SECTION(symbol, "section name") [;]	This pragma allocates space for a function in a given memory segment
DATA_SECTION (symbol, "section name") [;]	This pragma allocates space for a data variable in a given memory segment
MUST_ITERATE (min, max, multiple) [;]	This pragma gives the optimizer part of the compiler information on the number of times a loop will repeat
UNROLL (n) [;]	This pragma, when specified to the optimizer, tells the compiler how many times to unroll a loop.

An example of a pragma to specify the loop count is shown in Figure 12-43. The first code snippet does not have the pragma inserted and is less efficient than the second snippet which has the pragma for loop count inserted just before the loop in the source code.

Inefficient loop code

```
    C Code                             Compiler output

    int sum(const short *a, int n)     _sum
{                                      MOV #0, AR1
    int sum = 0;                       BCC L2,To<=#0
    int i;                             SUB #1, To, AR2
    for (i=0; i<n; i++)                MOV AR2,CSR
    {                                  RPT CSR
    sum += a[i];                        ADD *AR0+, AR1,AR1
    }                                  MOV AR1,T0
    return sum;                        return
}
```

```
Efficient loop code

        C Code                                      Compiler output

        int sum(const short *a, int n)              _sum
    {                                               SUB #1, T0, AR2
                                                    MOV AR2, CSR
        int sum = 0;                                MOV #0, AR1
        int i;                                      RPT CSR
        #pragma MUST_ITERATE(1)                     ADD *AR0+,AR1,AR1
        for (i=0; i<n; i++)                         MOV AR1,T0
        {                                           return
        sum += a[i];
        }
        return sum;
    }
```

```
voidfirFilter(short *x, int f, short *y, int N, int M, QScale)
{ int i, j, sum;
    #pragmaUNROLL(2)  ◄─────────── │ Unroll outer loop │
    for (j = 0; j < M; j++) {
        sum = 0;
        #pragmaUNROLL(2)  ◄────── │ Unroll inner loop │
        for (i = 0; i < N; i++)
                sum += x[i + j] *filterCoeff[f][i];
        y[j] = sum >>QScale;
        y[j] &= 0xfffe;
}}
```

Figure 12-43: Example of Using Pragmas to Improve the Efficiency of the Code.

12.13.1 Intrinsics

Modern optimizing compilers have special functions called **intrinsics** that map directly to inlined DSP instructions. Intrinsics help to optimize code quickly. They are called in the same way as a function call. Usually intrinsics are specified with some leading indicator such as the underscore (_).

As an example, if a developer were to write a routine to perform saturated addition in a higher level language such as C, it would look similar to the code in Figure 12-44. The

resulting assembly language for this routine is shown in Figure 12-45. This is quite
messy and inefficient. As an alternative, the developer could write a simple routine
calling a built in saturated add routine (Figure 12-46) which is much easier and produces
cleaner and more efficient assembly code (Figure 12-47). Figure 12-48 shows some of
the available intrinsics for the TMS320C55 DSP. Many modern DSPs support intrinsic
libraries of this type.

Saturated add is a process by which two operands are added together and, if the result is an
overflow, the result is set to the maximum positive value. This is useful in certain multimedia
applications where it is more desirable to have a result that is max positive instead of an
overflow which effectively becomes a negative number which looks undesirable in an image,
for example.

```
Int saturated_add(int a, int b)
{
    int result;

    result = a + b;

// check to see if a and b have the same sign

    if (((a^b) & 0x8000) == 0)
    {
        // if a and b have the same sign, check for
           underflow or overflow
        if ((result ^ a) & 0x8000)
        {
            // if the result has a different sign than a then
            underflow or overflow has
            // occurred. If a is negative, set result to
            max negative
            // If a is positive, set result to max positive
            result = ( a < 0) ? 0x8000 : 0x7FFF;
    }
}
    return result;
```

Figure 12-44: C Code to Perform Saturated Add.

```
Saturated_add:
        SP = SP - #1
                                    ; End Prolog Code
        AR1 = T1                    ;  |5|
        AR1 = AR1 + T0              ;  |5|
        T1 = T1 ^ T0                ;  |7|
        AR2 = T1 & #0x8000          ;  |7|
        if (AR2!=#0) goto L2        ;  |7|
                                    ; branch occurs ;  |7|
        AR2 = T0                    ;  |7|
        AR2 = AR2 ^ AR1             ;  |7|
        AR2 = AR2 & #0x8000         ;  |7|
        if (AR2==#0) goto L2        ;  |7|
                                    ; branch occurs ;  |7|
        if (T0<#0) goto L1          ;  |11|
                                    ; branch occurs ;  |11|
        T0 = #32767                 ;  |11|
        goto L3                     ;  |11|
                                    ; branch occurs ;  |11|
L1:
        AR1 = #-32768               ;  |11|
L2:
        T0 = AR1                    ;  |14|
L3:
                                    ; Begin Epilog Code
        SP = SP + #1                ;  |14|
return                              ;  |14|
                                    ; return occurs ;  |14|
```

Figure 12-45: TMS320C55 DSP Assembly Code for the Saturated Add Routine.

```
int sadd(int a, int b)
{
return _sadd(a,b);
}
```

Figure 12-46: TMS320C55 DSP Code for the Saturated Add Routine Using a Single Call to an Intrinsic.

```
Saturated_add:
    SP = SP - #1
                                    ; End Prolog Code

    bit(ST3, #ST3_SATA) = #1
    T0 = T0 + T1                    ; |3|

                                    ; Begin Epilog Code
    SP = SP + #1                    ; |3|
    bit(ST3, #ST3_SATA) = #0
    return                          ; |3|
                                    ; return occurs ; |3|
```

Figure 12-47: TMS320C55 DSP Assembly Code for the Saturated Add Routine Using a Single Call to an Intrinsic.

Intrinsic	Description
`int _sadd(int src1, int src2);`	Adds two 16–bit integers, with SATA set, producing a saturated 16–bit result.
`int _smpy(int src1, int src2);`	Multiplies src1 and src2, and shifts the result left by 1. Produces a saturated 16–bit result. (SATD and FRCT set).
`int _abss(int src);`	Creates a saturated 16–bit absolute value. _abss(0x8000) => 0x7FFF (SATA set)
`int _smpyr(int src1, int src2);`	Multiplies src1 and src2, shifts the result left by 1, and rounds by adding 2 15 to the result. (SATDand FRCT set)
`int _norm(int src);`	Produces the number of left shifts needed to normalize src.
`int _sshl(int src1, int src2);`	Shifts src1 left by src2 and produces a 16-bit result. The result is saturated if src2 is less than or equal to 8. (SATD set)
`long _lshrs(long src1, int src2);`	Shifts src1 right by src2 and produces a 32-bit result. Produces a saturated 32–bit result. (SATD set)
`long _laddc(long src1, int src2);`	Adds src1, src2, and Carry bit and produces a 32-bit result.
`long _lsubc(long src1, int src2);`	Optimizing C Code Subtracts src2 and logical inverse of sign bit from src1, and produces a 32-bit result.

Figure 12-48: Some Intrinsics for the TMS320C55 DSP (courtesy of Texas Instruments).

12.13.2 Keywords

Keywords are type modifiers that give the compiler information about how a variable is used. These can be very helpful in helping the optimizer part of the compiler make optimization decisions. Some common keywords in DSP compilers are:

- **Const**: This keyword defines a variable or pointer as having a constant value. The compiler can allocate the variable or pointer into a special data section which can be placed in ROM. This keyword will also provide information to the compiler that allows it to make more aggressive optimization decisions.

- **Interrupt**: This keyword will force the compiler to save and restore context and enable interrupts on exit from a particular pipelined loop or function.

- **Ioport**: This defines a variable as being in I/O space (this keyword is only used with global or static variables).

- **On-chip**: Using this keyword with a variable or structure will guarantee that that memory location is on-chip.

- **Restrict**: This keyword tells the compiler that only this pointer will access the memory location it points to (i.e., no aliasing of this location). This allows the compiler to perform optimization techniques such as software pipelining.

- **Volatile**: This keyword tells the compiler that this memory location may be changed without compiler's knowledge. Therefore the memory location should not be stored in a temporary register and, instead, be read from memory before each use.

12.13.3 Inlining

For small infrequently called functions it may make sense to paste them directly into code. This eliminates overhead associated with register storage and parameter passing. Inlining uses more program space, but speeds up execution, sometimes significantly. When functions are inlined, the optimizer can optimize the function and the surrounding code in new context. There are two types of inlining: static inlining and normal inlining. With static inlining the function being inlined is only placed in the code where it will be used. Normal inlining also has a function definition which allows the function to be called. The compiler, if specified, will automatically inline functions if the size is small enough. Inlining can also be definition controlled, where the programmer specifies which functions to inline.

Reducing Stack Access Time

When using a real-time operating system (RTOS) for task driven systems, there is overhead to consider that increases with the number of tasks in the system. The overhead in a task switch (or mailbox pend or post, semaphore operation, and so forth) can vary based on where the operating system structures are located. If the structures are in off-chip memory, the access time to perform the operation can be much longer than if the structure was in on-chip memory. The same holds true for the task stack space. If this is in off-chip memory, the performance suffers proportionally to the number of times the stack has to be accessed.

One solution is to allocate the stack in on-chip memory. If the stack is small enough, this may be a viable thing to do. But if there are many tasks in the system, there will not be enough on-chip memory to store all of the task stacks. However, special code can be written to move the stack on chip when it is needed the most. Before the task (or function) is complete, the stack can be moved back off chip. Figure 12-49 shows the code to do this. Figure 12-50 is a diagrammatic explanation of the steps to perform this operation. The steps are as follows:

1. Compile the C code to get an .asm file.
2. Modify the .asm file with the code in the example.
3. Assemble the new .asm file.
4. Link the system.

You need to be careful when doing this. This type of modification should not be done in a function that calls other functions. Also, interrupts should be disabled when performing this type of operation. Finally, the programmer needs to ensure that the secondary stack in on-chip memory does not grow too large, overwriting other data in on-chip memory!

```
SSP            .set         0x80001000
SSP2           .set         0x80000FFC

               MVK          SSP, A0
||             MVK          SSP2,B0

               MVKH                  SSP,A0
||             MVKH                  SSP2,B0

               STW  .D1     SP, *A0
||             MV    .L2    B0, SP
```

Figure 12-49: Modifying the Stack Pointer to Point to On-Chip Memory.

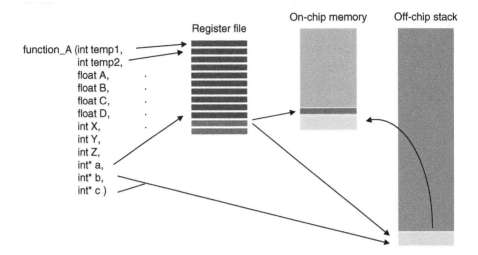

Figure 12-50: Moving the Stack Pointer On Chip to Increase Performance in a DSP Application.

12.13.4   Compilers Helping Out the Programmer

Compilers do their best to optimize applications based on the wishes of their programmers. Compilers also produce output that documents the decisions they were able to make or not make based on the specific source code provided to them by the programmer (see Figure 12-51). By analyzing this output, the programmer can understand the specific constraints and decisions and make appropriate adjustments in the source code to improve the performance of the compiler. In other words, if the

```
;*------------------------------------------------------------ *
;* SOFTWARE PIPELINE INFORMATION
;*
;*    Known Minimum Trip Count         : 1
;*    Known Max Trip Count Factor       : 1
;*    Loop Carried Dependency Bound(^) : 0
;*    Unpartitioned Resource Bound      : 1
;*    Partitioned Resource Bound(*)     : 1
;*    Resource Partition:
```

Figure 12-51: Compiler output can be used to diagnose compiler optimization results.

```
;*                                 A-side   B-side
;*   .L units                        0        0
;*   .S units                        0        1*
;*   .D units                        1*       1*
;*   .M units                        1*       0      Key Information for Loop
;*   .X cross paths                  1*       0
;*   .T address paths                1*       1*       ii = 1 (iteration
;*   Long read paths                 0        0        interval = 1 cycle)
;*   Long write paths                0        0        Means: Single Cycle
;*   Logical ops (.LS)               0        0        Inner Loop
;*   Addition ops (.LSD)             1        1      (.L or .S unit)
;*   Bound(.L .S .LS)                0        1*     (.L or .S or .D unit)
;*   Bound(.L .S .D .LS .LSD)        1*       1*     B side .M unit not used
;*                                                    Means: Only one MPY per
;*                                                    cycle
;*   Searching for software pipeline schedule at ...
;*     ii = 1 Schedule found with 8 iterations in parallel
;*   done
;*
;*   Collapsed epilog stages         : 7
;*   Prolog not entirely removed
;*   Collapsed prolog stages         : 2
;*
;*   Minimum required memory pad      : 14 bytes
;*
;*   Minimum safe trip count          : 1
;*------------------------------------------------------------- *
```

Figure 12-51: Continued

programmer understands the thought process of the compilation process, they are then able to re-orient the application to be more consistent with that thought process. The information in Figure 12-51 is an example of output generated by the compiler that can be analyzed by the programmer. This output can come in various forms, a simple text output file or a more fancy user interface to guide the process of analyzing the compiled output.

12.13.5 Summary of Coding Guidelines

Here is a summary of guidelines that the DSP programmer can use to produce the most highly optimized code for the DSP application. Many of these recommendations are general to all DSP compilers. The recommendation is to develop a list like the one below for whatever DSP device and compiler is being used. It will provide a useful reference for the DSP programming team during the software development phase of a DSP project.

General programming guidelines

1. Avoid removing registers for C compiler usage. Otherwise valuable compiler resources are being thrown away. There are some cases when it makes sense to use these resources. For example, it is acceptable to preserve a register for an interrupt routine.

2. To optimize functions selectively, place in separate files. This lets the programmer adjust the level of optimization on a file-specific basis.

3. Use the least possible number of *volatile* variables. The compiler cannot allocate registers for these, and also can't inline when variables are declared with the volatile keyword.

Variable declaration

1. Local variables/pointers are preferred instead of globals. The compiler uses stack-relative addressing for globals, which is not as efficient. If the programmer will frequently be using a global variable in a function, it is better to assign the global to a local variable and then use it.

2. Declare globals in file where they are used the most.

3. Allocate most often used elements of a structure, array, or bit-field in the first element, lowest address, or LSB; this eliminates the need for extra bytes to specify the address offset.

4. Use unsigned variables instead of *int* wherever possible; this provides a larger dynamic range and gives extra information to the compiler for optimization.

Variable declaration (data types)

1. Pay attention to data type significance. The better the information provided to the compiler, the better the efficiency of the resulting code.

2. Only use casting if absolutely necessary. Casting can use extra cycles, and can invoke wrong RTS functions if done wrong.

3. Avoid common mistakes in data type assumptions. Avoid code that assumes *int* and *long* are the same type. Also, use *int* for fixed-point arithmetic, since *long*

requires a call to a library which is less efficient. Also avoid code that assumes *char* is 8 bits or *long long* is 64 bits for the same reasons.

4. May be more convenient to define your own data types. *Int16* for 16-bit integer (*int*) and *Int32* for 32-bit integer (*long*). Experiment and see what is best for your DSP device and application.

Initialization of variables

1. Initialize global variables with constants at load time. This eliminates the need to have code copy values over at run-time.

2. When assigning the same values to global variables, rearrange code if it makes sense to do so. For example use a = b = c = 3; instead of a = 3; b = 3; c = 3). The first uses a register to store the same value to all, the second produces 3 separate long stores:

```
MOV #3, AR1          MOV #3, *(#_a)
MOV AR1, *(#_a)      MOV #3, *(#_b)
MOV AR1, *(#_b)      MOV #3, *(#_c)
MOV AR1, *(#_c)
(17 bytes)           (18 bytes)
```

3. Memory alignment requirements and stack management.

4. Group all like data declarations together. The compiler will usually align a 32-bit data on even boundary, so it will pad an extra 16-bit word in if needed.

5. Use the .align linker directive to guarantee stack alignment on even address. Since the compiler needs 32-bit data aligned on an even boundary, it starts the stack on an even boundary.

Loops

1. Split up loops comprised of two unrelated operations.

2. Avoid function calls and control statements inside loops; the compiler needs to preserve loop context in case of a call to a function. By taking function calls and control statements outside a loop if possible, the compiler can take advantage of the special looping optimizations in the hardware (for example the localrepeat and blockrepeat in the TI DSP) to further optimize the loop (Figure 12-52).

```
for (expression)
{
        Do A
        Call X  ◄────  All function calls in inner loops
                       must be inlined into the calling
        Do C           function!!
}
```

Figure 12-52: Do not call functions in inner loops of performance critical code.

3. Keep loop code small to enable compiler use of local repeat optimizations.

4. Avoid deeply nested loops; more deeply nested loops use less efficient types of looping.

5. Use an *int* or *unsigned int* instead of *long* for loop counter. DSP hardware generally uses a 16-bit register for a counter.

6. Use pragmas (if available) to give the compiler better information about loop counts.

Control code

1. The DSP compiler may generate similar code for nested if-then-else and switch-case statements if the number of cases is less than eight. If greater than eight, the compiler will generate a .switch label section.

2. For highly dense compare code, use switch statements instead of if-then-else.

3. Place the most common case at the start, since the compiler checks in order.

4. For single conditionals, it is always best to test against 0 instead of !0. For example, 'if (a==0)' produces more efficient code than 'if (a!=1)'.

Functions

1. When a function is only called by other functions in same file, make it a *static* function. This will allow the compiler to inline functions better.

2. When a global variable is only used by functions in the same file, make it a *static* variable.

3. Group minor functions in a single file with functions that use them. This makes file-level optimization better.

4. Too many parameters in function calls become inefficient. Once DSP registers are used up, the rest of the parameters go on the stack. Accessing variables from the stack is very inefficient.

5. Parameters that are used frequently in the subroutine should be passed in registers.

Intrinsics

1. There are some instructions on a DSP that the C compiler cannot implement efficiently. For example, the saturation function is hard to implement using standard instructions on many DSPs. To saturate manually requires a lot of code (check sign bits, determine proper limit, etc). DSPs that support specific instrinsics like saturate will allow the DSP to execute the function much more efficiently.

When developing an application, it is very easy (and sometimes even required) to use generic routines to do various computations. Many times the application developer does not realize how much overhead can be involved in using these generic routines. Often times a more generalized version of an algorithm or function is used because of simple availability instead of creating a more specialized version that better fits the specific need. Creating large numbers of specialized routines is generally a bad programming style as well as a maintenance headache. But strategic use of specialized routines can greatly improve performance in high performance code segments.

Use libraries

1. Some optimizations are more macro or global level optimizations. These optimizations are performed at the algorithm level. This is somewhat unique to DSP where there are many common routines such as FFT, FIR filters, IIR filters, and so on. Eventually, just about every DSP developer will be required to use one of these functions in the development of a DSP application. For common functions used in DSP, vendors have developed highly efficient implementations of these algorithms that can be easily reused. Many of these algorithms are implemented in C and are tested against standards and well documented. Examples include:

   • FFT

   • Filtering and convolution

   • Adaptive filtering

   • Correlation

- Trigonometric (i.e., sine)

- Math (, max, log, div)

- Matrix computations

Although many of these algorithms are very common routine in DSP, they can be complex to implement. Writing one from scratch would require an in-depth knowledge of how the algorithm works (for example an FFT), in-depth knowledge of the DSP architecture in order to optimize the algorithm, possibly expertise at assembly coding, which is hard to find, and time to get everything working right and optimized.

12.14   Profile-Based Compilation

Because there is a trade-off between code size and higher performance, it is often desirable to compile some functions in a DSP application for performance and others for code size. In fact, the ideal code size and performance for your application needs may be some combination of the different levels of optimization across all of the application functions. The challenge is in determining which functions need which options. In an application with 100 or more functions, each with five possible options, the number of option combinations starts to explode exponentially. Because of this, manual experimentation can take weeks of effort, and the DSP developer may rarely arrive at a solution close to optimal for the particular application needs. Profile-based compilation is one available technique which helps to solve this challenge by automating the entire process.

Profile-based compilation will automatically build and profile multiple compiler option sets. For example, this technique can build the entire application with the highest level of optimization and then profile each function to obtain its resulting code size and cycle count. This process is then repeated using the other compiler options at the remaining code-size reduction levels. The result is a set of different code size and cycle count data points for each function in the application. That data can then be plotted to show the most interesting combinations of functions and compiler options (Figure 12-53). The ideal location for the application is always at the origin of the graph in the lower left hand corner where cycle count and code size are both minimized.

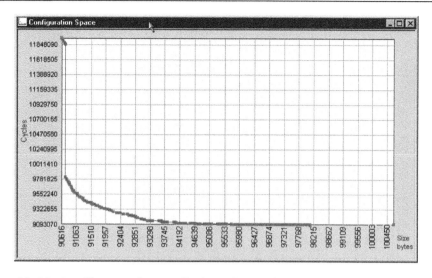

Figure 12-53: Profile-Based Compilation Shows the Various Trade-offs between Code Size and Performance.

Advantages

An advantage of a profile-based environment is the ability for the application developer to select a profile by selecting the appropriate point on the curve, depending on the overall system needs. This automatic approach saves many hours of experimenting manually. The ability to display profiling information (cycle count by module or function, for example) allows the developer to see the specifics of each piece of the application and work on that section independently, if desired (Figure 12-54).

Issues with Debugging Optimized Code

One word of caution: Do not optimize programs that you intend to debug with a symbolic debugger. The compiler optimizer rearranges assembler-language instructions which makes it difficult to map individual instructions to a line of source code. If compiling with optimization options, be aware that this rearrangement may give the appearance that the source-level statements are executed in the wrong order when using a symbolic debugger.

The DSP programmer can ask the compiler to generate symbolic debugging information for use during a debug session. Most DSP compilers have an option for doing this. DSP

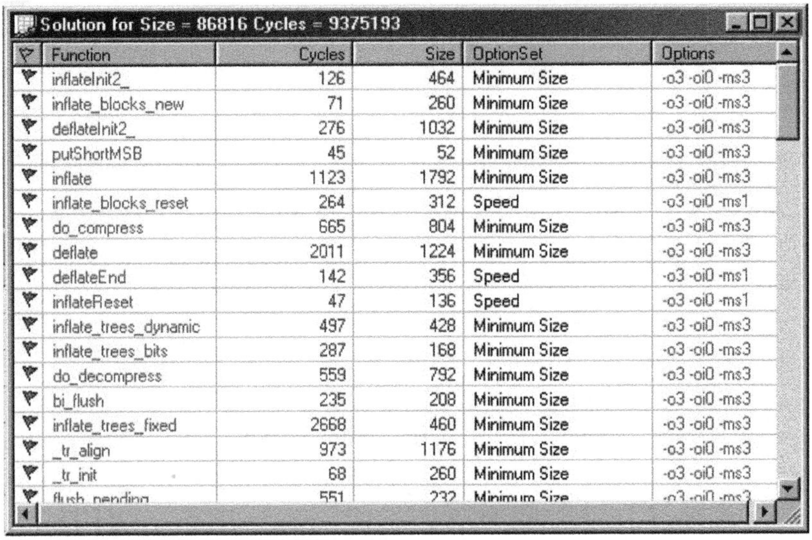
Solution for Size = 86816 Cycles = 9375193

Function	Cycles	Size	OptionSet	Options
inflateInit2_	126	464	Minimum Size	-o3 -oi0 -ms3
inflate_blocks_new	71	260	Minimum Size	-o3 -oi0 -ms3
deflateInit2_	276	1032	Minimum Size	-o3 -oi0 -ms3
putShortMSB	45	52	Minimum Size	-o3 -oi0 -ms3
inflate	1123	1792	Minimum Size	-o3 -oi0 -ms3
inflate_blocks_reset	264	312	Speed	-o3 -oi0 -ms1
do_compress	665	804	Minimum Size	-o3 -oi0 -ms3
deflate	2011	1224	Minimum Size	-o3 -oi0 -ms3
deflateEnd	142	356	Speed	-o3 -oi0 -ms1
inflateReset	47	136	Speed	-o3 -oi0 -ms1
inflate_trees_dynamic	497	428	Minimum Size	-o3 -oi0 -ms3
inflate_trees_bits	287	168	Minimum Size	-o3 -oi0 -ms3
do_decompress	559	792	Minimum Size	-o3 -oi0 -ms3
bi_flush	235	208	Minimum Size	-o3 -oi0 -ms3
inflate_trees_fixed	2668	460	Minimum Size	-o3 -oi0 -ms3
_tr_align	973	1176	Minimum Size	-o3 -oi0 -ms3
_tr_init	68	260	Minimum Size	-o3 -oi0 -ms3
flush_pending	551	232	Minimum Size	-o3 -oi0 -ms3

Figure 12-54: Profiling Information for DSP Functions.

compilers have directives that will generate symbolic debugging directives used by C source-level debugger. The downside to this is that it forces the compiler to disable many optimizations. The compiler will turn on the maximum optimization compatible with debugging. The best solution for debugging DSP code, however, is first to verify the program's correctness and then start turning on optimizations to improve performance.

12.14.1 Summary of the Code Optimization Process

We now expand our definition of the code optimization process first discussed at the beginning of the chapter. Figure 12-55 shows an expanded software development process for code optimization.

Although this process may vary based on the application, this process is a general flow for all DSP applications. There are 21 steps in this process that will be summarized below.

- **Step 1**: This step involves understanding the key performance scenarios for the application. A performance scenario is a path through the DSP application that will stress the available resources in the DSP. This could be performance, memory, and/or power. Once these key scenarios are understood, the

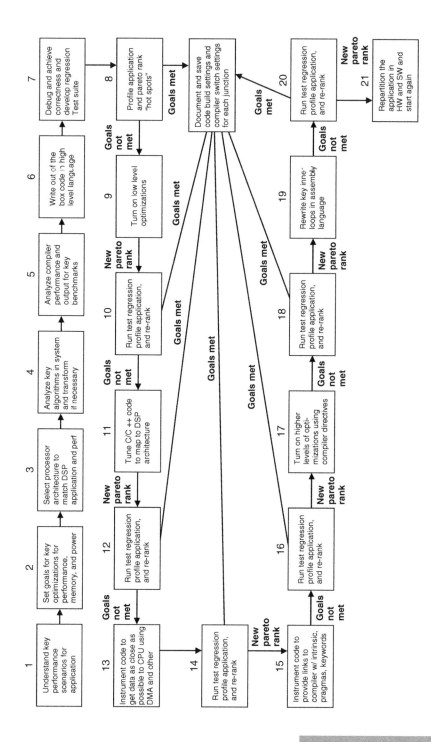

1 Understand key performance scenarios for application

2 Set goals for key optimizations for performance, memory, and power

3 Select processor architecture to match DSP application and perf

4 Analyze key algorithms in system and transform if necessary

5 Analyze compiler performance and output for key benchmarks

6 Write out of the box code in high level language

7 Debug and achieve correctness and develop regression Test suite

8 Profile application and pareto rank "hot spots"

Goals met

Document and save code build settings and compiler switch settings for each junction

Goals not met

9 Turn on low level optimizations

New pareto rank

10 Run test regression profile application, and re-rank

Goals not met

11 Tune C/C ++ code to map to DSP architecture

New pareto rank

12 Run test regression profile application, and re-rank

Goals not met

13 Instrument code to get data as close as possible to CPU using DMA and other

14 Run test regression profile application, and re-rank

New pareto rank

15 Instrument code to provide links to compiler w/ intrinsic, pragmas, keywords

Goals not met

16 Run test regression profile application, and re-rank

New pareto rank

17 Turn on higher levels of opti- mizations using compiler directives

Goals not met

18 Run test regression profile application, and re-rank

New pareto rank

19 Rewrite key inner- loops in assembly language

Goals not met

20 Run test regression profile application, and re-rank

Goals met

New pareto rank

21 Repartition the application in HW and SW and start again

Goals met

Goals met

Goals met

Goals met

Figure 12-55: The Expanded Code Optimization Process for DSP.

optimization process can focus on these "worst case" performance paths. If the developer can reach performance goals in these conditions, all other scenarios should meet their goals as well.

- **Step 2**: Once the key performance scenarios are understood, the developer then selects the key goals for each resource being optimized. One goal may be to consume no more than 75% of processor throughput or memory, for example. Once these goals are established, there is something to measure progress towards as well as stopping criteria. Most DSP developers optimize until they reach a certain goal and then stop.

- **Step 3**: Once these goals are selected, the developer, if not done already, selects the DSP to meet the goals of the application. At this point, no code is run, but the processor is analyzed to determine whether it can meet the goals through various modeling approaches.

- **Step 4**: This step involves analyzing key algorithms in the system and making any algorithm transformations necessary to improve the efficiency of the algorithm. This may be in terms of performance, memory, or power. An example of this is selecting a fast Fourier transform instead of a slower discrete Fourier transform.

- **Step 5**: This step involves doing a detailed analysis of the key algorithms in the system. These are the algorithms that will run most frequently in the system or otherwise consume the most resources. These algorithms should be benchmarked in detail, sometimes even to the point of writing these key algorithms in the target language and measuring the efficiency. Given that most of the application cycles may be consumed here, the developer must have detailed data for these key benchmarks. Alternatively, the developer can use industry benchmark data if there are algorithms that are very similar to the ones being used in the application. Examples of these industry benchmarks include the Embedded Processor Consortium at eembc.org and Berkeley Design Technology at bdti.com.

- **Step 6**: This step involves writing "out of the box" C/C++ code which is simply code with no architecture specific transformations done to "tune" the code to the target. This is the simplest and most portable structure for the code. This is the desired format if possible, and code should only be modified if the performance goals are not met. The developer should not undergo a thought process of "make it as fast as possible." Symptoms of this thought process include excessive optimization and premature optimization. Excessive optimization is when a developer keeps optimizing code even after the performance goals for the application have been met. Premature optimization is when the developer begins optimizing the application before understanding the key areas that should be optimized (following the 80/20 rule). Excessive and premature optimization are dangerous because these consume project resources, delay releases, and compromise good software designs without directly improving performance.

- **Step 7**: This is the "make it work right before you make it work fast" approach. Before starting to optimize the application which could potentially break a working application because of the complexity involved, the developer must make the application work correctly. This is most easily done by turning off optimizations and debugging the application using a standard debugger until the application is working. To get a working application, the developer must also create a test regression that is run to confirm that the application is indeed working. Part of this regression should be used for all future optimizations to ensure that the application is still working correctly after each successive round of optimization.

- **Step 8**: This step involves running the application and collecting profiling information for each function of the application. This profiling information could be cycles consumed for each function, memory used, or power consumed. This data can then be pareto ranked to determine the biggest contributors to the performance bottlenecks. The developer can then focus on these key parts of the application for further optimization. If the goals are met, then no optimizations are needed. If the goals are not met, the developer moves on to step 9.

- **Step 9**: In this step, the developer turns on basic compiler optimizations. These include many of the machine independent optimizations that compilers are good at finding. The developer can select options to reduce cycles (increase performance) or to reduce memory size. Power can also be considered during this phase.

- **Steps 10, 12, 14, 16, and 18**: These steps involve re-running the test regression for the application and measuring performance against goals. If the goals are met, then the developer is through. If not, the developer reprofiles the application and establishes a new pareto rank of top performance, memory, or power bottlenecks.

- **Step 13**: This step involves restructuring or tuning the C/C++ code to map the code more efficiently to the DSP architecture. Examples of this were discussed earlier in the chapter. Restructuring C or C++ code can lead to significant performance improvements but makes the code potentially less portable and readable. The developer should first attempt this where there are the most significant performance bottlenecks in the system.

- **Step 15**: In this step, the C/C++ code is instrumented with very specific information to help the compiler make more aggressive optimizations based on these "hints" from the developer. Three common forms of this instrumentation include using special instructions with intrinsics, pragmas, and keywords.

- **Step 17**: If the compiler supports multiple levels of optimization, the developer should proceed to turn on higher levels of optimization to allow the compiler to be more aggressive in searching for code transformations to yield more performance gains. These higher levels of optimization are advanced and will cause the compiler to run longer searching for these optimizations. Also, there is the chance that a more aggressive optimization will do something to break the code or otherwise cause it to behave incorrectly. This is where a regression test suite that is run periodically is so important.

- **Step 19**: As final resort, the developer rewrites the key performance bottlenecks in assembly language if the result can yield performance improvements above what the compiler can produce. This is a final step, since writing assembly language reduces portability, increases maintainability, decreases readability, and has other side effects. For advanced architectures, assembly language may mean writing highly complex and parallel code that runs on multiple independent execution units. This can be very difficult to learn to do well and generally there are not many DSP developers that can become expert assembly language programmers without a significant ramp time.

- **Step 21**: Each of the optimization steps described above are iterative. The developer may iterate through these phases multiple times before moving on to the next phase. If the performance goals are not being met by this phase, the developer needs to consider repartitioning the hardware and software for the system. These decisions are painful and costly but they must be considered when goals are not being met by the end of this optimization process.

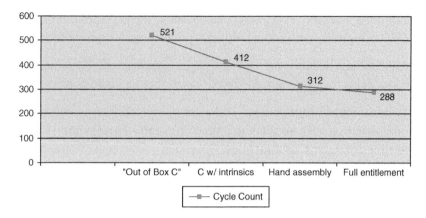

Figure 12-56: Achieving Full Entitlement. Most of the significant performance improvements are achieved early in the optimization process.

As shown in Figure 12-56, when moving through the code optimization process, improvements become harder and harder to achieve. When optimizing generic "out of the box" C or C++ code, just a few optimizations can lead to significant performance improvements. Once the code is tuned to the DSP architecture and the right hints are given to the compiler in the form of intrinsics, pragmas, and keywords, additional performance gains are difficult to achieve. The developer must know where to look for these additional improvements. That's why profiling and measuring are so important. Even assembly language programming cannot always yield full entitlement from the DSP device. The developer ultimately has to decide how far down the curve in Figure 12-56 the effort should go, as the cost/benefit ratio gets less justifiable given the time required to perform the optimizations at these lower levels of the curve.

12.15 Summary

Coding for speed requires the programmer to match the expression of the application's algorithm to the particular resources and capabilities of the processor. Key among these fit issues is how data is staged for the various operations. The iterative nature of DSP algorithms also makes loop efficiency critically important. A full understanding of when to unroll loops and when and how to pipeline loops will be essential if you are to write high-performance DSP code—even if you rely on the compiler to draft most of such code.

Whereas hand-tuned assembly code was once common for DSP programmers, modern optimizing DSP compilers, with some help, can produce very high-performance code.

Embedded real-time applications are an exercise in optimization. There are three main optimization strategies that the embedded DSP developer needs to consider:

- **DSP architecture optimization**: DSPs are optimized microprocessors that perform signal processing functions very efficiently by providing hardware support for common DSP functions;

- **DSP algorithm optimization**: Choosing the right implementation technique for standard and often used DSP algorithms can have a significant impact on system performance, and

- **DSP compiler optimization**: DSP compilers are tools that help the embedded programmer exploit the DSP architecture by mapping code onto the resources in such a way as to utilize as much of the processing resources as possible, gaining the highest level of architecture entitlement as possible.

References

TMS320C62XX Programmers Guide. Texas Instruments, 1997.

Hennesey, John L., and Patterson, David A. *Computer Architecture: A Quantitative Approach*. Palo Alto, CA: Morgan Kaufmann Publishers, Inc., 1990.

Kernighan, Brian W., and Pike, Rob. *The Practice of Programming*. Addison Wesley, 1999.

TMS320C55x_DSP_Programmer's_Guide.

TMS320C55xx Optimizing C/C++ Compiler User's Guide (SPRU281C)TMS320C55x. DSP Programmer's Guide (SPRU376A).

Generating Efficient Code with TMS320 DSPs: Style Guidelines (SPRA366).

How to Write Multiplies Correctly in C Code (SPRA683).

TMS320C55x DSP Library Programmer's Reference (SPRU422).

CHAPTER 13

Embedded Processors

Peter Wilson

In the olden days, boards were crammed with lots of packages of "glue logic," usually TTL ICs that performed all of the various functions not realizable by the processor and related components. Glue logic could be simple decoders to enable I/O and memory, or even complex state machines that relieved the processor of tough computational tasks.

Then the PAL arrived, probably reaching perfection in the 22V10 device. PALs had lots of burnable (later, electrically reprogrammable) fuses. Working with Boolean algebra, the designer could create most any combinatorial bit of logic from the inputs, and in many devices then create one or more registers, whose content could also be used as inputs. PALs greatly lowered glue logic's chip count.

The logical next step was the FPGA, which today comes in a mind-boggling variety of flavors. FPGAs contain vast numbers of mostly uncommitted logic elements. They're so versatile that today it's common to plop an entire microprocessor into the FPGA's design. A lot of vendors offer both hard and soft cores, predesigned micros that are either core components of the FPGA, or IP one adds to the device's logic via a few mouse clicks inside the vendor's design software. The most popular at this writing are the Micro- and Pico-Blaze products from Xilinix, and Altera's NIOS-II offerings. It's astonishing just how little effort is required to design and implement a system using these products. The first time I tried the NIOS-II product I had a working embedded app in under 15 minutes.

Peter Wilson could have titled his chapter "Processor Design Patterns for FPGAs," for, like the design patterns now becoming common in software engineering, the hardware developer's toolkit is being enriched by IP blocks that perform certain functions. Indeed, it's reasonable to take the analogy further: with SystemC, Verilog, VHDL, and similar hardware descriptor languages now the norm, hardware designers work just like software engineers, using similar tools, techniques, and processes. Unfortunately, too many on the hardware side *haven't stolen best practices from their software brethren. A prime example is configuration management.*

Pete could have also titled this chapter "An Introduction to VHDL." For in the course of just a handful of pages he takes one through all of the VHDL needed to implement an entire processor! The initiated may be surprised just how straightforward the process is, and how much it mirrors C programming.

Is the 10-instruction processor Pete outlines practical? In a truly tiny application where transistor counts mediate all decisions, maybe so. Decades ago, when hardware cost money, I saw a functioning machine with merely 3 opcodes! But the process he shows is both fascinating and useful. If you haven't delved deeply into CPU design, here's your Baedeker. If you're using an off-the-shelf soft core, Pete's descriptions will give you great insight into what's going on in the system.

—Jack Ganssle

13.1   Introduction

This application example chapter concentrates on the key topic of Integrating Processors onto Field Programmable Gate Array (FPGA) designs. This ranges from simple 8-bit microprocessors up to large IP processor cores that require an element of hardware–software co-design involved. This chapter will take the reader through the basics of implementing a behavioral-based microprocessor for evaluation of algorithms, through to the practicalities of structurally correct models that can be synthesized and implemented on an FPGA.

One of the major challenges facing hardware designers in the 21st century is the problem of hardware–software co-design. This has moved on from a basic partitioning mechanism based on standard hardware architectures to the current situation where the algorithm itself can be optimized at a compilation level for performance or power by implementing appropriately at different levels with hardware or software as required.

This aspect suits FPGAs perfectly, as they can handle fixed hardware architecture that runs software compiled onto memory, they can implement optimal hardware running at much faster rates than a software equivalent could, and there is now the option of configurable hardware that can adapt to the changing requirements of a modified environment.

13.2   A Simple Embedded Processor

13.2.1   Embedded Processor Architecture

A useful example of an embedded processor is to consider a generic microcontroller in the context of an FPGA platform. Take a simple example of a generic 8-bit microcontroller shown in Figure 13-1.

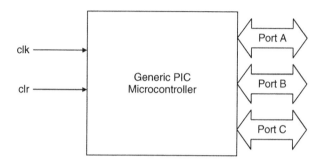

Figure 13-1: Simple Microcontroller.

As can be seen from Figure 13-1, the microcontroller is a 'general purpose microprocessor,' with a simple clock (clk) and reset (clr), and three 8-bit ports (A, B, and C). Within the microcontroller itself, there needs to be the following basic elements:

1. A control unit: This is required to manage the clock and reset of the processor, manage the data flow and instruction set flow, and control the port interfaces. There will also need to be a Program Counter (PC).

2. An Arithmetic Logic Unit (ALU): A PIC will need to be able to carry out at least some rudimentary processing—carried out in the ALU.

3. An address bus.

4. A data bus.

5. Internal registers.

6. An instruction decoder.

7. A Read Only Memory (ROM) to hold the program.

While each of these individual elements (1–6) can be implemented simply enough using a standard FPGA, the ROM presents a specific difficulty. If we implement a ROM as a set of registers, then obviously this will be hugely inefficient in an FPGA architecture. However, in most modern FPGA platforms, there are blocks of Random Access Memory (RAM) on the FPGA that can be accessed and it makes a lot of sense to design a RAM block for use as a ROM by initializing it with the ROM values on reset and then using that to run the program.

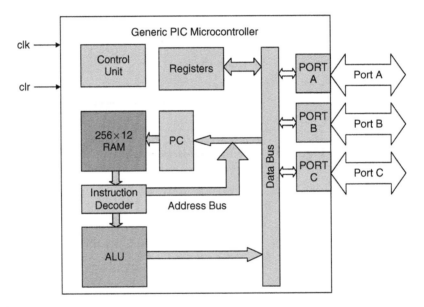

Figure 13-2: Embedded Microcontroller Architecture.

This aspect of the embedded core raises an important issue, which is the reduction in efficiency of using embedded rather than dedicated cores. There is usually a compromise involved and in this case it is that the ROM needs to be implemented in a different manner, in this case with a hardware penalty. The second issue is what type of memory core to use? In an FPGA RAM, the memory can usually be organized in a variety of configurations to vary the depth (number of memory addresses required) and the width (width of the data bus). For example a 512 address RAM block with an 8-bit address width would be equivalent to a 256 address RAM block with a 16-bit address width.

If the equivalent ROM is, say 12 bits wide and 256, then we can use the 256×16 RAM block and ignore the top four bits. The resulting embedded processor architecture could be of the form shown in Figure 13-2.

13.2.2 Basic Instructions

When we program a microprocessor of any type, there are three different ways of representing the code that will run on the processor. These are machine code (1's and 0's), assembler (low-level instructions such as LOAD, STORE, ...) and high-level code (such as C, Fortran, or Pascal). Regardless of the language used, the code will always be compiled or assembled into machine code at the lowest level for programming into memory. High-level code (e.g., C) is compiled and assembler code is assembled (as the name suggests) into machine code for the specific platform.

Clearly a detailed explanation of a compiler is beyond the scope of this book, but the same basic process can be seen in an assembler and this is useful to discuss in this context.

Every processor has a basic 'Instruction Set,' which is simply the list of functions that can be run in a program on the processor. Take the simple example of the following pseudocode expression:

$$b = a + 2$$

In this example, we are taking the variable a and adding the integer value 2 to it, and then storing the result in the variable b. In a processor, the use of a variable is simply a memory location that stores the value, and so to load a variable we use an assembler command as follows:

LOAD a

What is actually going on here? Whenever we retrieve a variable value from memory, the implication is that we are going to put the value of the variable in the register called the accumulator (ACC). The command 'LOAD a' could be expressed in natural language as 'LOAD the value of the memory location denoted by a into the accumulator register ACC'.

The next stage of the process is to add the integer value 2 to the accumulator. This is a simple matter, as instead of an address, the value is simply added to the current value stored in the accumulator. The assembly language command would be something like:

ADD #x02

Notice that we have used the x to denote a hexadecimal number. If we wished to add a variable, say called c, then the command would be the same, except that it would use the address c instead of the absolute number. The command would therefore be:

ADD c

Now we have the value of a +2 stored in the accumulator register (ACC). This could be stored in a memory location, or put onto a port (e.g., PORT A). It is useful to notice that for a number we use the key character # to indicate that we are adding the value and not using the argument as the address.

In the pseudocode example, we are storing the result of the addition in the variable called b, so the command would be something like this:

STORE b

While this is superficially a complete definition of the instruction set requirements, there is one specific design detail that has to be decided on for any processor. This is the number of instructions and the data bus size. If we have a set of instructions with the number of instructions denoted by I, then the number of bits in the opcode (n) must conform to the following rule:

$$2^n \leq I$$

In other words, the number of bits provides the number of unique different codes that can be defined, and this defines the size of the instruction set possible. For example, if $n = 3$, then with three bits there are eight possible unique opcodes, and so the maximum size of the instruction set is eight.

13.2.3   Fetch Execute Cycle

The standard method of executing a program in a processor is to store the program in memory and then follow a strict sequence of events to carry out the instructions. The first stage is to use the PC to increment the program line, this then calls up the next command from memory in the correct order, and then the instruction can be loaded into the appropriate register for execution. This is called the *fetch execute cycle*.

What is happening at this point? First the contents of the PC is loaded into the Memory Address Register (MAR). The data in the memory location are then retrieved and loaded into the Memory Data Register (MDR). The contents of the MDR can then be transferred into the Instruction Register (IR). In a basic processor, the PC can then be incremented by one (or in fact this could take place immediately after the PC has been loaded into the MDR).

Once the opcode (and arguments if appropriate) are loaded, then the instruction can be executed. Essentially, each instruction has its own state machine and control path, which is linked to the IR and a sequencer that defines all the control signals required to move the data correctly around the memory and registers for that instruction. We will discuss registers in the next section, but in addition to the C, IR, and accumulator (ACC) mentioned already, we require two emory registers as a minimum, the MDR and MAR.

For example, consider the simple command LOAD a, from the previous example. What is required to actually execute this instruction? First, the opcode is decoded and this defines that the command is a 'LOAD' command. The next stage is to identify the address. As the command has not used the # symbol to denote an absolute address, this is stored in the variable *a*. The next stage, therefore is to load the value in location *a* into the MDR, by setting MAR = a and then retrieving the value of a from the RAM. This value is then transferred to the accumulator (ACC).

13.2.4   *Embedded Processor Register Allocation*

The design of the registers partly depends on whether we wish to 'clone' a PIC device or create a modified version that has more custom behavior. In either case there are some mandatory registers that must be defined as part of the design. We can assume that we need an accumulator (ACC), a Program Counter (PC), and the three input/output ports (PORTA, PORTB, PORTC). Also, we can define the IR, MAR, and MDR.

In addition to the data for the ports, we need to have a definition of the port direction and this requires three more registers for managing the tristate buffers into the data bus to and from the ports (DIRA, DIRB, DIRC). In addition to this, we can define a number (essentially arbitrary) of registers for general purpose usage. In the general case the naming, order, and numbering of registers does not matter; however, if we intend to use a specific device as a template, and perhaps use the same bit code, then it is vital that the registers are configured in exactly the same way as the original device and in the same order.

In this example, we do not have a base device to worry about, and so we can define the general purpose registers (24 in all) with the names REG0 to REG23. In conjunction with the general purpose registers, we need to have a small decoder to select the correct register and put the contents onto the data bus (F).

13.2.5   A Basic Instruction Set

In order for the device to operate as a processor, we must define some basic instructions in the form of an instruction set. For this simple example we can define some very basic instructions that will carry out basic program elements, ALU functions, memory functions. These are summarized in Table 13-1.

Table 13-1

Command	Description
LOAD arg	This command loads an argument into the accumulator. If the argument has the prefix # then it is the absolute number, otherwise it is the address and this is taken from the relevant memory address. Examples: LOAD #01 LOAD abc
STORE arg	This command stores an argument from the accumulator into memory. If the argument has the prefix # then it is the absolute address, otherwise it is the address and this is taken from the relevant memory address. Examples: STORE #01 STORE abc
ADD arg	This command adds an argument to the accumulator. If the argument has the prefix # then it is the absolute number, otherwise it is the address and this is taken from the relevant memory address. Examples: ADD #01 ADD abc
NOT	This command carries out the NOT function on the accumulator.
AND arg	This command ands an argument with the accumulator. If the argument has the prefix # then it is the absolute number, otherwise it is the address and this is taken from the relevant memory address. Examples: AND #01 AND abc

Table 13-1 *(Continued)*

Command	Description
OR arg	This command ors an argument with the accumulator. If the argument has the prefix # then it is the absolute number, otherwise it is the address and this is taken from the relevant memory address. Examples: OR #01 OR abc
XOR arg	This command xors an argument with the accumulator. If the argument has the prefix # then it is the absolute number, otherwise it is the address and this is taken from the relevant memory address. Examples: XOR #01 XOR abc
INC	This command carries out an increment by one on the accumulator.
SUB arg	This command subtracts an argument from the accumulator. If the argument has the prefix # then it is the absolute number, otherwise it is the address and this is taken from the relevant memory address. Examples: SUB #01 SUB abc
BRANCH arg	This command allows the program to branch to a specific point in the program. This may be very useful for looping and program flow. If the argument has the prefix # then it is the absolute number, otherwise it is the address and this is taken from the relevant memory address. Examples: BRANCH #01 BRANCH abc

In this simple instruction set, there are 10 separate instructions. This implies that we need at least 4 bits to describe each of the instructions given in the table above. Given that we wish to have 8 bits for each data word, we need to have the ability to store the program memory in a ROM that has words of at least 12 bits wide. In order to cater for a greater number of instructions, and also to handle the situation for specification of different addressing modes (such as the difference between absolute numbers and variables), we can therefore suggest a 16-bit system for the program memory.

Notice that at this stage there are no definitions for port interfaces or registers. We can extend the model to handle this behavior later.

13.2.6   Structural or Behavioral?

So far in the design of the simple microprocessor, we have not specified details beyond a fairly abstract structural description of the processor in terms of registers and busses. At this stage we have a decision about the implementation of the design with regard to the program and architecture.

One option is to take a program (written in assembly language) and simply convert this into a state machine that can easily be implemented in a VHDL model for testing out the algorithm. Using this approach, the program can be very simply modified and recompiled based on simple rules that restrict the code to the use of registers and techniques applicable to the processor in question. This can be useful for investigating and developing algorithms, but is more ideal than the final implementation as there will be control signals and delays due to memory access in a processor plus memory configuration that will be better in a dedicated hardware design.

Another option is to develop a simple model of the processor that does have some of the features of the final implementation of the processor, but still uses an assembly language description of the model to test. This has advantages in that no compilation to machine code is required, but there are still not the detailed hardware characteristics of the final processor architecture that may cause practical issues on final implementation.

The third option is to develop the model of the processor structurally and then the machine code can be read in directly from the ROM. This is an excellent approach that is very useful for checking both the program and the possible quirks of the hardware/software combination as the architecture of the model reflects directly the structure of the model to be implemented on the FPGA.

13.2.7 Machine Code Instruction Set

In order to create a suitable instruction set for decoding instructions for our processor, the assembly language instruction set needs to have an equivalent machine code instruction set that can be decoded by the sequencer in the processor. The resulting opcode/instruction is given in Table 13-2.

Table 13-2

Command	Opcode (Binary)
LOAD arg	0000
STORE arg	0001
ADD arg	0010
NOT	0011
AND arg	0100
OR arg	0101
XOR arg	0110
INC	0111
SUB arg	1000
BRANCH arg	1001

13.2.8 Structural Elements of the Microprocessor

Taking the abstract design of the microprocessor given in Figure 13-2 we can redraw with the exact registers and bus configuration as shown in the structural diagram in Figure 13-3. Using this model we can create separate VHDL models for each of the blocks that are connected to the internal bus and then design the control block to handle all the relevant sequencing and control flags to each of the blocks in turn.

Before this can be started, however, it makes sense to define the basic criteria of the models and the first is to define the basic type. In any digital model it is sensible to ensure that data can be passed between standard models and so in this case we shall use the std_logic_1164 library that is the standard for digital models.

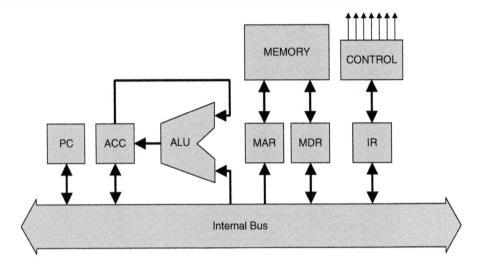

Figure 13-3: Structural Model of the Microprocessor.

In order to use this library, each signal shall be defined as of the basic type std_logic and also the library ieee.std_logic_1164.all shall be declared in the header of each of the models in the processor.

Finally, each block in the processor shall be defined as a separate block for implementation in VHDL.

13.2.9 Processor Functions Package

In order to simplify the VHDL for each of the individual blocks, a set of standard functions have been defined in a package call processor_functions. This is used to define useful types and functions for this set of models. The VHDL for the package is given below:

```
Library ieee;
Use ieee.std_logic_1164.all;

Package processor_functions is
      Type opcode is (load, store, add, not, and,
        or, xor, inc, sub, branch);
      Function Decode (word : std_logic_vector)
        return opcode;
```

```
      Constant n : integer := 16;
      Constant oplen : integer := 4;
      Type memory_array is array (0 to 2**(n-oplen-1)
        of Std_logic_vector(n-1 downto 0);
      Constant reg_zero : unsigned (n-1 downto 0) :=
          (others => '0');
End package processor_functions;

Package body processor_functions is
      Function Decode (word : std_logic_vector) return
        opcode is
          Variable opcode_out : opcode;

      Begin

          Case word(n-1 downto n-oplen-1) is
                  When "0000" => opcode_out : = load;
                  When "0001" => opcode_out : = store;
                  When "0010" => opcode_out : = add;
                  When "0011" => opcode_out : = not;
                  When "0100" => opcode_out : = and;
                  When "0101" => opcode_out : = or;
                  When "0110" => opcode_out : = xor;
                  When "0111" => opcode_out : = inc;
                  When "1000" => opcode_out : = sub;
                  When "1001" => opcode_out : = branch;
                  When others => null;
          End case;
          Return opcode_out;
      End function decode;
End package body processor_functions;
```

13.2.10 The PC

The PC needs to have the system clock and reset connections, the system bus (defined as inout so as to be readable and writable by the PC register block). In addition, there are several control signals required for correct operation. The first is the signal to increment the PC (PC_inc), the second is the control signal load the PC with a specified value (PC_load) and the final is the signal to make the register contents visible on the internal bus (PC_valid). This signal ensures that the value of the PC register will appear to be high impedance ('Z') when the register is not required on the processor bus. The system bus (PC_bus) is defined as a std_logic_vector, with direction inout to ensure the ability to read and write. The resulting VHDL entity is given below:

```
library ieee;
use ieee.std_logic_1164.all;
entity pc is
      Port (
            Clk : IN std_logic;
            Nrst : IN std_logic;
            PC_inc : IN std_logic;
            PC_load : IN std_logic;
            PC_valid : IN std_logic;
            PC_bus : INOUT std_logic_vector(n-1 downto 0)
      );
End entity PC;
```

The architecture for the PC must handle all of the various configurations of the PC control signals and also the communication of the data into and from the internal bus correctly. The PC model has an asynchronous part and a synchronous section. If the PC_valid goes low at any time, the value of the PC_bus signal should be set to 'Z' across all of its bits. Also, if the reset signal goes low, then the PC should reset to zero.

The synchronous part of the model is the increment and load functionality. When the clk rising edge occurs, then the two signals PC_load and PC_inc are used to define the function of the counter. The precedence is that if the increment function is high, then regardless of the load function, then the counter will increment. If the increment function (PC_inc) is low, then the PC will load the current value on the bus, if and only if the PC_load signal is also high.

The resulting VHDL is given below:

```
architecture RTL of PC is
      signal counter : unsigned (n-1 downto 0);
begin

      PC_bus <= std_logic_vector(counter)
                  when PC_valid = '1' else
                    (others => 'Z');
      process (clk, nrst) is
```

```
      begin
          if nrst = '0' then
                      count <= 0;
          elsif rising_edge(clk) then
                  if PC_inc = '1' then
                        count <= count + 1;
                  else
                        if PC_load = '1' then
                              count <= unsigned(PC_bus);
                        end if;
                  end if;
          end if;
      end process;
end architecture RTL;
```

13.2.11   The IR

The IR has the same clock and reset signals as the PC, and also the same interface to the
bus (IR_bus) defined as a std_logic_vector of type INOUT. The IR also has two further
control signals, the first being the command to load the IR (IR_load), and the second
being to load the required address onto the system bus (IR_address). The final
connection is the decoded opcode that is to be sent to the system controller. This is
defined as a simple unsigned integer value with the same size as the basic system bus.
The basic VHDL for the entity of the IR is given below:

```
library ieee;
use ieee.std_logic_1164.all;
use work.processor_functions.all;
entity ir is
    Port (
          Clk : IN std_logic;
          Nrst : IN std_logic;
          IR_load : IN std_logic;
          IR_valid : IN std_logic;
          IR_address : IN std_logic;
          IR_opcode : OUT opcode;
          IR_bus : INOUT std_logic_vector(n-1 downto 0)
    );
End entity IR;
```

The function of the IR is to decode the opcode in binary form and then pass to the control block. If the IR_valid is low, the bus value should be set to 'Z' for all bits. If the reset signal (nsrt) is low, then the register value internally should be set to all 0's.

On the rising edge of the clock, the value on the bus shall be sent to the internal register and the output opcode shall be decoded asynchronously when the value in the IR changes.

The resulting VHDL architecture is given below:

```
architecture RTL of IR is
      signal IR_internal : std_logic_vector (n-1 downto 0);
begin
      IR_bus <= IR_internal
          when IR_valid = '1' else (others => 'Z');
      IR_opcode <= Decode(IR_internal);
      process (clk, nrst) is
      begin
            if nrst = '0' then
                IR_internal <= (others => '0');
            elsif rising_edge(clk) then
                if IR_load = '1' then
                      IR_internal <= IR_bus;
                end if;
            end if;
      end process;
end architecture RTL;
```

In this VHDL, notice that we have used the predefined function Decode from the processor_functions package previously defined. This will look at the top four bits of the address given to the IR and decode the relevant opcode for passing to the controller.

13.2.12   *The Arithmetic and Logic Unit*

The Arithmetic and Logic Unit (ALU) has the same clock and reset signals as the PC, and also the same interface to the bus (ALU_bus) defined as a std_logic_vector of type INOUT. The ALU also has three further control signals, which can be decoded to map to the eight individual functions required of the ALU. The ALU also contains the Accumulator (ACC) which is a std_logic_vector of the size defined for the system bus width. There is also a single-bit output ALU_zero which goes high when all the bits in the accumulator are zero.

The basic VHDL for the entity of the ALU is given below:

```
library ieee;
use ieee.std_logic_1164.all;
use work.processor_functions.all;
entity alu is
      Port (
            Clk : IN std_logic;
            Nrst : IN std_logic;
            ALU_cmd : IN std_logic_vector(2 downto 0);
            ALU_zero : OUT std_logic;
            ALU_valid : IN std_logic;
            ALU_bus : INOUT std_logic_vector(n-1 downto 0)
      );
End entity alu;
```

The function of the ALU is to decode the ALU_cmd in binary form and then carry out the relevant function on the data on the bus, and the current data in the accumulator. If the ALU_valid is low, the bus value should be set to 'Z' for all bits. If the reset signal (nsrt) is low, then the register value internally should be set to all 0's.

On the rising edge of the clock, the value on the bus shall be sent to the internal register and the command shall be decoded.

The resulting VHDL architecture is given below:

```
architecture RTL of ALU is
      signal ACC : std_logic_vector (n-1 downto 0);
begin
      ALU_bus <= ACC
          when ACC_valid = '1' else (others => 'Z');
      ALU_zero <= '1' when acc = reg_zero else '0';
      process (clk, nrst) is
      begin
            if nrst = '0' then
                  ACC <= (others => '0');
            elsif rising_edge(clk) then
```

```
                    case ACC_cmd is
                    -- Load the Bus value into the
                         accumulator
                    when "000" => ACC <= ALU_bus;
                    -- Add the ACC to the Bus value
                    When "001" => ACC <= add(ACC,ALU_bus);
                    -- NOT the Bus value
                    When "010" => ACC <= NOT ALU_bus;
                    -- OR the ACC to the Bus value
                    When "011" => ACC <= ACC or ALU_bus;
                    -- AND the ACC to the Bus value
                    When "100" => ACC <= ACC and ALU_bus;
                    -- XOR the ACC to the Bus value
                    When "101" => ACC <= ACC xor ALU_bus;
                    -- Increment ACC
                    When "110" => ACC <= ACC + 1;
                    -- Store the ACC value
                    When "111" => ALU_bus <= ACC;
              end if;
         end process;
end architecture RTL;
```

13.2.13 The Memory

The processor requires a RAM memory, with an address register (MAR) and a data register (MDR). There therefore needs to be a load signal for each of these registers: MDR_load and MAR_load. As it is a memory, there also needs to be an enable signal (M_en), and also a signal denote Read or Write modes (M_rw). Finally, the connection to the system bus is a standard *inout* vector as has been defined for the other registers in the microprocessor.

The basic VHDL for the entity of the memory block is given below:

```
library ieee;
use ieee.std_logic_1164.all;
use work.processor_functions.all;
entity memory is
      Port (
            Clk : IN std_logic;
            Nrst : IN std_logic;
```

```
               MDR_load : IN std_logic;
               MAR_load : IN std_logic;
               MAR_valid : IN std_logic;
               M_en : IN std_logic;
               M_rw : IN std_logic;
               MEM_bus : INOUT std_logic_vector(n-1
                 downto 0)
         );
End entity memory;
```

The memory block has three aspects. The first is the function that the memory address is loaded into the MAR. The second function is either reading from or writing to the memory using the MDR. The final function, or aspect of the memory is to store the actual program that the processor will run. In the VHDL model, we will achieve this by using a constant array to store the program values.

The resulting basic VHDL architecture is given below:

```
architecture RTL of memory is
      signal mdr : std_logic_vector(wordlen-1 downto 0);
      signal mar : unsigned(wordlen-oplen-1 downto 0);
                  begin
      MEM_bus <= mdr
           when MEM_valid = '1' else (others => 'Z');
      process (clk, nrst) is
           variable contents : memory_array;
           constant program : contents :=
           (
                0 => "0000000000000011",
                1 => "0010000000000100",
                2 => "0001000000000101",
                3 => "0000000000001100",
                4 => "0000000000000011",
                5 => "0000000000000000" ,
                Others => (others => '0')
           );
begin
           if nrst = '0' then
                mdr <= (others => '0');
                mdr <= (others => '0');
```

```
            contents := program;
      elsif rising_edge(clk) then
            if MAR_load = '1' then
                  mar <= unsigned(MEM_bus(n-oplen-
                                    1 downto 0));
            elsif MDR_load = '1' then
                  mdr <= MEM_bus;
            elsif MEM_en = '1' then
                  if MEM_rw = '0' then
                        mdr <= contents(to_integer
                                    (mar));
                  else
                        mem(to_integer(mar))
                              := mdr;
                  end if;
            end if;
      end if;
   end process;
end architecture RTL;
```

We can look at some of the VHDL in a bit more detail and explain what is going on at this stage. There are two internal signals to the block, mdr and mar (the data and address, respectively). The first aspect to notice is that we have defined the MAR as an unsigned rather than as a std_logic_vector. We have done this to make indexing direct. The MDR remains as a std_logic_vector. We can use an integer directly, but an unsigned translates easily into an std_logic_vector.

```
signal mdr : std_logic_vector(wordlen-1 downto 0);
signal mar : unsigned(wordlen-oplen-1 downto 0);
```

The second aspect is to look at the actual program itself. We clearly have the possibility of a large array of addresses, but in this case we are defining a simple three-line program:

$$c = a + b$$

The binary code is shown below:

```
0 => "0000000000000011",
1 => "0010000000000100",
2 => "0001000000000101",
3 => "0000000000001100",
4 => "0000000000000011",
5 => "0000000000000000" ,
Others => (others => '0')
```

For example, consider the line of the declared value for address 0. The 16 bits are defined as 0000000000000011. If we split this into the opcode and data parts we get the following:

```
Opcode    0000
```

```
Data      000000000011 (3)
```

In other words this means LOAD the variable from address 3. Similarly, the second line is ADD from 4, and finally the third command is STORE in 5. In addresses 3, 4, and 5, the three data variables are stored.

13.2.14 *Microcontroller: Controller*

The operation of the processor is controlled in detail by the sequencer, or controller block. The function of this part of the processor is to take the current PC address, look up the relevant instruction from memory, move the data around as required, setting up all the relevant control signals at the right time, with the right values.

As a result, the controller must have the clock and reset signals (as for the other blocks in the design), a connection to the global bus, and finally all the relevant control signals must be output. An example entity of a controller is given below:

```
library ieee;
use ieee.std_logic_1164.all;
use work.processor_functions.all;
entity controller is
        generic (
                n : integer := 16
        );
        Port (
                Clk : IN std_logic;
```

```
          Nrst : IN std_logic;
          IR_load : OUT std_logic;
          IR_valid : OUT std_logic;
          IR_address : OUT std_logic;
          PC_inc : OUT std_logic;
          PC_load : OUT std_logic;
          PC_valid : OUT std_logic;
          MDR_load : OUT std_logic;
          MAR_load : OUT std_logic;
          MAR_valid : OUT std_logic;
          M_en : OUT std_logic;
          M_rw : OUT std_logic;
          ALU_cmd : OUT std_logic_vector(2 downto 0);
          CONTROL_bus : INOUT std_logic_vector(n-1
            downto 0)
      };
End entity controller;
```

Using this entity, the control signals for each separate block are then defined, and these can be used to carry out the functionality requested by the program. The architecture for the controller is then defined as a basic state machine to drive the correct signals. The basic state machine for the processor is defined in Figure 13-4.

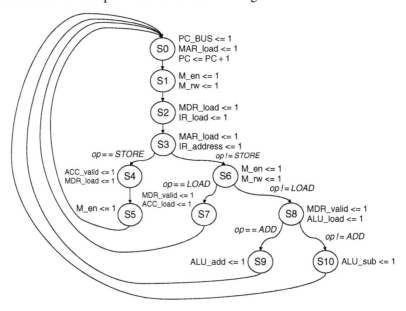

Figure 13-4: Basic Processor Controller State Machine.

We can implement this using a basic VHDL architecture that implements each state using a new state type and a case statement to manage the flow of the state machine. The basic VHDL architecture is shown below and it includes the basic synchronous machine control section (reset and clock) the management of the next stage logic:

```
architecture RTL of controller is
      type states is
       (s0,s1,s2,s3,s4,s5,s6,s7,s8,s9,s10);
      signal current_state, next_state : states;
begin
          state_sequence: process (clk, nrst) is
               if nrst = '0' then
                      current_state <= s0;
               else
                      if rising_edge(clk) then
                          current_state <=
                              next_state;
                      end if;
               end if;
          end process state_sequence;

          state_machine : process ( present_state,
            opcode ) is
              -- state machine goes here
          End process state_machine;
end architecture;
```

You can see from this VHDL that the first process (state_sequence) manages the transition of the current_state to the next_state and also the reset condition. Notice that this is a synchronous machine and as such waits for the rising_edge of the clock, and that the reset is asynchronous. The second process (state_machine) waits for a change in the state or the opcode and this is used to manage the transition to the next state, although the actual transition itself is managed by the state_sequence process. This process is given in the VHDL below:

```vhdl
state_machine : process ( present_state,
  opcode ) is
begin
    -- Reset all the control signals
    IR_load <= '0';
    IR_valid <= '0';
    IR_address <= '0';
    PC_inc <= '0';
    PC_load <= '0';
    PC_valid <= '0';
    MDR_load <= '0';
    MAR_load <= '0';
    MAR_valid <= '0';
    M_en <= '0';
    M_rw <= '0';
    Case current_state is
    When s0 =>
        PC_valid <= '1'; MAR_load <= '1';
        PC_inc <= '1'; PC_load <= '1';
        Next_state <= s1;
    When s1 =>
        M_en <='1'; M_rw <= '1';
        Next_state <= s2;
    When s2 =>
        MDR_valid <= '1'; IR_load <= '1';
        Next_state <= s3;
    When s3 =>
        MAR_load <= '1'; IR_address <= '1';
        If opcode = STORE then
                Next_state <= s4;
        else
                Next_state <=s6;
        End if;
    When s4 =>
        MDR_load <= '1'; ACC_valid <= '1';
        Next_state <= s5;
    When s5 =>
        M_en <= '1';
        Next_state <= s0;
    When s6 =>
        M_en <= '1'; M_rw <= '1';
```

```
            If opcode = LOAD then
                   Next_state <= s7;
            else
                   Next_state <= s8;
            End if;
      When s7 =>
            MDR_valid <= '1'; ACC_load <= '1';
            Next_state <= s0;
      When s8 =>
            M_en<='1'; M_rw <= '1';
            If opcode = ADD then
                   Next_state <= s9;
            else
                   Next_state <= s10;
            End if;
      When s9 =>
            ALU_add <= '1';
            Next_state <= s0;
      When s10 =>
            ALU_sub <= '1';
            Next_state <= s0;
             End case;
End process state_machine;
```

13.2.15 Summary of a Simple Microprocessor

Now that the important elements of the processor have been defined, it is a simple
matter to instantiate them in a basic VHDL netlist and create a microprocessor using
these building blocks. It is also a simple matter to modify the functionality of the
processor by changing the address/data bus widths or extend the instruction set.

13.3 Soft Core Processors on an FPGA

While the previous example of a simple microprocessor is useful as a design exercise
and helpful to gain understanding about how microprocessors operate, in practice most
FPGA vendors provide standard processor cores as part of an embedded development kit
that includes compilers and other libraries. For example, this could be the Microblaze
core from Xilinx or the NIOS core supplied by Altera. In all these cases the basic idea is
the same, that a standard configurable core can be instantiated in the design and code
compiled using a standard compiler and downloaded to the processor core in question.

Each soft core is different and rather than describe the details of a particular case, the reader is encouraged to experiment with the offerings from the FPGA vendors to see which suits their application the best.

In any soft core development system there are several key functions that are required to make the process easy to implement. The first is the system building function. This enables a core to be designed into a hardware system that includes memory modules, control functions, Direct Memory Access (DMA) functions, data interfaces and interrupts. The second is the choice of processor types to implement. A basic NIOS II or similar embedded core will typically have a performance in the region of 100–200 MIPS, and the processor design tools will allow the size of the core to be traded off with the hardware resources available and the performance required.

13.4   Summary

The topic of embedded processors on FPGAs would be suitable for a complete book in itself. In this chapter the basic techniques have been described for implementing a simple processor directly on the FPGA and the approach for implementing soft cores on FPGAs have been introduced.

Index

Saturated add function
  code, 498
  TMS320C55 DSP assembly
    code, 510
  intrinsic, 510
  using DSP intrinsics, 498
Scanning, 406, 412
Seebeck voltage, 423
Sensor, actuator, and control
    applications and circuits
    (hard tasks), 157–160
  actual E-2 system layout, 159
  E2BUS PC-host interface; *see*
    E2BUS PC-host interface
  host-to-module
    communications protocol,
    165–168
  basic function
    prototypes, 167
  SPI clock and data
    signals, 166
  RS-422–compatible indicator
    panel, 202–224
  simplified system layout, 159
  speed-controlled DC motor
    with Tach feedback and
    thermal cutoff, 184–193
  stepper motor controller,
    168–184
  circuit, 169
  two-axis attitude sensor using
    MEMS accelerometer,
    193–202
Set point gain; *see* 'Rabit', term
Setting time; *see* Conversion time
Simple microprocessor, 553
Single analog input channel, 397
Snooze function, 88
  flow chart of, 82
Sobel edge detector, 337, 343
Software (firmware musings),
    269–270
Software application
    examples, 331
  automotive driver
    assistance, 332

automotive safety systems,
    333, 335
  basic camera placement
    regions, 334–335
  collision avoidance and
    adaptive cruise control,
    333–335
  image acquisition, 337–338
  lane departure: system
    example, 335–337
  smart airbags, 333
baseline JPEG compression
    overview
  coding the AC coefficients
    (run-length
    coding), 353
  coding the DC
    coefficients, 352
  discrete cosine transform
    (DCT), 349–350
  entropy encoding, 353–354
  format conversion, 359–360
  frame types, 358–359
  Huffman coding, 354–355
  JPEG file interchange
    format (JFIF), 355
  motion estima-
    tion/compensation,
    357–358
  MPEG-2 encoder
    frameworks, 360–364
  MPEG-2 encoding, 355–356
  preprocessing, 347–348
  quantization, 350–351
  spatial filtering, 348–349
  zigzag sorting; *see* Zigzag
    sorting
code optimization study using
    open-source algorithms
  ogg, 365–366
  open source, 364–365
  speex, 367–368
  vorbis, 366–367
memory and data
    movement, 338
  decision making, 346–347
  edge detection, 342–343

image filtering, 340–342
  lane tracking, 346
  projection correction,
    338–340
  straight line detection:
    Hough transform,
    343–346
optimizing vorbis and speex
    on blackfin, 368
  assembly optimization, 373
  compiler optimization,
    368–370
  data management, 372–373
  instruction execution, 372
  system optimization,
    370–371
Software configuration
    management
    (SCM/CM), 226
  versioning, 226
Software controllers, 35–37
Software counter, 383
Software instrumentation,
    61–62
  low-intrusion, 61–62
Software instrumentation,
    embedded software, testing,
    61–62
Software pipelining, 461,
    484–486
  example, 486–487
  compiler-generated pipeline,
    488–490
  minimally parallel
    implementation, 488
  serial implementation, 487
  five-stage instruction
    pipeline, 462
  five-stage pipe, 485
Software testing
  choosing test cases, 54
  coverage tests; *see*
    White-box tests
  functional tests; *see*
    Black-box testing
  ideal time to stop test, 53
  ideal time to test

Printed and bound by CPI Group (UK) Ltd, Croydon, CR0 4YY

03/10/2024

01040331-0004